W9-BIG-784

INTRODUCTION TO FISH PHYSIOLOGY

Dr. Lynwood S. Smith

Front cover: *Trichogaster leeri* (pearl gourami), photo courtesy of Wardley Products Company, Inc.

Back cover: *Salvelinus fontinalis* (brook trout), photo courtesy of Dr. D. Terver, Nancy Aquarium, France.

Title page: *Malapterurus electricus* (electric catfish), photo by Gerhard Budich.

ISBN 0-87666-549-0

t.f.h.

© 1982 T.F.H. Publications, Inc.

Distributed in the U.S. by T.F.H. Publications, Inc., 211 West Sylvania Avenue, PO Box 427, Neptune, NJ 07753; in England by T.F.H. (Gt. Britain) Ltd., 13 Nutley Lane, Reigate, Surrey; in Canada to the pet trade by Rolf C. Hagen Ltd., 3225 Sartelon Street, Montreal 382, Quebec; in Canada to the book trade by H & L Pet Supplies, Inc., 27 Kingston Crescent, Kitchener, Ontario N28 2T6; in Southeast Asia by Y.W. Ong, 9 Lorong 36 Geylang, Singapore 14; in Australia and the South Pacific by Pet Imports Pty. Ltd., P.O. Box 149, Brookvale 2100, N.S.W. Australia; in South Africa by Valid Agencies, P.O. Box 51901, Randburg 2125 South Africa. Published by T.F.H. Publications, Inc., Ltd., the British Crown Colony of Hong Kong.

Contents

Preface:

This book describes in general how fish function, particularly at the level of organs and organ systems. In many ways, fish are typical vertebrates differing only slightly from humans. In other ways, their aquatic environment imposes strict requirements or offers unique opportunities which have resulted in some most unusual functions having no counterpart in man. I have tried to present both such kinds of physiological functions in simple, direct language accompanied by examples of typical experimental data. Most of these examples come from experiments on salmon and trout because salmonids are by far the most studied fish. Readers desiring further information on any of these examples should use the author(s) and year given at the end of the Figure or Table caption to find the full bibliographic listing at the end of the book and then go to the original source at a library. Since fish physiology is a rapidly-growing subject, readers can pursue a wealth of materials which are beyond the scope of this book.

Readers should have a minimum of difficulty with technical vocabulary in this book, although technical words could not be avoided completely. A collegiate or other large dictionary should suffice to define the occasional unfamiliar word. If basic concepts need further clarification, readers should consult a textbook on vertebrate or even biomedical physiology—fish mostly function similarly to other vertebrates. Sadly, such books will also probably impress you how much is known about mammals and how comparatively little is known about fish.

I hope that readers find this to be a satisfying and useful book. The intended audience includes college students, professional fisheries people of all kinds, and interested laymen, particularly pet fish hobbyists and growers. If this book provides some insight into the functional opportunities and limitations of the many lifestyles of fishes, then it will have been successful.

Lynwood S. Smith
Seattle, Washington

Chapter I: INTRODUCTION

A. COMPARISON BETWEEN AQUATIC AND TERRESTRIAL LIFE

Fish are typical vertebrates in many respects, and most people already know something about the physiological functioning of familiar vertebrates—dogs, cats, farm animals, themselves. While fish have many functions which are comparable to those of land vertebrates, others of their functions relate to their living in water. You may have some reasonable insight into what it would be like to live a dog's life, but it is quite another thing to imagine what it could be like to live like a fish. Their aquatic realm is too different from our terrestrial one for there to be relationships between them.

The physical and chemical properties of water dominate the lives of aquatic organisms in ways largely unknown to most land mammals. The density of water slows movement and demands streamlining for those species which choose to move rapidly, but at the same time it makes possible neutral buoyancy and almost effortless control of posture. Most land vertebrates function on the surface of their environment, while fish can operate in three dimensions with different problems of orientation and location. Air-breathing animals (including some fish) have an abundant supply of oxygen available, while water-breathers have only about 5 percent or less as much oxygen available in water, even when it is saturated. (Table I-1). Fish thus have a double problem of pumping a heavy respiratory medium over their gills and then having to pump relatively large amounts of it because of the low oxygen content. The site necessary for making adequate exchange of respiratory gases is also necessarily an excellent site of the transfer of water, ions, and heat. This leaves the fish in a compromising situation in which optimum respiration is not possible because of the associated problems in osmoregulation.

At standard temperature and pressure (STP = 0° C and 760 mm HG)
1 liter O_2 weighs 1,428 mg.

Air is approximately 20% O_2 = 285 mg O_2/liter air.

AT STP, 1 liter of air-saturated water contains 14.6 mg O_2 or 5.1% as much as air.

Solubility of O_2 in water decreases at higher temperatures. Water saturated with air contains:

At 10° C	11.3 mg O_2/liter
At 20° C	9.1 mg O_2/liter
At 30° C	7.5 mg O_2/liter

For O_2 dissolved in water, 10 mg O_2 = 7 ml O_2 at any temperature.

Table I-1. Some comparisons between the oxygen content of water and air. (See also Table IV-1.)

Further, it is impossible for the average fish to maintain a body temperature different from that of the environment. Tunas, for example, can do this only by the possession of a heat exchanger system in their muscle comparable to that in the gills; most of the heat in the blood leaving the muscle is transferred to the arterial blood entering the muscle by an elaborate counter current heat exchanger. All other fish have a body temperature which strictly follows that of the environment, for better or for worse. The temperature stability of water protects fish from rapid (e.g.—diurnal) temperature changes and is generally advantageous in this regard in temperate climates. If the water eventually becomes too warm for survival, however, the fish has no way to avoid it unless able to leave for cooler places—deeper water, for example. Fish are similarly exposed intimately to everything which is dissolved in water, good or bad.

Most sensory organs of fish are not greatly different from those of land vertebrates, but the resulting sensations which they perceive underwater are probably quite different in several respects from what we sense. Most fish can probably see as acutely and over about the same color range that we do. However, visibility in water is a few hundred meters at best and a small fraction of a centimeter at worst. In humans experiencing little or no vision, maintaining a balance is often a problem—our equilibrium is predominantly visual. In fish there appear to be no such problems, and they orient effectively even in poor visibility. In humans, taste and smell, pressure and hearing are distinct senses. In fish, each of these pairs probably constitutes a single, continuous spectrum. Some fish even have one sense that that humans do not—detection of low level electrical fields.

Thus, while we can have some generally useful insight into what a fish's life may be like, it is very difficult to apply one's intuition to the details. At the same time, it is important that a fish physiologist try to do so. It is an important part of many physiological experiments that data be obtained from "happy" fish rather than distressed ones. Recognizing the difference between these two extremes of behavior becomes an intangible factor in research in fish physiology. One must try to think like a fish.

B. SWIMMING MECHANICS

There are two related problems which a fish must solve to swim efficiently. One is to produce a minimum of energy-robbing turbulence while moving through the water. The other is to convert muscular contraction into some kind of effective propulsive motion. Although the two problems are somewhat interrelated, let's look at turbulence problems first.

The general idea of streamlining is recognized by most people. A streamlined object is smoothly-rounded and spindle-shaped with no sudden changes in outline. In fish, for which swimming is important, however, the minor details also become important. For a fish of any given weight, a long, slim fish goes through the water with less effort than a short, fat fish. For this reason, fast-swimming fish have very compact viscera and many have no swim bladder, both acting to minimize the fish's cross sectional area and the accompanying form (pressure) drag which is proportional to the cross sectional area. If the maximum diameter of the fish exceeds about 15 percent of the body length, hydrodynamic data suggest that the flow over the body is probably turbulent downstream from the thickest part of the body. Most fast-swimming fish have the widest part of their body as far posterior as possible so as to delay the initiation of turbulence as long as possible. Thus fish like sculpins, ling cod or catfish will never be famous for their swimming prowess—their head is the widest part of their body and produces lots of turbulence.

The amount of surface area of a fish also influences the amount of drag to be overcome during swimming. A body which is flattened has much more surface area than the one which is round in cross section. Large fins also add dramatically to surface area. Spines which project at right angles into the flow of water create huge amounts of drag. When form drag and friction (surface area) drag are combined, drag increases by perhaps as much as the cube of the velocity. In practical terms, this means that for a fish to double its swimming velocity it must increase its energy output by about five or six times. Such numbers must be taken with a certain amount of caution because they are extrapolated from data on plaster models of fish tested in wind tunnels. Tests on living, swimming fish have been made only a few times and have given variable results. However, it appears probable that a living, swimming fish has less drag than a dead, limp one being towed through the water or a model fish in a wind tunnel. A more complete discussion of the factors appears in Webb, 1975.

Most fish swim by moving the caudal fin back and forth by alternate contractions of the muscles on each side of the body. A few fish use other fins for propulsion, often the dorsal or the pectorals. When fin motions are analyzed in slow-motion movie films, the motions are seen to be quite complex and to vary widely according to the body shape and muscle arrangements of the fish. The wave-like nature of the swimming motions is best seen in long, slim fish such as eels, but also occurs in short fish. This wave of muscular contraction begins just behind the head, alternately in each side of the body, and progresses posteriorly as an S-shaped wave whose amplitude increases as it travels posteriorly. The wave has a component of its motion which pushes backward against the water, thus propelling the fish forward. The propulsion wave travels down the fish about 15 percent faster than the fish travels forward—i.e., there is about 15 percent slippage. The side-to-side component of the propulsion wave carries the caudal fin from side to side, also providing for-

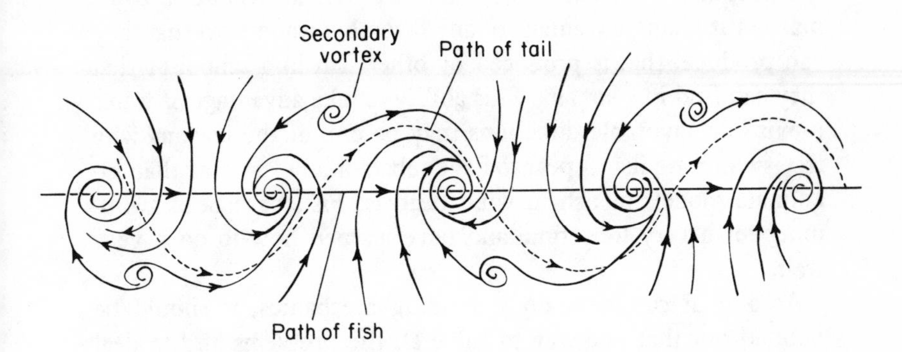

Fig. I-1. Schematic representation of the wake vortices produced by a fast-swimming fish. Swirls (vortices) develop alternately along each side of the body and then move into a relatively straight line behind the fish. Energy savings may occur as the fish's tail passes in front of each swirl, traveling in the same direction as the water in the swirl. (Redrawn from Rosen (1959), from Wedemeyer, et al., 1976, by permission of TFH Publications.)

ward propulsion in somewhat the same fashion as the use of a sculling oar. The relative importance of body undulation versus caudal fin oscillation for propulsion varies from perhaps 80 percent body and 20 percent caudal for eel-like fish to 20 percent body and 80 percent caudal for swimmers like tunas or swordfish. Most fish are intermediate between these extremes.

It should not be overlooked that fins do a great many things for fish besides propel them. They are used for equilibrium (upright balance), braking, turning, walking and climbing. Some are modified for making sounds, providing a spiny defense or acting as feelers. Even the adipose fin of salmonids and catfish probably serves some significant but presently unknown function at some time in the life cycle and is not simply a vestigial remnant of a fin that was once useful.

The water flow near a fish also turns out to be quite a bit more complex than might be supposed. There are small whirlpools (vortices) formed on either side of a fish behind the head as the propulsive waves originate. The vortices grow and are left behind the fish as a swirling wake, not unlike the whirlpools left by each stroke of an oar (Fig. I-1). There is also crossflow above and below the caudal peduncle which complicates any analysis of the propulsive pressure on the tail. In general, fish seem to be masters at taking advantage of any favorable water flow that they can produce, that is produced by other fish in a school or that they can find in a stream. This ability to take advantage of water moving in favorable directions may be one of the reasons why free-swimming fish appear to have substantially less drag than inanimate models of fish in wind tunnels. Fish have been swimming for a very long time and have learned how to do it very well.

As a final comment on swimming mechanics, it should be pointed out that one way to solve all the problems and to deal with the energy needed for swimming is to not swim. A number of fish have adopted lifestyles which require little swimming. They can be sedentary and filter food out of their respiratory water or develop means of attracting food to come to them rather than chasing it. They can escape predators by camouflage rather than speed or can develop bony plates or spines for defense. Strong-swimming, often predatory fish have sophisticated streamlining and gimmicks which are stimulating to an engineer and satisfying to analyze, but streamlining isn't the only way to be a fish.

C. COMPOSITION OF A FISH

1. Gross Anatomy

Salmonids are relatively primitive among the approximately 25,000 species of teleost fishes without being aberrant or archaic. They can be used as being illustrative of a number of general anatomical features which are typical of fish. The general body shape is characteristic of swimming, rather than sedentary fish. The caudal fin is relatively large and there is more muscle in the posterior part of the body to operate the caudal fin than in slowly swimming, hovering or sedentary fish such as in the cod-like fish or the flatfish. In salmonids the pectoral fins are placed below the gill covers and the pelvic fins are below the dorsal fin. In contrast, the more advanced perciform fishes have their pectorals near the midline and the pelvic fins ventral and sometimes even anterior to the pectorals. Muscle is the largest single tissue in the body, with dark muscle being found laterally on either side of large amounts of white muscle. Dark muscle is associated with sustained swimming, white muscle with burst (non-sustainable, emergency) swimming. Muscle is arranged on the axial skeleton in slanting layers shaped in somewhat of a W-shape whose significance is unknown. Contraction of the muscles shortens each side of the body alternately during swimming and produces a compression load on the vertebral column.

The visceral organs of salmonids are compactly arranged and typical of most vertebrates, but with certain notable exceptions. The digestive tract is relatively short, but its surface area is greatly expanded just below the pylorus by the presence of pyloric caecae. The number of pyloric caecae is of taxonomic significance in distinguishing the five species of Pacific salmon. The kidney of salmonids is formed by the fusion of a pair of embryonic kidneys, producing a median structure which runs most of the length of the visceral cavity. The anterior portion is hormonal in function and not excretory, while the posterior portion is called an opisthonephros rather than a metanephros as in the higher vertebrates. The whole kidney also functions in the formation of blood cells (there is no bone marrow as in mammals). The reproductive organs originate anteriorly and dorsally in the visceral cavity in typical vertebrate fashion, but the urogenital ducts are quite confusing in their derivation and in their termination at the urogenital papilla just posterior to the anal opening. There is no definitive study as of the time of this writing on how to properly designate the reproductive and excretory ducts according to their embryological derivation. My policy is to use noncommittal names such as urinary duct and vent and not to worry about whether calling the wide place in a urinary duct a true bladder or not. The functional capacities of these structures are obvious and undisputed. There are certain typical vertebrate organs which appear to be absent because they are diffusely scattered. The pancreas is scattered along the mesenteries of the intestine, and the thyroid gland consists of clusters of follicles scattered in the muscles of the isthmus around the ventral aorta.

A number of salmonid structures are typical of vertebrates as a whole. The swim bladder and its pneumatic duct are

Organ	Wet Weight (as%Body Weight)
Liver	1.22
Spleen	0.13
Intestine	4.69[a]
Heart	1.22
Swim Bladder	0.22
Kidney	0.86
Muscle	55.8
Skin	8.68
Axial Skeleton	13.5[b]
Gills, Gill Bars	2.76
Head	11.83
Total	99.91[c]

[a]Includes visceral fat.

[b]Includes vertebral column and fins with associated bones; excludes head skeleton.

[c]These fish were killed and stored on ice for up to two days before weighing. Weight losses during this storage were adjusted for. Weights of individual organs usually totaled between 90-95% of the total body weight as a result of evaporation during dissection as well as removal of blood from organs such as spleen and heart. Evaporation was particularly noticeable on skin and fins. However, all organ weights were increased proportionately so that the total percentage approximated 100%.

Table I-2. Organ and tissue weights of smolting coho salmon, *Oncorhynchus kisutch*, 130-140 mm fork length, 24-30 g total weight. (Unpublished data, L.S. Smith, 1975.)

homologous to the trachea and lungs of the tetrapod vertebrates. The nervous system and its supporting skeleton are quite typical, with 10 cranial nerves and all of the standard lobes in their normal position and enclosed in a typical cranium. Even in the flatfish where there has been a 90° change in body orientation and the eyes and nostrils have migrated to one side of the head, only the nerve endings and sensory organs have migrated—the brain is still in its original position, along with the semicircular canals and otoliths, i.e., on its side in the adult fish.

Thus there are few surprises in fish anatomy to anyone familiar with other vertebrates, but there are many interesting adaptations of the basic vertebrate body plan.

2. Organ Weights

Before beginning the detailed discussions of the functions of individual organs of fish, it is useful to consider the relationships of these organs and tissues to the whole fish. Data to explore this area are relatively scarce for fish, although readily available for many experimental animals. Table I-2 presents data for coho salmon smolts *(Oncorhynchus kisutch).* The major point of this table is to provide some perspective about the importance of organs. Some active, vital organs such as the heart are quite small in terms of the total body weight. Other organs such as the skin, which is comparatively large, often are ignored by physiologists. The quantity of muscle is perhaps surprisingly large compared to land vertebrates, but presumably relates to the fact that almost all of the muscle can be used for swimming and very little for posture. Further, the fish does not have to support the weight of the muscle with its skeleton because the muscle mass has almost neutral buoyancy in water. The relative weights of the viscera are quite variable in many fish because of the tendency to store fat there during non-reproductive periods.

3. Body Composition and Fluid Compartments

Figure I-2 is a simplistic representation of the gross chemical

Fig. I-2. Diagrammatic representation of the tissue composition and fluid compartments of a typical teleost fish. Ranges of values show approximate variability to be expected among different species or in the same individuals under different circumstances.

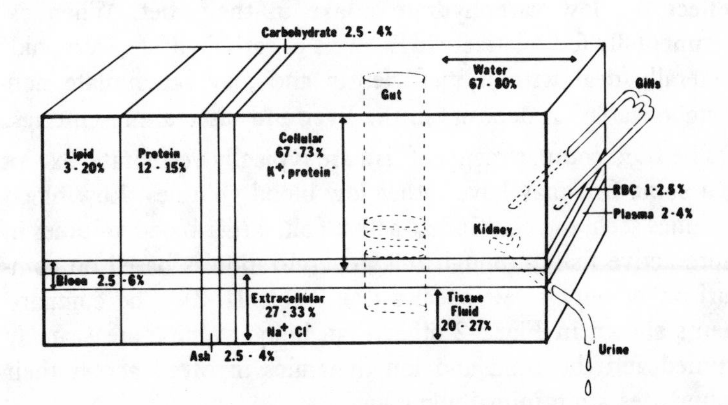

composition of the average salmonid which is probably also approximately correct for many other teleost groups. The figure also shows the three major fluid compartments inside the body. The composition of fish is quantitatively similar to that of many land vertebrates, but with certain major exceptions. The lipid content of salmon has been observed to be as high as 40 percent (by weight) of the muscle in premigratory coho and as low as one percent in post-spawning salmon. The lipid is stored as polyunsaturated oils and used for muscle metabolism during the nonfeeding period of migration. As the fat stores become depleted, muscle proteins are also consumed. Both fat and protein are replaced by water, presumably to maintain the fish's streamlined shape for maximal hydrodynamic efficiency. Smaller changes in muscle composition occur with changes in diet. There is usually an inverse relationship between lipid and water regardless of the cause—each replacing the other—so that long-term, major changes in body composition may show little change in total body weight.

On the other hand, there may also be major short-term changes in body water content which take only a few minutes or a few hours during respiratory or osmoregulatory stress without any change in lipid content. Rainbow trout put into about a 5 percent salt solution can lose up to 6 percent of their body weight in two minutes, for example.

Compared to many vertebrates, fish are relatively low in glycogen and other carbohydrates. In wild fish, this is thought to reflect the low carbohydrate intake in their diet. When experimentally fed relatively high levels of carbohydrate, salmonids generally deal with it inefficiently and may accumulate nonmetabolizable fat deposits in the liver and visceral mesenteries.

The fluid compartments of fish are typically vertebrate, except that some fish may have rather low blood volumes. Low blood volumes seem to occur in less active fish, larger blood volumes in more active fish, although this generalization is based on comparison of only a few species (see Table III-3). The compartments shown in Fig. I-2 other than blood volume are scarcely studied and the fluid and ion dynamics involved across their boundaries are minimally known.

Chapter II: OSMOREGULATION

A. OVERVIEW

1. Evolutionary Aspects

The basic problem of osmoregulation is common to all organisms—as soon as the internal concentration of anything exceeds that of the same substance in the aquatic environment, a diffusion gradient occurs. And when a cell membrane is placed across this diffusion pathway so that some substances pass through the membrane faster than others with various concentration gradients, then osmotic pressure occurs. There is thus no organism which lives in freshwater without carrying on osmoregulation to maintain an osmotic pressure differential. In seawater, some organisms resemble the salt mixture of seawater very closely and spend little effort to control their internal environment. All the remainder devote a significant portion of their basal metabolic rate of energy consumption to retain their internal salts and other dissolved materials at concentrations different than those in their environment.

The blood of hagfish closely resembles seawater, at least in its salt composition. Only a few ions differ, principally Mg^{++}. In the strict sense of the word, hagfish do not osmoregulate, but only ion regulate since the total osmotic pressure of their plasma is almost identical to that of seawater.

Both lampreys and elasmobranchs osmoregulate; that is, the concentrations of most substances in their blood differ from seawater and the total osmotic pressure (total number of particles in solution) is different from that in the environment. The elasmobranchs accomplish this in a way quite different from teleosts, and the process will be described briefly. The concentrations of salts in shark blood are approximately like that in teleosts—collectively equivalent to about 1% NaCl. However, the kidney tubules of elasmobranchs retain urea until the concentration in the blood reaches about 5%. Since it does not readily diffuse across the gills, it stays in the blood and raises the total osmotic pressure of the blood to a level greater than that in seawater. This means that water moves from the environment into elasmobranchs without any effort on their part except to retain the urea. However, there is still about a 2:1 concentration gradient into the elasmobranchs for NaCl, so salt diffuses inward as well as water. The rectal gland, a small finger-like structure in the posterior visceral cavity, actively secretes the excess NaCl. Thus, elasmobranch osmoregulation is not as free of metabolic costs as might appear at first. Some sharks migrate from the sea into Lake Nicaragua and have reduced levels of urea in their blood while in the lake. Other details of their freshwater osmoregulation are under study.

The general features of teleost osmoregulation are diagrammed in Fig. II-1. There is no accumulation of urea, and the total osmotic pressure (osmolarity) of teleost blood is greater than that of freshwater and less than that of seawater. Thus they osmoregulate in either environment.

Marine teleosts passively lose water through any permeable surface and gain salt. Their primary means available by which to replace the lost water is to drink seawater. The gut actively takes up the monovalent ions (Na^+, K^+, Cl^-) and most of the water into the blood, leaving most of the divalent ions (Mg^{++}, Ca^{++}, SO_4^{--}) in the gut as a rectal fluid having about the same osmolarity as the blood; the kidneys excrete the small amounts of divalent ions reaching the blood. Special chloride cells in the gills actively secrete the Cl^- and Na^+ ions from the blood back into the environment. The concentration differential across the gill membranes of the various charged particles also results in a potential difference of about 20-25 millivolts (mV) (10 mV in salmonids), with the inside of the fish being positive. This enhances the removal across the gills of positive ions from the fish.

Freshwater teleosts are almost in exactly the reverse situation of the marine teleosts—most drink little or no water since large amounts of water diffuse inward across the gill membranes, all of

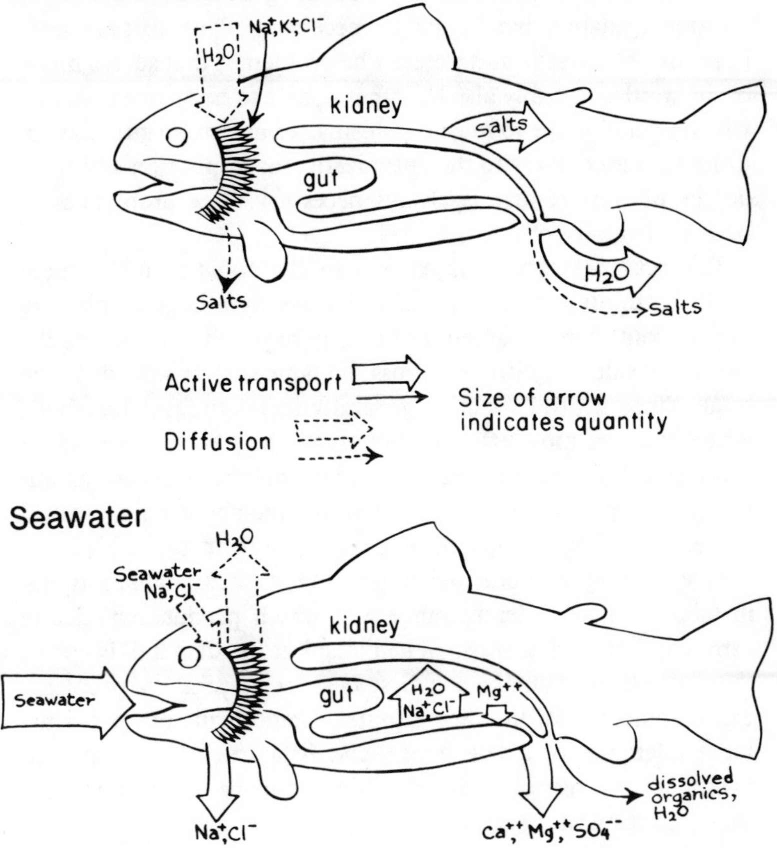

Fig. II-1. The general pattern of water and salt movement in osmoregulation of fresh and saltwater teleost fishes, not including sharks and rays. (From Wedemeyer, et al., 1976, by permission of TFH Publications.)

them produce large amounts of urine to get rid of the excess water and they use metabolic energy to reabsorb as many salts as possible from their urine and also to absorb salts from their environment. Again, there is an electrical differential between inside and outside, but this time it is -30 to -40 mV (less in high-Ca^{++} water), which enhances the transport of positive ions across the gills.

2. Basic Problems Met By Osmoregulation

Osmoregulation would not be necessary if membranes were impermeable to salts and water. The problem is that such a membrane would probably also be impermeable to oxygen, a situation which is intolerable to most organisms. Thus osmoregulation optimization is contrary to the optimization of respiration and some degree of compromise is always necessary. The problem is to balance a variety of needs.

The relative scarcity of oxygen in the aquatic environment could be minimized as a problem by having a huge respiratory surface, but then the amount of energy required to deal with the water and salt that diffused across the same surface would also be huge. Since natural selection generally seems to favor metabolic efficiency, the most efficient osmoregulatory surface would be the smallest or least permeable one, but this would severely limit the oxygen inflow and the rate at which metabolic energy could be released. The compromise is somewhere in the middle, of course, giving a moderate respiratory capacity which is adjustable over a moderate range and which produces moderate osmoregulatory demands. When respiratory demand is great, those needs are usually met first because respiratory problems can become crucial in a few minutes. Osmoregulatory problems seem tolerable for a few hours, and fish usually seem able to postpone dealing with the osmotic problems until after they solve the respiratory ones.

B. OSMOREGULATORY FUNCTIONS IN THE GILLS: CHLORIDE CELLS

1. Anatomy

The chloride cells of teleosts were given their name on the basis of their location at the base of the gill filaments, their staining properties and their ultrastructure. They were first noticed because their acidophilic staining qualities were similar to the acid-secreting cells of the gastric lining. Later studies with the electron microscope showed abundant membranes and mitochondria inside the cells which make major structural changes during migrations between freshwater and seawater, suggesting some kind of secretory or transport activity which changes during migration. Functional and biochemical studies have further indicated that gills of teleosts are the site of one or more ion pumps for chloride, sodium and potassium ions. Because of their location, there has been no possibility of isolating these cells for direct confirmation of their ion transporting role, but the many indirect studies mostly agree on the connection between ion transport functions and these acidophilic cells. The recent discovery of chloride cells in the epidermis lining the inside of the gill covers of killifish *(Fundulus)* permits isolated preparations to be made and should soon answer many of the present questions.

2. Ion Pumps: Maetz's Model

The chloride cell model proposed by Maetz and his colleagues is shown in Fig. II-2. The basic feature of the freshwater model, for which there is the most experimental evidence, is a coupled exchange of Na^+ for NH_4^+ or H^+ and Cl^- for HCO_3^- in which Na^+ and Cl^- are transported against the concentration gradient. The system consumes energy which is released by a Na^+-K^+ activated ATPase (adenosin triphosphotase) which is located on intracellular membranes. The chloride pump has been well identified by biochemical means and its structure has been linked circumstantially to intracellular membranes.

The kind of evidence which led Maetz and others to postulate the ion coupling model involved the change in concentration of various ions in the fish's environment and blood or the transfer of radio-sodium, etc., from fish to environment or environment to fish. Only in the latter was a direct measurement of the actual ion flux possible. Maetz found that addition of NH_4^+ to the fish's environment slowed the rate of excretion of ammonium ions, as would be expected because of the decreased diffusion gradient, but that the uptake rate of Na^+ also decreased at the same time. Early work did not show the same proportion of change in both ions, but later work considering both ammonium and hydrogen ions showed that the sum of the two was nearly proportional to the Na^+ transport. Similar approximate correlations were found with Cl^- and HCO_3^-.

In goldfish, Maetz found that Na^+ and Cl^- transport were not

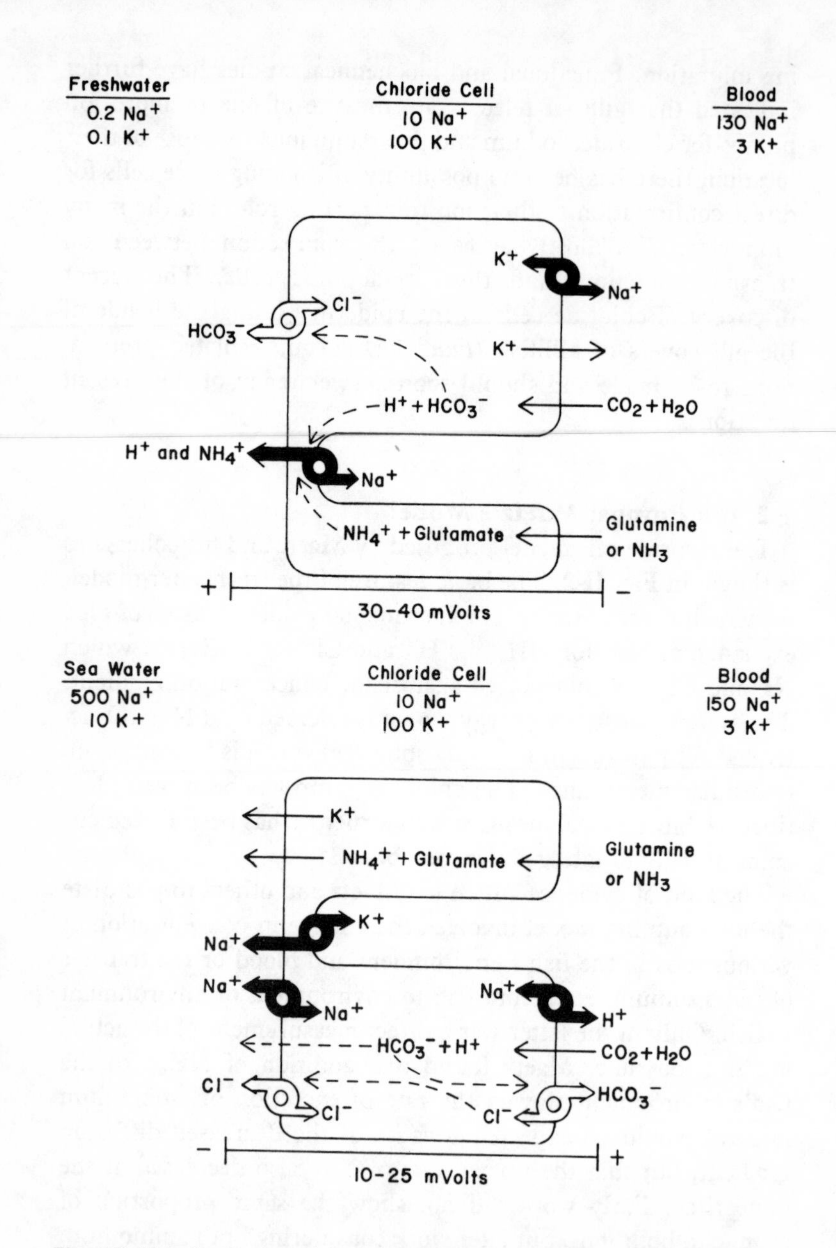

Fig. II-2. Hypothetical model of chloride cells in freshwater and seawater. Ion exchange mechanisms are shown as arrows coupled to a circle. Open arrows indicate active transport processes; solid arrows are considered as passive diffusion by some authors, as active transport by others. Units on ion concentrations are millimoles/liter. (Redrawn from Maetz, 1971, with modifications and additions.)

linked. They could even move in opposite directions for several days at a time. This is not entirely unreasonable, since Cl^- might be required to produce gastric HCl without any direct need for Na^+, for example, but the results were still somewhat startling.

After the ion exchange model was first proposed, the ammonia level in the blood was measured and found to be extremely low. This was not surprising, even though ammonia is the major excretory product from protein catabolism, because ammonia is highly toxic and could not be tolerated at any significant concentration. When the ammonia concentration was combined with typical blood flow past the chloride cells, there was not enough ammonia being delivered to the gills to account for the amount of ammonia and ammonium ion which was being excreted. However, the amino acid glutamine is common in fish plasma and the enzyme glutaminase, which removes the $-NH_2^-$ (amino) group from glutamine as NH_3, was found in the gills. Carbonic anhydrase is also present to provide the H^+ (from HCO_3^-) to make NH_3 into NH_4^+. The major ions are thus accounted for under most circumstances by the model.

The ion exchange model for chloride cells of marine fish is less well established than for freshwater fish. The major difference is that Na^+, NH_4^+, Cl^- and HCO_3^- are all moving outward and are not available for exchange. Maetz found Na^+ and Cl^- to be going both inward and outward and hypothesized an exchange diffusion (no net transport) as shown in the lower half of Fig. II-2. There also appeared to be some inward transport of K^+, possibly in exchange for Na^+, in eels *(Anguilla).*

3. Transepithelial Potential: Kirschner's Model

An electrical phenomenon associated with ion transport was described by Kirschner and several co-workers in 1974. He found that a rainbow trout in seawater had a potential difference between the inside and outside of the fish of about +10 mV (fish positive to its environment) or more. When the fish's environment was changed rapidly to freshwater, the potential changed just as rapidly to −30 mV (fish negative to its environment). The potential followed the environmental salinity in either direction in a few minutes (Fig. II-3). Nearly the same electrical changes

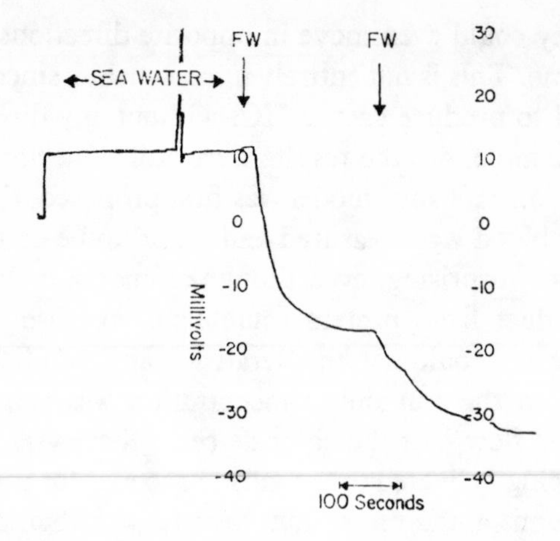

Fig. II-3. The effect on the transepithelial potential (TEP) of changing the fluids perfusing the gills of a steelhead trout from seawater to freshwater. At the first arrow the recirculation reservoir was changed from seawater to one containing 1 mM NaCl (freshwater). At the second arrow, the reservoir was changed again with the same solution. A third change (not shown) brought the TEP down to − 43 mV. (From Greenwald, et al., 1974, with permission of author and the publisher.)

occurred if the fish's environment was changed from freshwater to an NaCl solution having the same strength as sea water (Fig. II-4). Kirschner called this electrical gradient a transepithelial potential (TEP).

The TEP is produced by the active transport of Cl^- across the gill epithelium. Since chloride ions carry the same electrical charge as an electron, when they are transported actively (meaning that energy is used by the transport process and involves hypothetical molecular machinery called an ion pump) against the diffusion gradient, there is a piling up of electrical charges as well as Cl^- on one side of the membrane. An important feature of the system is that the membrane through which the chloride ions were transported is not very permeable to the ions so that they don't leak back into the cell as fast as they are pumped out.

The total system also transfers Na^+ and K^+, but little or no active transport is needed for these ions. The polarity of the TEP is such that positive ions move in the same directions as the actively transported negative ions, but without requiring any additional effort. The positive ions must lag behind somewhat, or they would neutralize the TEP which attracts them. According to Kirschner, the permeability of the positive ions is quite different from that for Cl^-. The relative permeabilities for Na^+:K^+:Cl^- are 1:10:0.3, respectively. The ion concentrations and the resulting TEP agree within about 5%, according to Kirschner's calculations.

The two models of ion transport agree and disagree in several respects. Both identify an active transport system (ion pump) for Cl^- and agree on the general degree of permeability of gill membranes to each of the monovalent ions—Na^+, K^+, and Cl^-. Both models indicate the presence of Na^+-K^+ activated ATPase and

Fig II-4. The effect of NaCl on the transepithelial potential (TEP) across the gills of steelhead trout. The TEP in freshwater was nearly −40mV. At the first arrow, the gills were exposed to 500 mM NaCl (approximately equivalent to seawater) which produced a TEP of +8 mV. At the second arrow, the NaCl was replaced by freshwater. The experiment demonstrates that most of the TEP can be produced solely by the NaCl gradient between blood and water at the gills. (From Greenwald, et al., 1974, with permission of the author and the publisher.)

J^{Na}_{out} AND TEP IN SW AND FW

Medium	J^{Na}_{out}	TEP
	μeq (100 g)⁻¹ h⁻¹	*mV*
SW	155±16 (16)	+10.4±0.9 (15)
FW	39.6±5.8 (16)	−34.9±1.4 (15)

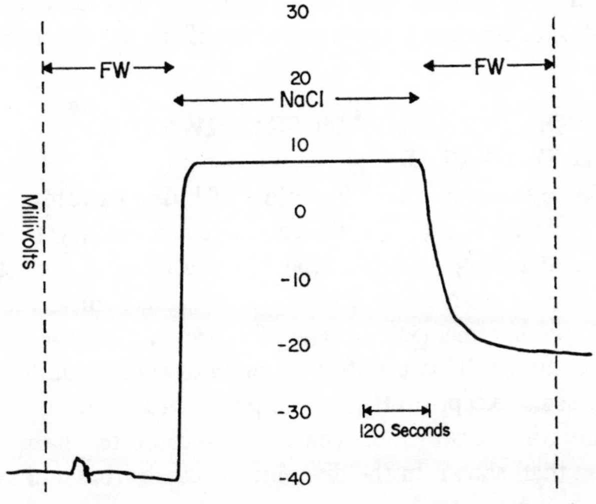

carbonic anhydrase enzymes. The two models also differ in several significant respects. In Maetz's ion exchange model, ion exchange is not entirely necessary but often occurs. In the TEP, no ion exchange or Na^+-K^+ pump are needed since the electrical gradient accounts for all the transport of positive ions. The ion exchange model assumes that the gill membranes are impermeable to divalent ions, particular Mg^{++}, Ca^{++}, and SO_4^{--}, while the TEP experiments demonstrated the passage of such ions as well as an organic molecule, choline. What appears to be a major point of difference between the two models—the change-over between freshwater and marine osmoregulation in anadromous fish—probably is not. In the ion exchange model, it takes several days to complete the transfer process which involves considerable increase in RNA-DNA activity as the intracellular structures for ion transport are torn down and rebuilt so as to function in the opposite direction. The TEP changes almost instantly as the salinity changes, but the ion fluxes do not reach a new equilibrium for several days.

At the present time there is insufficient information available to resolve all of the apparent differences between the two models, but it seems that each model represents part of the osmoregulatory functions of gills. Further experiments are needed to monitor the flow of a large number of substances at the same time as the electrical potentials are measured. Then new models can be proposed which incorporate both TEP and ionic data and also take into account present problems of species differences.

C. KIDNEY FUNCTION IN FRESHWATER AND SEAWATER

1. Basic anatomy and function of kidney tubules

The basic structure of a typical vertebrate kidney tubule is represented in Fig. II-5. Arterial blood arrives at a cluster of capillaries, the glomerulus, where plasma is mechanically filtered through the basement membrane of Bowman's capsule into the proximal tubule. This ultrafiltrate contains everything found in blood plasma except proteins. The proximal and distal tubules, which are considerably more complex and contorted in the actual situation than shown in the diagram, reabsorb from and secrete

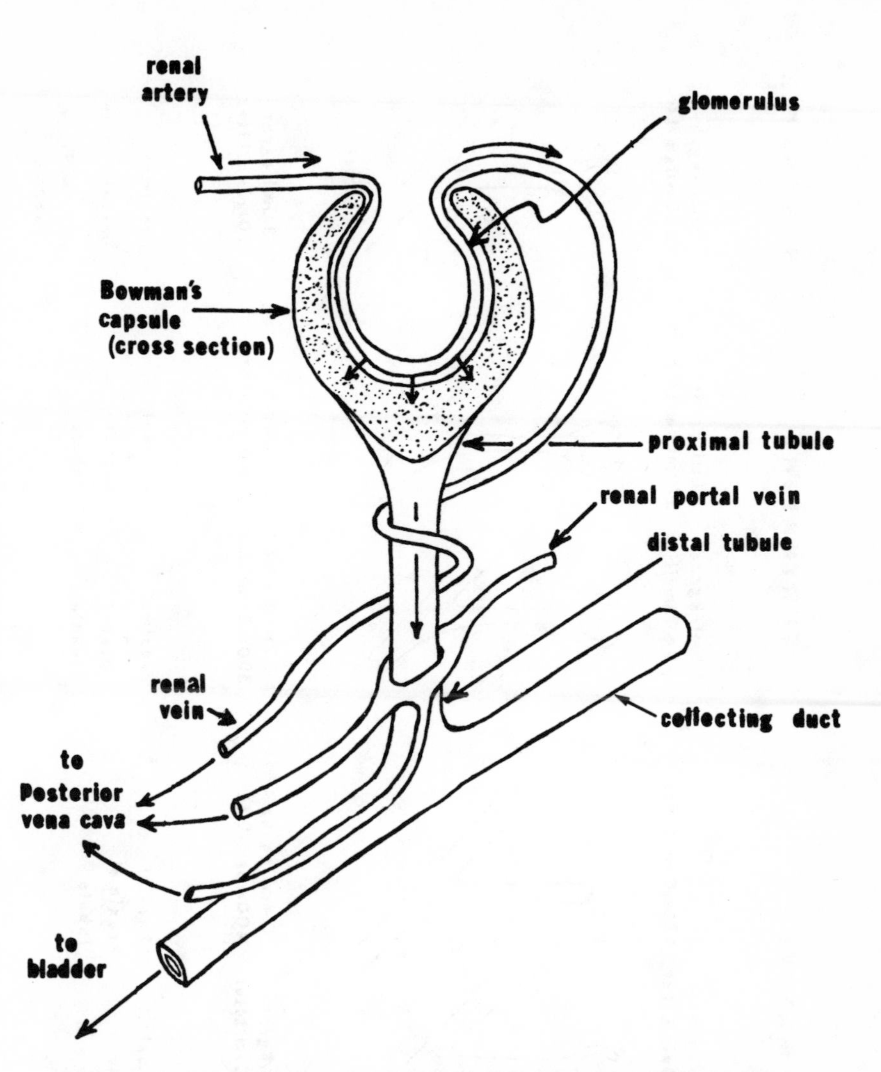

Fig. II-5. Diagram showing the functional elements of a typical teleost nephron (kidney unit). Arterial blood passes through a glomerulus where some of the plasma phosphorus, is filtered through the capillary walls into the Bowman's capsule and then into the proximal tubule. Both reabsorption and secretion by cells in the tubule modify the composition of the filtrate to determine the final composition of the urine which enters the collecting duct. Venous blood (renal vein, renal portal vein) carries away reabsorbed materials to prevent excessive concentration gradients from developing and thereby increasing the osmotic work required for producing urine. Note: The tubule has been greatly shortened for ease of representation and the detailed pattern of branching of the renal and renal portal veins is unknown.

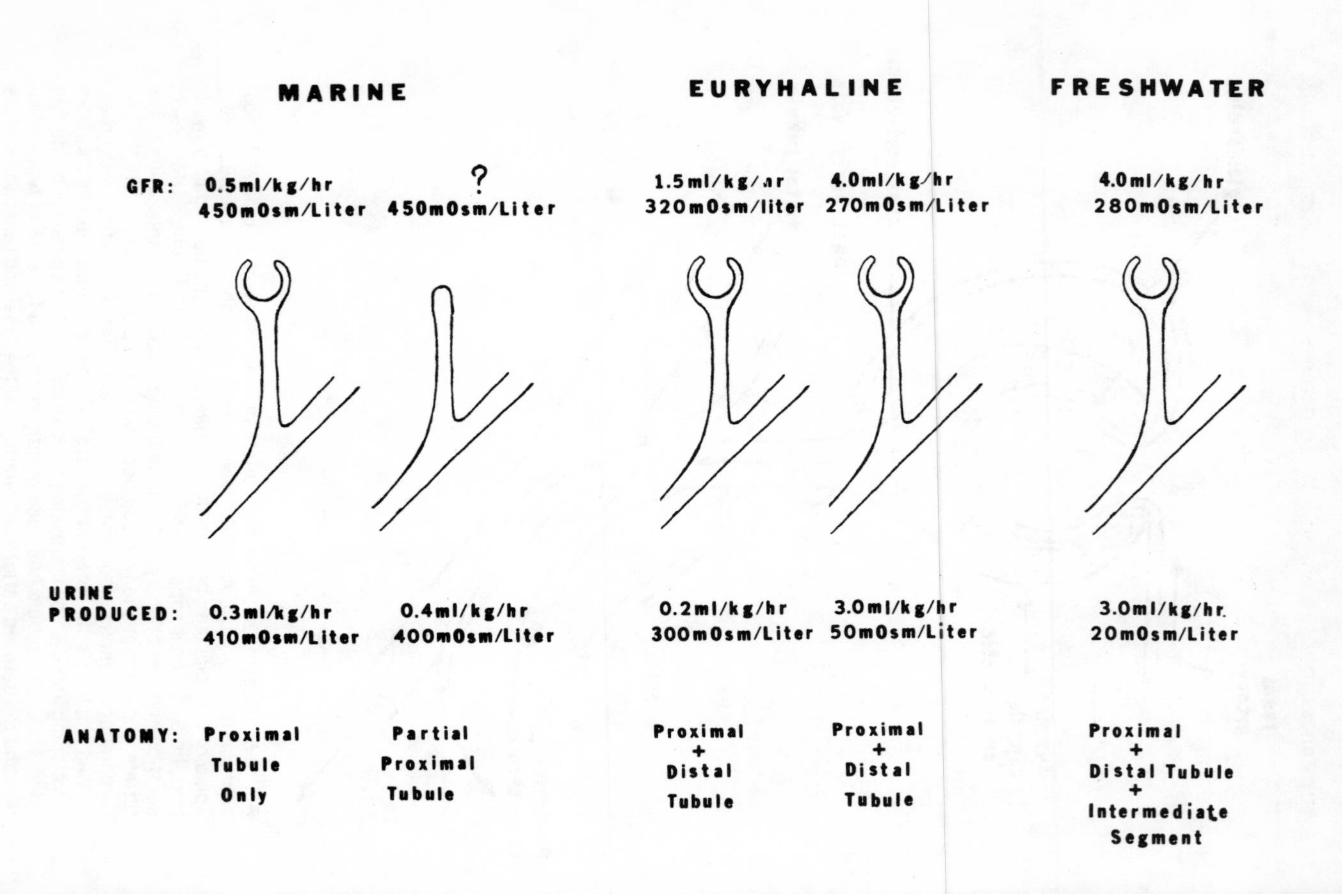
MARINE
EURYHALINE
FRESHWATER
GFR:
0.5ml/kg/hr
450mOsm/Liter
?
450mOsm/Liter
1.5ml/kg/hr
320mOsm/liter
4.0ml/kg/hr
270mOsm/Liter
4.0ml/kg/hr
280mOsm/Liter
URINE PRODUCED:
0.3ml/kg/hr
410mOsm/Liter
0.4ml/kg/hr
400mOsm/Liter
0.2ml/kg/hr
300mOsm/Liter
3.0ml/kg/hr
50mOsm/Liter
3.0ml/kg/hr.
20mOsm/Liter
ANATOMY:
Proximal Tubule Only
Partial Proximal Tubule
Proximal + Distal Tubule
Proximal + Distal Tubule
Proximal + Distal Tubule + Intermediate Segment

substances into the ultrafiltrate to determine the composition of the eventual urine. This occurs partly by active secretion and partly by diffusion. The vessel draining the glomerulus probably continues its course in the vicinity of the tubule before joining the posterior vena cava. The renal portal vein (from the caudal vein and probably others) also contributes to producing a large low pressure blood supply through the kidney. Both vessels carry away the tubular reabsorption products and minimize diffusion back into the tubule.

The basic idea of this system is a relatively fail-safe one in which everything smaller than proteins is filtered and will potentially be voided in the urine. Routinely useful substances are reabsorbed into the blood, usually with upper limitations above which any excess goes into the urine. This is simpler than having a secretory mechanism for every possible molecule which the fish might encounter and eventually get into its body. Further modification of the urine also takes place in the collecting ducts and urinary bladder. In the bladder of seawater fish, salts are actively absorbed and water follows osmotically, probably gaining water for the fish at less energy cost than moving seawater across the gut wall.

2. Comparison of blood and urine in freshwater and seawater fish.

A generalized picture of teleost kidney function appears in Fig. II-6 as a series of kidney tubules showing the quantity and osmolarity of blood and urine in various environments. For the moment, we will examine only the total number of particles in

Opposite page: Fig. II-6. Generalized representation of the typical input and output of kidneys of teleost fishes in freshwater and seawater. Note that euryhaline (including anadromous) fishes do not maintain as high a gradient between their blood and the environment as do the strictly freshwater and marine fishes. Note: Although the tubules are represented diagrammatically as the same in all cases, the actual tubule components listed below each unit show the differences in histological composition of each tubule. (Redrawn and simplified from Hickman and Trump, 1969, with permission of Academic Press.)

solution (osmolarity) and volume of fluid being processed by the kidney. In marine teleosts, the blood is slightly more concentrated than in freshwater fish and little plasma is filtered (although the quantity may be underestimated by present methods). The urine is also relatively concentrated, although still more dilute than the blood. The reason for the small amount of urine being produced is thus at least partly related to water conservation—i.e., the kidney is an osmoregulatory liability. The more active it becomes in seawater, the more water would be lost and would have to be replaced from some other source. In contrast, the freshwater kidney is just the opposite. While the blood is slightly more dilute than in the marine teleosts, there is nearly ten times as much of it filtered and the urine is mostly water—the function of the kidney is to conserve salt and excrete water.

In general, then, there are no marine fish which can steadily produce urine which is more concentrated than their blood plasma and no freshwater fish which can produce salt-free urine. Marine fish die from water loss when they are prevented from drinking. If you put a freshwater fish into distilled water and changed the water frequently, the fish's blood would slowly become more dilute until it died because of the continued loss of salts through the kidney. Thus one of the major steps in evolution of land vertebrates was to develop a kidney capable of conserving water sufficiently well to provide reasonable independence from a constant source of water. This problem was also solved by fish which spend considerable periods out of water—the walking catfish, the climbing perch and the African lungfish. There are a number of other physiological problems—respiration, drying and vision, for example—which also had to be solved in the transition to land, as well as osmoregulation.

Euryhaline fish, including the anadromous salmonids, possess a kidney which is intermediate between the strictly freshwater and strictly marine teleosts. The blood picture is similar to that already presented, although perhaps slightly more dilute in seawater than the marine fish, but the urine is not as concentrated in seawater as marine fish or as dilute in freshwater as in freshwater fish. The kidney of the euryhaline fish maintains less of a concentration gradient between blood and urine than either the strictly freshwater or marine types. The greater rate of loss of salts or water in the urine means greater quantities involved in active transport, but the lesser gradients between fish and environment involved could also mean less total effort to maintain them so the end result may not be too different bioenergetically than either the strictly marine or the strictly freshwater osmoregulation.

3. Glomerular and Aglomerular Kidney Tubules

Considering that glomerular ultrafiltration in marine teleosts represents a potential hazard for water loss, it is perhaps not surprising that a kidney should evolve which has no glomerulus. In an aglomerular kidney there is no ultrafiltration and all of the formation of urine is carried out by the tubules. A wide variety of marine fishes are aglomerular, including the pipefish, seahorses, midshipman, goosefish and at least some sculpins, puffers, clingfish, sea poachers, porcupinefish and toadfish. It was thought for some years that an aglomerular kidney would not allow an aglomerular fish to survive in freshwater. However, it is now known that there are two such exceptional species, both in the pipefish family. The details of how they accomplish this feat are scanty since the reports cover mostly the histological structure of the kidney tubules. There are also several known instances of kidneys which have mixtures of glomerular and aglomerular tubules in the same kidney.

The functional characteristics of a marine aglomerular kidney are shown in Fig. II-6 as being approximately comparable to that of the marine glomerular kidney except that it lacks any glomerular filtration rate and usually lacks a distal tubule. One hypothesis is that the upper portion of the tubule may secrete a hyperosmotic solution into the tubule so that water and other substances diffuse into the tubule. Much is yet to be learned about aglomerular kidney function, especially in freshwater.

A complete kidney tubule consists of a glomerulus, neck, proximal segment with up to three subdivisions, intermediate segment, distal segment, collecting tubule and collecting duct, all of which can transport dissolved substances. The intermediate segment occurs only in freshwater fish and in land vertebrates, where it is credited with allowing land vertebrates to retain water and produce a urine which is hyperosmotic to blood. Marine teleosts lack a distal tubule, and aglomerular marine teleosts lack the glomerulus, neck and upper proximal tubule as well. In freshwater glomerular kidneys, virtually all portions of the tubule reabsorb dissolved minerals—ions, glucose, etc.—but the distal tubule is where urine becomes dilute via absorption of salt without water. Only creatine and some organic acids are secreted by the proximal tubule into the urine. In seawater, an important function of the proximal tubule is excretion of Mg^{++} and other divalent ions. In euryhaline and anadromous fish, the direction of transport of divalent ions and waters reverses as the fish moves between freshwater and marine environments. Readers desiring further details can consult Hickman and Trump, 1969.

4. Bladder Function

In most teleosts there is some kind of enlargement of the urinary ducts used for storing urine and therefore serving as a bladder. In the catfish family Ictaluridae the bladder is a distinct pouch off to one side of the urinary duct, while in the salmonids it is only a wide space in the urinary duct. Regardless of the anatomical confusion over the embryological origin and exact naming of these urogenital ducts and pouches, this text will call any structure which stores urine a bladder.

The urinary bladder appears to serve as an osmoregulatory organ as well as a storage organ. There is a transepithelial potential (TEP) between the inside and outside of the bladder. Recent work on the urinary bladder of marine fish showed that NaCl was absorbed by active transport and water accompanied it. Saving this water and excreting the salt across the gills was considered more energy efficient than absorbing the same amount of water and salt from the gut.

To mention a behavioral rather than a physiological matter, I found that the urine of juvenile (20-30 cm) salmonids in seawater had a pungent, somewhat oily odor. Having previously wondered why a fish would need to store urine at all, the thought occurred that such a penetrating odor could be readily detected by a predator and easily followed if there were a constant emission of urine. I never had the opportunity to test whether a predator could follow a urine trail or not or whether these fish actually produced an odor trail.

5. Diuresis: Causes and Consequences

Diuresis refers to any excessive production of urine. In humans, caffeine and alcohol both produce diuresis by influencing the hormone secretion which regulates the water reabsorbing ability of the collecting tubules. While such a mechanism seems not to have been looked for in fish, fish are clearly subject to periods of diuresis, some fish being affected more severely than others. In our laboratory, adult coho salmon in freshwater produced increased amounts of urine for two to three hours following anaesthesia and surgical installation of dorsal aortic and urinary catheters. This period corresponded to the period of elevated respiration while the oxygen debt incurred during surgery was repaid. Homer Smith and his students, who pioneered in the study of teleost kidney physiology from the mid-1920's to the mid-1950's, found an instance of diuresis in the goosefish *(Lophius piscatorius)*. When captured and injured, even in a minor way such as loss of a few scales, urine production increased to the point where the fish died in as little as 24 hours. In this marine fish, the urine both increased in volume and became more dilute, leading to severe dehydration. Diuresis in freshwater (as in the adult coho) was not so obvious because urine production was already high, but the deleterious factor was the increased salt loss. Homer Smith named this phenomenon laboratory diuresis because of his continued difficulties in dealing with it in doing laboratory experiments on the goosefish. The term laboratory diuresis is not commonly heard today because there is little work presently being done on the goosefish and no other species examined so far seems to have such a severe diuretic response to stress.

A case of the milder form of diuresis in rainbow trout appears in Table II-1A. The diuretic stimulus was a 30-minute period of hypoxia, during and after which the fish maximized its respiratory exchange at the expense of taking on a considerable load of

A

Chemical constituent	Control	Post-hypoxic stress		
	4 hours	1 hour	4 hours	20 hours
Urine flow ml/kg/day	91.9 (19)[2] ± 24.7 [1]	129.9 (17) ± 13.6	75.8 (17) ± 24.3	89.5 (15) ± 18.1
pH	7.08 (13) ± 0.24	6.82 (10) ± 0.19	6.55 (10) ± 0.21	7.14 (8) ± 0.16
Total CO_2 vol. %	9.3 (13) ± 4.6	12.6 (10) ± 5.6	5.4 (10) ± 1.6	13.1 (8) ± 4.0
Lactic acid mg/100 ml	0.8 (10) ± 0.9	19.4 (7) ± 13.6	44.4 (8) ± 52.2	— ( ) ±
Na mg/l	223.4 (18) ± 109.3	593.1 (15) ± 138.8	346.4 (16) ± 306.6	208.7 (15) ± 103.5
K mg/l	67.3 (18) ± 17.4	137.9 (15) ± 34.5	94.2 (16) ± 22.8	64.3 (15) ± 17.6
Ca mg/l	44.6 (18) ± 16.4	38.9 (15) ± 13.6	53.2 (16) ± 10.4	55.2 (15) ± 16.6
Mg mg/l	59.8 (18) ± 25.5	77.6 (15) ± 22.8	75.4 (16) ± 22.2	43.4 (15) ± 19.2
Cl meq/l	9.0 (18) ± 5.2	25.3 (13) ± 17.2	15.7 (13) ± 10.7	8.5 (15) ± 3.5
PO_4 mM/l	11.1 (18) ± 11.7	21.3 (16) ± 9.5	26.2 (17) ± 19.7	5.0 (15) ± 5.7

[1]Standard deviation.
[2]Number of samples analyzed is in parentheses.

Table II-1. Two examples of diuresis in rainbow trout. A. Quantity and chemical composition of the urine of rainbow trout subjected to a 30 min. period of hypoxia at 12°C. Numbers in parentheses indicate the number of fish sampled. Note that the units for several of the ions are mg/l rather than the more commonly used m Eq/l. (From Hunn, 1969, with permission of author and publisher.)

Table II-1 (cont.). B. Effect of short-term vs. long-term hypoxia (3.0 or 4.5 mgO_2/liter) on hematocrit and urine formation rate (UFR) of rainbow trout. (From Swift and Lloyd, 1974, with permission from *J. Fish Biol.* 6: 379-387. Copyright by Academic Press, Inc. [London] Ltd.)

B

Expt No.	Duration of hypoxia (h)	Temp. (°C)	Mean UFR (ml/kg/h ± S.D.) Pre-test	Test	Post-test	No. of fish	Haematocrit (%) Control	Test
	Short term							
1	3	13	2.1 ± 0.8	4.7 ± 1.4	—	8	32 (8)*	46 (8)
2	4	12	2.6 ± 0.9	6.0 ± 1.9	—	9	—	—
3	4	13	2.9 ± 0.7	5.1 ± 1.2	—	5	40 (8)	54 (10)
4	4	16	3.3 ± 0.6	5.5 ± 0.9	—	8	— —	— —
5	5	13-5	2.6 ± 0.6	4.5 ± 0.9	1.9 ± 0.3† 2.6 ± 0.5	8	39 (19)	49 (20)
	Long term							
6	24	17	2.6 ± 0.5	5.3 ± 0.8‡ 3.3 ± 1.3	3.3 ± 1.3	6	—	—
7	24	17	2.7 ± 1.1	6.2 ± 2.0‡ 3.2 ± 1.2	3.2 ± 1.2	7	—	—
8	24	17	—	—	—	—	33 (8)	40 (6)
9	24	18	—	—	—	—	33 (4)	39 (7)

*Number of fish used to determine haematocrit.
†Immediate post-stress value (up to 10 h).
‡Initial test rate.

water. Not only did the urine flow and salt content increase somewhat, but the total salt loss (urine flow x salt content) was even more significant. However, such losses are still not normally lethal, as they were in the goosefish. In Pacific salmon following 30 minutes of anaesthesia and surgery, the resulting diuresis lasted only for about two hours.

6. Change of Kidney Function in Anadromous and Euryhaline Fishes

Anadromous and euryhaline fishes are hypoosmotic to seawater and hyperosmotic to freshwater. Their total osmotic picture must change from one of excreting water and conserving salts in freshwater to excreting salts and conserving water in a marine environment. In an estuarine environment having approximately the same salt concentration as in the fish's blood plasma, all osmoregulatory organs reduce their activity and urine production may cease. While it has been suggested that a transitional period in estuarine waters was needed or at least desirable for migratory salmonids, there seemed to be no distress in coho salmon smolts transferred directly into seawater. Rainbow trout can be adapted to seawater only gradually over several days.

In contrast to the complete change in direction of gill osmoregulation, the change of kidney function during the transfer between freshwater and marine environments is primarily one of quantity and not direction. The major changes in kidney function of adult coho salmon as they enter freshwater are shown in Table II-2. Urine production increased dramatically (60-fold in this fish, but 8- to 12-fold is more common) as a result of increased glomerular filtration rate (GFR) and reduced reabsorption of water. Note that while the *concentration* of all the ions in the urine decreased, the *rate* of excretion of several of them actually was greater in freshwater than in seawater because of the great increase in quantity of urine being produced. A most important function in urine production in seawater is the excretion of Mg^{++}. The relatively large amounts of Mg^{++} ion in the urine decrease rapidly upon entry into freshwater mostly because there is little Mg^{++} in most freshwater systems, although some excretion usually continues throughout the freshwater residence. The

Parameter	Salt water Mean	Salt water S.D.	Salt water *n*	Fresh water Mean	Fresh water S.D.	Fresh water *n*
Flow rates (ml/kg × hr)]						
Urine rate	0.487	0.216	13	1.81	1.14	37
Glomerular filtration rate	1.05	0.443	9	2.57	1.05	25
Urine ion concentrations (m-equiv/l.)						
Sodium	23.6	7.55	13	16.5	7.37	25
Potassium	1.62	0.115	13	1.47	0.517	25
Calcium	5.71	0.911	13	3.96	3.12	25
Magnesium	208	19.1	13	70.5	97.6	25
Chloride	177	11.4	13	37.8	55.5	36
Plasma ion concentrations (m-equiv/l.)						
Sodium	181	2.81	11	154	4.62	34
Potassium	2.70	0.155	11	2.36	0.375	34
Calcium	1.53	0.281	11	1.53	0.273	34
Magnesium	1.27	0.179	11	0.944	0.144	34
Chloride	155	1.95	11	134	5.06	34
Excretion rates [μ-equiv/(kg × hr)]						
Sodium	11.0	4.34	13	22.1	20.3	25
Potassium	0.791	0.370	13	1.75	1.16	25
Calcium	2.72	1.07	13	3.10	0.967	25
Magnesium	100	39.4	13	35.9	50.9	25
Chloride	85.6	35.3	13	33.8	29.8	36
Filtered ion load [μ-equiv/kg × hr)]						
Sodium	162	70.4	9	336	132	25
Potassium	2.54	1.09	9	5.24	1.71	25
Calcium	0.466	0.191	9	1.16	0.448	25
Magnesium	0.944	0.286	9	1.72	0.787	25
Chloride	158	68.2	9	330	130	25

Table II-2. Comparisons of kidney functions for one adult coho salmon in seawater and freshwater. (From Miles, 1971, with permission of publisher.)

Mg^{++} data in Table II-2 is still much higher than typical for freshwater fishes. There has been some suggestion that the salmonid kidney is functionally aglomerular in seawater. It can be seen that there definitely is a GFR in seawater, although there is no way to know from these data whether this is a combination of a few normal glomeruli and many non-filtering glomeruli or whether it represents a reduced GFR for all the glomeruli.

Equivalent detail is not available for smolting salmon, but the process appears to be the reverse of that in the adult. Upon entering seawater, the GFR and urine flow decrease and Mg^{++} content of the urine increases. The transition in both directions for anguillid eels is also similar although the exact rates and concentrations may differ somewhat. A number of other fish make similar migrations, sometimes several times a year. These include cutthroat trout, sticklebacks, steelhead trout and a small Japanese salmonid, the ayu. Little information is available on kidney function in these fish.

Some euryhaline species which enter salt springs or inhabit evaporation ponds experience even greater changes in environmental salinity than anadromous fish and have correspondingly drastic changes in kidney (and gill) function. The killifish *Fundulus kansae* lives in salt springs up to 40°C. (104°F.), but still inhabits freshwater. Upon entering hypertonic saline waters, the GFR decreased from about 600 to about 40 ml/kg/day, urine flow decreased from about 250 to 10 ml/kg/day and the urine osmolarity rose until it was slightly greater than that of the blood serum. The latter ability is rare among fish, being found mostly in reptiles, birds and mammals. The fish urine is still hypo-osmotic to the environment so there is no net gain of water by the kidney, only less of a loss.

D. INTESTINAL FUNCTION IN OSMOREGULATION

Drinking seawater is the major source of water by which marine teleosts replace the water lost by diffusion across the gills and through the kidney. Possible losses through the skin, if any, are difficult to assess. Drinking also seems to occur in micro amounts in some freshwater fish, although the quantities must almost be totally insignificant in comparison to the diffusional influx across the gills and in some species (e.g., eels) across the skin. Further, the markers (radioisotopes and large, inert organic molecules such as polyethylene glycol) which are assumed not to permeate the gill membranes may do so to some small degree and thus give false indications of drinking in freshwater. Some water may also be ingested with food. Other than to mention this problem of measuring the drinking of freshwater, the remaining discussion will be devoted to the drinking of seawater.

1. Location of Intestinal Osmoregulation

Once having determined that seawater enters one end of the intestine and that a fairly concentrated solution of divalent ions leaves the other end, one soon wonders where this alteration of composition could occur without interfering with digestive processes. Seawater is alkaline—pH 8-8.5—while an active salmonid stomach is acid at about a pH of 3, and the activity of the pepsin enzyme ceases above a pH of about 4.5. Digestion and osmoregulatory processes are clearly contradictory as far as the stomach is concerned. While some possibility of alternation of digestion and osmoregulation might seem possible, my own (unpublished) observations on juvenile marine salmon with an abundant supply of amphipods show they never had an empty stomach. It seems much more likely that the nearness of the esophageal and intestinal openings in salmon stomachs is, in part, a way of getting seawater through the stomach with minimal contact to the food being digested in the pouch-like posterior portion of the stomach. By default, this places intestinal osmoregulation somewhere in the intestinal tract posterior to the stomach. Eels, on the other hand, may differ from salmon. Recent work on eels suggests that their long esophagus is permeable to monovalent ions but not to water, thus accounting for the apparent dilution of ingested seawater seen in the stomachs of several species of marine fish.

Working in various parts of Puget Sound, we saw several adult coho and chinook salmon which had apparently stopped feeding while still in hyperosmotic saltwater. Their stomachs were empty and shrunken to about a third of normal while the upper part of their intestines was barely 5-6 mm in diameter. The posterior of their intestine appeared normal—about 20 mm in diameter and with an abundant blood supply. The change from atrophied to normal status was abrupt—only a few millimeters of length for a 3-fold change in diameter. A University of Washington M.S. thesis project confirmed a similar configuration in estuarine adult steelhead trout. Since these fish were not feeding but necessarily still osmoregulating in the marine fashion, it appeared that the only likely activity in the posterior third of the intestine was osmoregulation.

2. Stimulus For Drinking

Much of the research on osmoregulatory drinking has been done in Japan on anguillid eels. Normal drinking rates are 20-40 µl/100g/hr in freshwater and more than 10 ml/100g/hr in seawater. Upon withdrawing about 30% of the blood volume, a freshwater eel drank at a rate of 0.5-1.0 ml/100g/hr for 15 hours and finally returned to a normal drinking rate after about 40 hours. Intravenous infusion of 0.9% NaCl (approximately isosmotic with blood) interrupted the drinking of freshwater in this eel for only a few minutes. Conversely, intravenous infusion of a large amount of 2% NaCl (hypertonic to blood) into a seawater eel stopped drinking for several hours. Larger volumes of hypertonic saline stopped drinking for longer times. Eels put into hypertonic sucrose or mannitol solutions lost 12% of their body weight in 12-16 hours and died from dehydration, but drank at less than 5% of the seawater drinking rate. The conclusion from this and other studies was that drinking was controlled by the presence of Cl^-. It was also shown in eels that distending the stomach with mannitol solution would slow down but not stop the drinking rate of seawater eels.

In seawater-adapted rainbow trout drinking rates were proportional to the external salinity. Drinking rates in one-third-, half- and full-strength seawater were 47, 95 and 129 ml/kg body wt/day, respectively. Using polyethylene glycol as a marker, the investigators showed that freshwater rainbow trout did not drink but could be stimulated to do so by adding sucrose (1 g/l) to their environment. In this dilute medium drinking was associated with an average weight gain of 9% of body weight. In one-third-strength seawater, which is almost isosmotic with the plasma, there would have been little diffusion across the gills. Thus, the drinking rate was a means to replace the water lost in the urine.

3. Ionic Uptake and Regulation by the Intestine, Production of Rectal Fluid

In rainbow trout adapted to increasing salinities, the osmolarity of the intestinal fluid remained constant, but the ion composition and fluid volume changed considerably (Table II-3). Na^+ and K^+ almost disappeared from the intestinal fluid, presumably having been increasingly transported into the blood. Mg^{++} and SO_4^{--} became more than twice as concentrated as in seawater, probably through a combination of the intestinal wall becoming nearly impermeable to these ions and the active transport of monovalent ions from the intestine into the blood. Part of the apparent concentration of the divalent ions came from the absorption of water by the intestine and reduction of the remaining water volume containing the divalent ions. The consistently high concentrations of Cl^- in the intestinal fluid were considered normal and have also been noted in eels. In one experiment 60-80% of the water was absorbed, leaving the divalent ions concentrated in the remaining 20-40%. The drinking rate associated with these changes was described in the previous paragraph.

The composition of intestinal fluid at various locations along the intestine appears in Fig. II-7. The location of the ion regulation is not obvious from these data. While the posterior and terminal portions of the intestine obviously produced major differences in the intestinal fluid as compared to that in the anterior intestine, it is also obvious that fluid in the anterior intestine was already rather different from seawater. It has been hypothesized by several authors that ingested seawater equilibrates with plasma in the upper intestine rather quickly and then active transport of ions takes place under isosmotic conditions. This is in contrast to uptake of hyperosmotic fluids from the intestine in eels (Skadhauge, 1969). A later explanation involved active secretion of salt (largely Cl^-) into intercellular spaces. Water then flows from the gut lumen into these intercellular channels along osmotic gradients and into the body fluid spaces at the basal ends of the intercellular channels.

Medium	Intestinal fluid (mM/l)						Osmolarity (mOsm/l)	pH
	Na^+	K^+	Ca^{2+}	Mg^{2+}	SO_4^{2-}	Cl^-		
FW*	170 ± 3 (4)†	4·0 ± 0·3 (4)	2·1 ± 0·1 (4)	0	0	70 ± 5 (4)	300 ± 3 (4)	8·5 ± 1·0 (4)
⅓ SW	140 ± 6 (11)	4·0 ± 0·4 (11)	3·0 ± 0·3 (11)	50 ± 4 (11)	30 ± 4 (8)	80 ± 5 (9)	310 ± 6 (10)	8·4 ± 0·1 (6)
½ SW	80 ± 7 (13)	3·0 ± 0·2 (13)	3·0 ± 0·2 (11)	90 ± 5 (11)	60 ± 6 (11)	70 ± 3 (13)	320 ± 6 (13)	8·4 ± 0·1 (7)
SW	20 ± 3 (4)	1·0 ± 0·1 (4)	2·2 ± 0·1 (4)	120 ± 2 (4)	110 ± 1 (4)	50 ± 2 (4)	300 ± 9 (4)	8·1 ± 0·1 (4)
	External media (mM/l)							
⅓ SW	146	3	3	17	10	170	330	7·8
½ SW	220	5	5	25	15	267	480	7·8
SW	450	10	9	50	30	530	960	8·1

* Sucrose added to the medium (1 g/l).

† Number in parentheses denotes sample size.

Table II-3. Composition of intestinal (rectal) fluid collected by catheter from rainbow trout *(Salmo gairdneri)* in fresh water and several strengths of sea water. The importance of the intestine in marine osmoregulation can be seen in the increasing concentrations of MG^{++} and SO_4^{--} and the decreasing amounts of monovalent ions appearing in rectal fluid with increasing salinity. (From Shehadah and Gordon, 1969, with permission of author and publisher.)

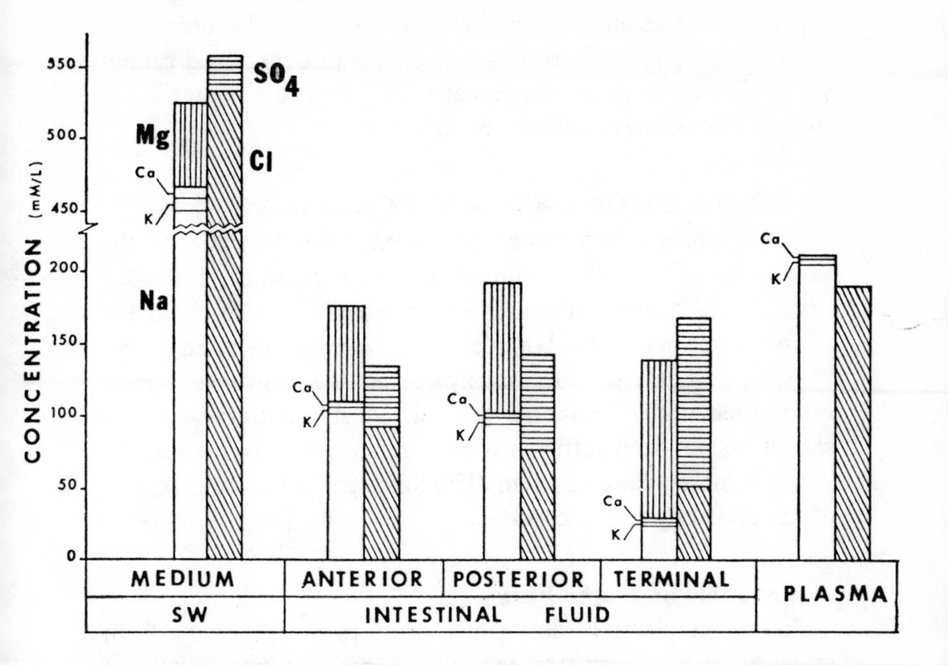

Fig. II-7. The osmoregulatory role of the intestine of rainbow trout showing changes in the ionic composition of ingested seawater as it passes through the intestine. Most of the NA^{+} and Cl^{-} are absorbed into the blood, with MG^{++} and SO_4^{--} accumulating in the intestine, probably through absorption of water. (From Shehadah and Gordon, 1967, with permission of M.S. Gordon and the publisher.)

Upon measuring the rate of ingestion of various ions and comparing this to the rate of excretion to balance the ion "budget," Hickman (1968c) found some discrepancies (also see the flounder data in the next section of this chapter). One of the larger discrepancies was for Ca^{++}. In the rainbow trout data above, most of the missing Ca^{++} was found as a $CaCO_3$ precipitate in the mucous tube which surrounded the fecal material. This appeared to be the route for excretion of about two thirds of the Ca^{++} ingested in seawater. Ca^{++} precipitate has also been seen in the urine of coho salmon, and a high content of Ca^{++} is found in the slime produced by hagfish. While it is acknowledged that calcium plays an important role in the osmotic integrity of membranes and is essential in electrical transmission by nerve and muscle membranes, these kinds of details just discussed cannot yet be fitted into those functional roles. See also Chapter X for further discussion of calcium metabolism.

E. INTEGRATION OF OSMOREGULATION

The osmoregulatory system is not an organ system with the same sense of obvious continuity that the nervous system is, for example. It is more of a happenstance collection of all of the various semi-permeable boundaries between the fish and its environment. These sites necessarily operate as a coordinated team, most often under hormonal control. Most osmoregulatory research has not been sufficiently comprehensive to encompass the whole osmoregulatory system. The following are some examples of the relatively few such cases.

1. Models of Water Flux

One of the pieces of data which are commonly missing from many osmoregulatory studies is the change in water content of fish. This is partly because water is so omnipresent and so readily movable that it is difficult to isolate which package (compartment) of water is being observed. On the other hand, water content of fish can be followed in terms of changes in body weight, with suitable controls, by assuming that short-term changes in body weight are mostly water—i.e., that changes in the ion content of the fish are insignificant. Wet weights can be compared to dry weights at the end of experiments, too, although such determinations are obviously limited to one time for any given fish.

A model for changes in water flux during entry into seawater for coho smolts is shown in Fig. II-8. This model is based on changes in ion concentrations and body weight. In freshwater, the urine output is approximately equal to the diffusional influx of water, mostly across the gills, but possibly may include other surfaces such as skin and intestine. Upon transfer into seawater, the water flux across the gills and other external surfaces reverses. The kidney, of course, is unable to reverse its water flux by producing hyperosmotic urine, so the best it can do is to

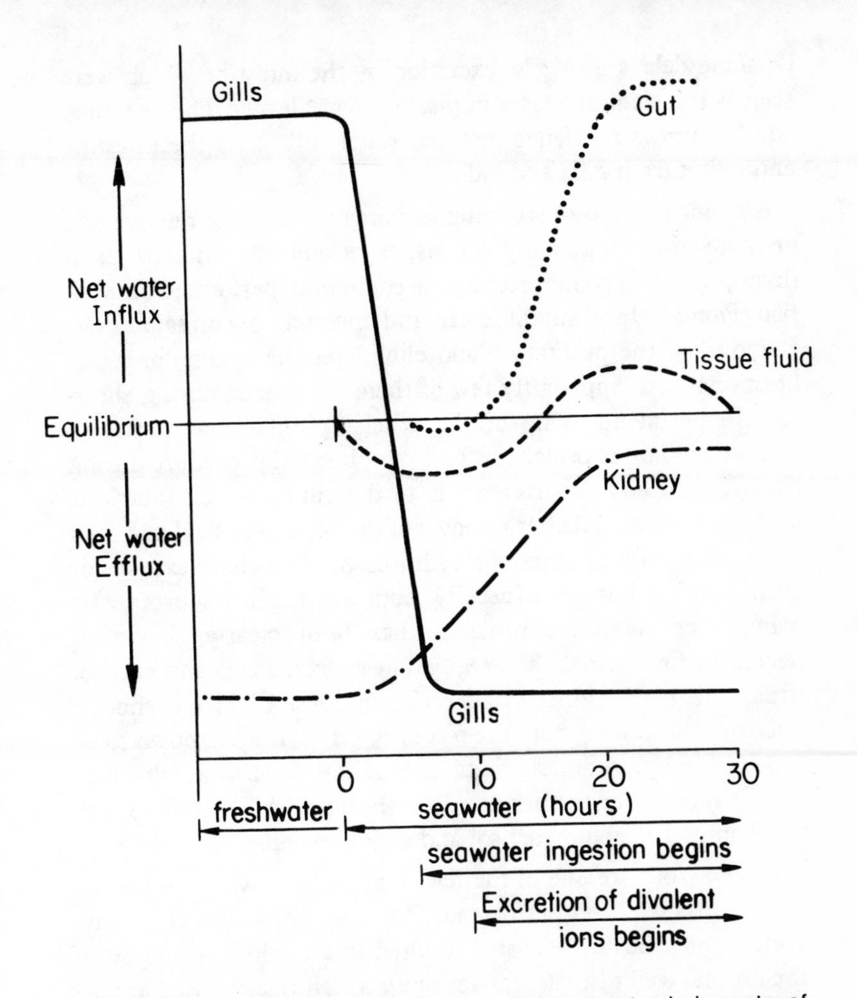

Fig. II-8. Theoretical changes in movement of plasma water during entry of coho smolts into seawater. (Redrawn by TFH from Miles and Smith, 1968, with permission of the journal.)

minimize its water loss. Thus there is a net water loss for at least the first 18 hours in seawater. The immediate water loss comes from the blood, but blood rapidly equilibrates with tissue fluid so that minimal changes in concentration of blood ions occur. Continued dehydration leads to initiation of the drinking of seawater (shown here as delayed after entry into seawater, but experimentally shown in eels to begin immediately). Drinking seawater eventually replaces the lost tissue fluid, although there appear to be some delays in Mg^{++} excretion by the intestine which were seen as transient increases in plasma Mg^{++} levels. The new final equilibrium is a balance between water influx via the intestine and efflux from gills and kidney.

A major method of studying osmoregulation is to remove one or more of the regulatory organs, especially the pituitary gland (hypophysis). Hypophysectomy is commonly performed on killifish *(Fundulus)* and anguillid eels and uncommonly on salmonids. Removal of the pituitary gland eliminates the source for many hormones, but apparently few of them are crucial during short-term experiments. One of the crucial pituitary hormones in osmoregulation is prolactin. The experimental procedure during hypophysectomy experiments is to determine which functions decrease or even fail after removal of the pituitary gland and then try to restore those functions by injection of pituitary extracts or even purified hormone (usually from mammalian sources). By using such methods, prolactin has been clearly shown as necessary for normal osmoregulation in freshwater and cortisol (from the head kidney, under control of ACTH (adrenocorticotropis hormone) from the pituitary gland) is also known to be involved in hydromineral regulation in both fresh- and saltwater. See Chapter X for further details of the endocrine system as well as Chapter XII about cortisol and stress responses.

Brown trout are one of the few salmonids on which hypophysectomies have been performed (Oduleye, 1975 a,b). Hypophysectomy of trout in seawater resulted in a 25-30% reduction of the normal water flux (turnover rate), which was restored to normal by prolactin injections. While the mechanism of this effect was unproved, a common explanation was that the permeability of the gills to water decreased. In eels and goldfish in seawater, hypophysectomy also is accompanied by reduced drinking and urine flow rates. There are conflicting results as to whether the kidney responds to prolactin or is controlled separately and simply follows the reduced water flux across the gills. Separate control of the kidney is also possible through the caudal neurosecretory system (urophysis).

Cortisol appears to produce a wider range of osmoregulatory effects than prolactin (which is already broad), which may relate more to ion regulation especially in seawater, than to water regulation in freshwater. In most cases prolactin and cortisol have similar or related effects. Calcium, on the other hand, is antagonistic to both hormones. Calcium in the water somehow decreases the permeability of cell membranes and improves the watertight integrity of the cement between cells (which is where considerable osmotic ultrafiltration and diffusion are thought to occur). In brown trout in freshwater, flux, drinking rate and urine flow all decreased markedly when fish were transferred from tap water (0.3 mM Ca^{++}) to tap water with added calcium (10 mM/liter). In experiments in Florida with several species of high-seas fish which would normally be killed by anything more than a 10% dilution of seawater, they survived up to a 90% dilution of seawater when crushed limestone ($CaCO_3$) was added to their aquarium water, presumably because the calcium reduced the loss of ions, the permeability and the influx of water. In the general control of permeability to water, calcium seems to be antagonistic to both cortisol and prolactin.

2. Ion Budget for Flounder

One of the more complete studies of osmoregulation was for the southern flounder of the U.S. Atlantic coast. This flounder makes seasonal migrations, being offshore in winter and inshore in summer. It sometimes comes into completely freshwater. A typical pattern of kidney function during the year is shown in Fig. II-9. The amounts and pathways of the major ions in the seawater which is swallowed are shown in Table II-4. The division between monovalent and divalent ions is clear. Monovalent ions were absorbed from the gut and excreted primarily by the gills, while divalent ions remained mostly in the rectal fluid. Any divalent ions that were absorbed from the gut were excreted by the kidney. The major controlling factor determining both urine volume and composition was the glomerular filtration rate (GFR). Although rather variable, these urine data could be grouped into higher and lower flow groups (Fig. II-10). It appeared that the kidney tubules processed the ultrafiltrate from the glomerulus in a variable fashion according to the GFR. Magnesium ion was an exception to this (Fig. II-11). Since the GFR increased only slightly during the infusion of Mg^{++} into

	Swallowed	Rectal excretion		Absorbed		Renal excretion		Extrarenal excretion	
	µmoles/h	µmoles/h	% of intake	µmoles/h	% of intake	µmoles/h	% of absorption	µmoles/h	% of absorption
Sodium	1956.1	23.31	1.2	1932.8	98.8	2.69	0.13	1930.1	99.87
Chloride	2281.6	140.30	6.1	2141.3	93.9	22.68	1.05	2118.6	98.95
Potassium	41.46	0.82	2.0	40.64	98.0	0.29	0.71	40.35	99.29
Calcium	42.63	13.43	31.5	29.20	68.5	3.33	11.40	25.87	88.60
Magnesium	226.30	200.00	84.5	26.23	15.5	26.23	100.00	—	—
Sulfate	128.24	116.99	88.7	11.24	11.3	11.24	100.00	—	—
	ml/h	ml/h		ml/h		ml/h		ml/h	
Water	4.57	1.11	24.2	3.46	75.8	0.179	5.2	3.28	94.8

NOTE: The quantities of electrolytes given under "extrarenal excretion" pertain only to ions absorbed from the gut. All values per 1 kg body weight.

Table II-4. Routes of absorption and excretion of the major ions in sea water swallowed by the southern flounder, *Paralichthys lethostigma.* The quantities of ions given under "extrarenal excretion" pertain only to ions absorbed from the gut. All values are per 1 kg of weight. (From Hickman, 1968c, with permission of author and National Research Council of Canada.)

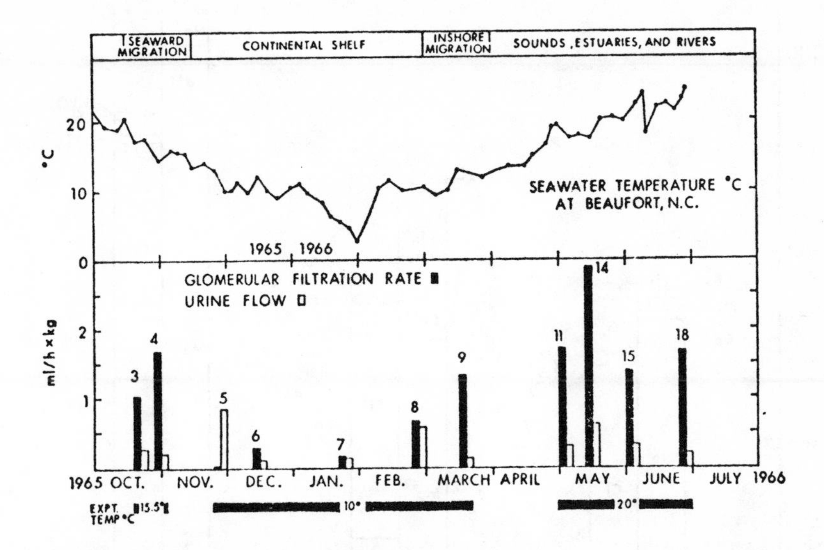

Fig. II-9. Average glomerular filtration rate (black bars) and urine flow rates (white bars) of 11 southern flounder studies at different times of the year. The upper graph shows the seawater temperature in the holding pen for the flounder. Experimental temperature appears at the bottom. The top line shows the time of migration and typical habitat of adult southern flounder during the year. (From Hickman, 1968a, with permission of the author and the National Research Council of Canada.)

the plasma while urine Mg^{++} increased greatly, the tubules must have actively secreted the magnesium from the blood into the urine. One of the insights gained from experiments such as this is the vital importance of maintaining low levels of Mg^{++} internally. The consequences of failing to do so were seen as one of the lethal effects of scale loss discussed in the chapter on applied physiology. It should also be pointed out that there are many marine animals whose internal Mg^{++} levels are similar to that of seawater. These are mostly invertebrates, but also include the primitive hagfish.

One of the more interesting aspects of the flounder study concerned Ca^{++}. Upon completing the analysis of the various fluid flows and ion concentrations, almost everything could be completely accounted for except calcium, of which about one third

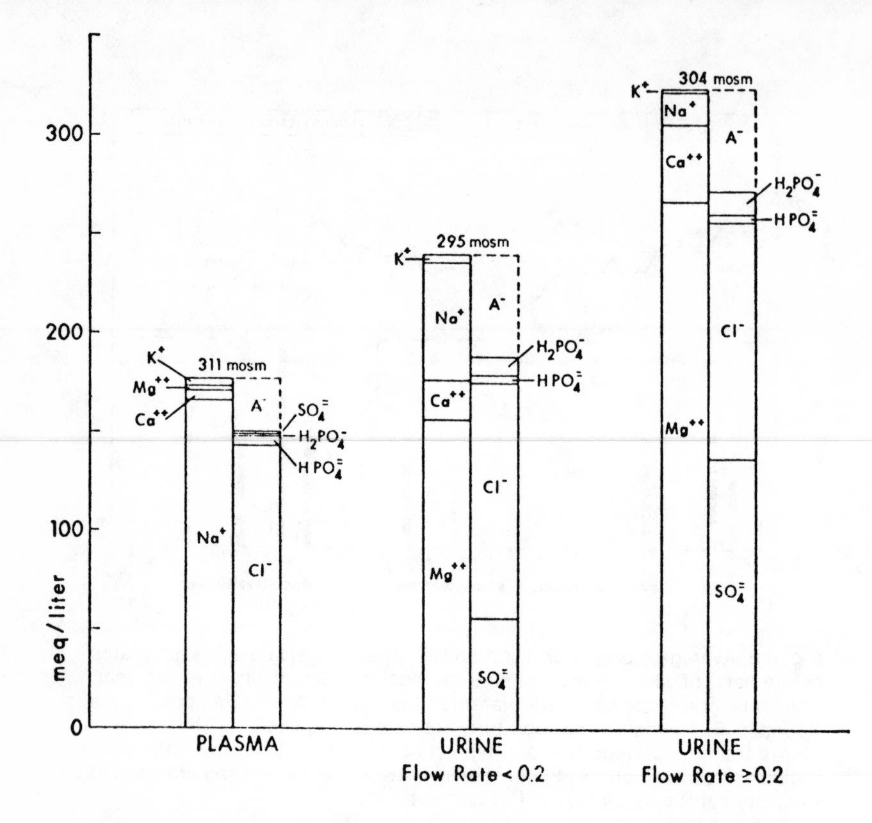

Fig. II-10. Comparison of the ionic composition of plasma and urine in the southern flounder. The composition of the plasma was stable throughout the year, but the urine composition differed between the low flow rates (winter) and high flow rates (summer) shown in Fig. II-9. (From Hickman, 1968b, with permission of author and National Research Council of Canada.)

disappeared. This was even after taking into account that major portions of the calcium were excreted in solid form in both urine and rectal fluid. There was also some question as to why the calcium remained as a fine powdery precipitate and was never seen to form calculi—the equivalent of human kidney stones. An obvious hypothesis for the disappearance of the calcium was that it was deposited in the skeleton. However, even skeletal deposition has its limits, especially in older fish whose growth rate is slow, and it was the older fish that were being used as experimental animals. The lost calcium still had not been found as of 1978.

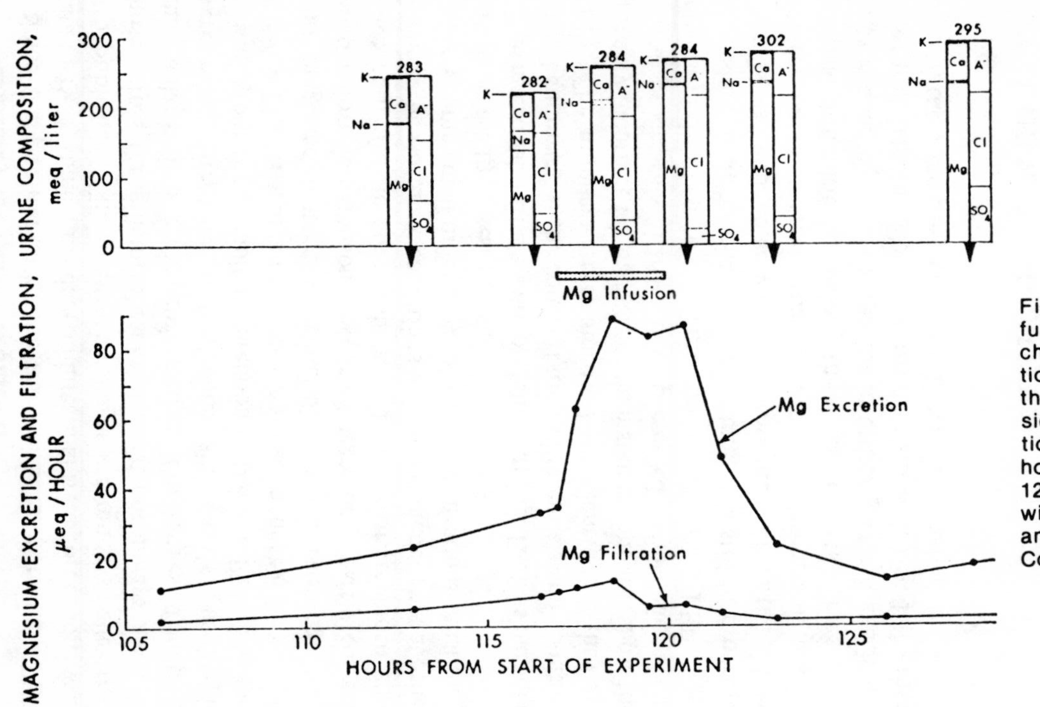

Fig. II-11. The effect of infusion of magnesium chloride on urine composition and excretion rates in the southern flounder. Infusion of a 224 m Eq/l solution of $MgCl_2$ began at hour 117 and ended at hour 120. (From Hickman, 1968b, with permission of author and National Research Council of Canada.)

3. Summary for Osmoregulation

Since the publication of the transepithelial potential (TEP) data in 1974, some TEP data on intestinal transport of ions were also published. The intestinal voltages are much smaller than those across the gill membranes and did not always have the same polarity in a given environment. The ranges of voltages were mostly negative in freshwater and mostly positive in seawater, but comparable in direction to the gill TEP. The fact that any TEP existed in the intestine suggests that earlier authors who suggested that seawater rapidly became isosmotic with blood before ion transport occurred were only partly correct. Since both marine and freshwater TEP ranges included zero voltages, the isosmotic conditions were not excluded, but at other voltages some small differential of ion concentrations must have existed across the intestinal wall.

The intestinal TEP data are presented in Fig. II-12 as part of a general summary of the concentration differentials and resulting TEPs. The numbers are rounded off sufficiently that they are probably incorrect for many specific instances but still show the general processes which we have presented. Knowledge about portions of the diagram is still incomplete. There is a TEP across the bladder wall, but the work describing it is still in progress. There is also undoubtedly a TEP across the walls of the kidney tubules, but the small size of tubules presents significant technical difficulties in making the measurements. The skin of fishes is a largely unknown factor in osmoregulation and is left out of this diagram. Compare this diagram with Fig. II-1—they both tell the same story but use different information.

F. ROLE OF WATER QUALITY

While a number of factors concerning water quality are discussed in Chapter XII, some of those concepts need to be discussed in the context of osmoregulation. Having read the many instances of regulation of the internal concentration of various dissolved materials by fish, the reader may be misled into thinking that only those kinds of materials diffuse into or out of fish. This is not true. Anything in solution potentially can enter fish and even some colloidal-sized particles have been seen to cross

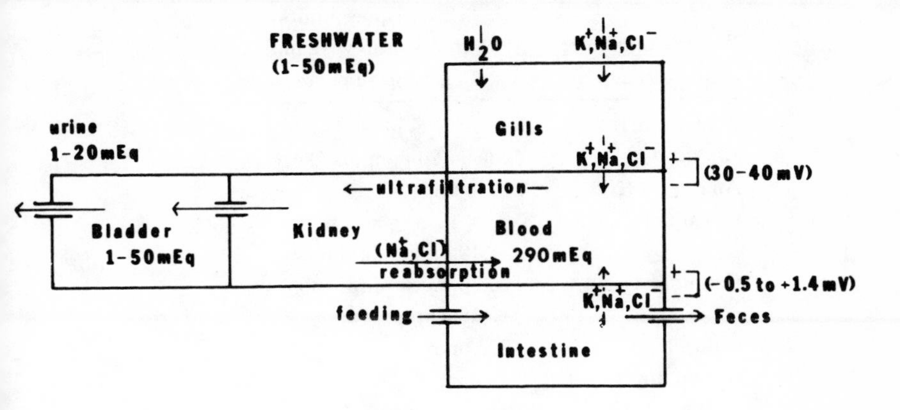

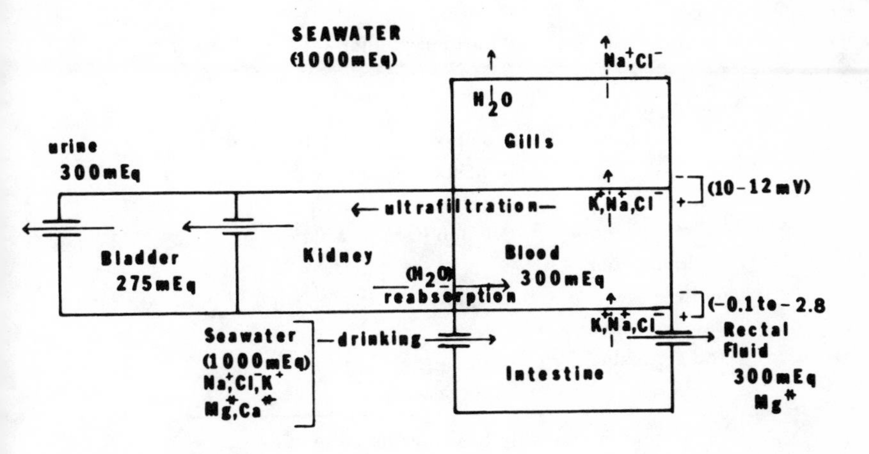

Fig. II-12. Summary of known and probable transepithelial potentials (TEP) in teleost fishes. The diagrams illustrate the major fluid compartments in fish, the concentration of some of the major ions or total ion concentrations in compartments, and the range of TEP observed across membranes between compartments. Fluid transported along a tube is represented by arrows passing through openings in compartments. Arrows crossing compartment boundaries represent transport through membranes. Data comes from a variety of salmonid and non-salmonid species.

Water chemistry characteristic	Upper limits (unless otherwise indicated) for continuous exposure
Acidity (pH)	6-9
Alkalinity	At least 20 ppm (as $CaCO_3$)
Ammonia (NH_3)	0.02 ppm [a]
Cadmium	0.0004 ppm in soft water [a] (<100 ppm alkalinity)
Cadmium	0.003 ppm in hard water (>100 ppm)
Chlorine	0.003 ppm
Chromium	0.03 ppm
Copper	0.006 ppm in soft water; 0.03 ppm in hard water
Hydrogen sulfide	0.002 ppm
Lead	0.03 ppm
Mercury (organic or inorganic)	0.2 parts per billion (ppb) maximum, 0.05 ppb average
Nitrogen (dissolved)	110% maximum total dissolved atmospheric gas pressure
Nitrite (NO_2^-)	100 ppb in soft water; 200 ppb in hard water
Polychlorinated biphenyls	0.002 ppm
Total suspended and settleable solids	80 ppm or less

[a] 0.005 ppm is probably healthier for salmonids.

Opposite page and above: Table II-5. Suggested water chemistry limits conducive to optimum health of warm- and cold-water fishes. (From Wedemeyer and Yasutake, 1977, with permission of author and publisher.)

pH	Water temperature C (F in parentheses)				
	5 (41)	10 (50)	15 (59)	20 (68)	25 (77)
6.5	0.04	0.06	0.09	0.13	0.18
6.7	0.06	0.09	0.14	0.20	0.28
7.0	0.12	0.19	0.27	0.40	0.55
7.3	0.25	0.37	0.54	0.79	1.10
7.5	0.39	0.59	0.85	1.25	1.73
7.7	0.62	0.92	1.35	1.96	2.72
8.0	1.22	1.82	2.65	3.83	5.28
8.3	2.41	3.58	5.16	7.36	10.00
8.5	3.77	5.55	7.98	11.18	14.97
8.7	5.85	8.53	12.02	16.63	21.82
9.0	11.02	15.68	21.42	28.47	35.76

pH	Water Temperature C (F in parentheses)				
	5 (41)	10 (50)	15 (59)	20 (68)	25 (77)
6.5	50	33.3	22.2	15.4	11.1
7.0	16.7	10.5	7.4	5.0	3.6
7.5	5.1	3.4	2.3	1.6	1.2
8.0	1.6	1.1	0.7	0.5	0.4
8.5	0.5	0.4	0.3	0.2	0.1
9.0	0.2	0.1	0.09	0.07	0.05

[a] Percentage $NH_3 = \frac{\text{ppm } NH_3 (100)}{\text{ppm } NH_3 + NH_4^+}$

through the gill tissues into the blood. Some substances pass across quickly, others very slowly, but all can eventually enter the fish. Thus fish are intimately associated with their environment because they have no way to avoid it. This is quite different from a person swimming in the same water as the fish (as long as the person drinks no water) because humans have no respiratory surface in contact with the water and have a relatively impermeable skin. The great degree to which people are isolated from their environment makes it difficult to appreciate the fish's situation.

Thus anyone concerned with the health of fish must be concerned with water quality. Some criteria for water quality are suggested in Table II-5. This mostly lists impurities or toxicants because these are the substances that most often cause problems in natural waters. However, there are undoubtedly some kinds of water within all of these suggested limits that will not grow fish or perhaps will grow some fish and not others. Distilled water, for example, contains no toxicants but has none of the minerals needed to keep fish alive, either. Thus the Table is not a complete description of the water quality requirements of fish.

In recirculating or even partly recirculating water systems, the ammonia which fish excrete becomes a crucial factor in water quality. The molecular form of ammonia, NH_3, is highly toxic (see Table II-5) while the ionic form, NH_4^+, is only slightly toxic. Depending on the pH of the water, much of the ammonia may become ionized immediately and prevent major problems of toxicity (Table II-6A). When recirculation involves a filter, bacteria there commonly convert either NH_3 or NH_4^+ to nitrite (NO_2^-). If nitrification is incomplete, the nitrite may also accumulate and reach toxic levels. Thus recirculating water systems can be difficult to understand, particularly when the commonly used chemical tests for ammonia measure total ammonia (NH_3 and NH_4^+). Then information such as that in Table II-6B is required to interpret the test results. See also Chapter XII for further discussion of ammonia and toxicants.

Chapter III: CIRCULATORY SYSTEM

A. OVERVIEW

1. Anatomy

While the circulatory systems of all vertebrates have the same general plan, each group of vertebrates has certain differences to fit their anatomy, physiology and environmental conditions. The major routes of blood vessels in salmonids are illustrated diagrammatically in Figure III-1. Fish also have a lymphatic system, but it is little known compared to that of land vertebrates.

Starting at the heart, there is only one way out—into the ventral aorta and the gills. After gas exchange takes place, however, there are multiple exit routes, including several small but vital ones. The coronary artery leaves the second gill arch and returns to the heart along the ventral side of the ventral aorta, providing oxygenated blood to the heart and to the thyroid follicles scattered around the ventral aorta. Arising from the first gill arch, a vessel goes to the pseudobranch and then to the choroid gland behind the eye before connecting to the venous system. The role of both of these organs is described below but probably involves control of ventilation and of gas secretion into the eye fluids, respectively. The branchial (recurrent) vein is a bypass from the gills directly back to the heart when not all of the cardiac output needs to go into the dorsal aorta and other efferent vessels. While not all of the implications of the recurrent vein are understood yet, it apparently may return a major part of the cardiac output directly back to the venous side of the heart when the fish is at rest.

The dorsal aorta is the major source of blood for most parts of the body. It supplies the head, the trunk muscles, the pectoral

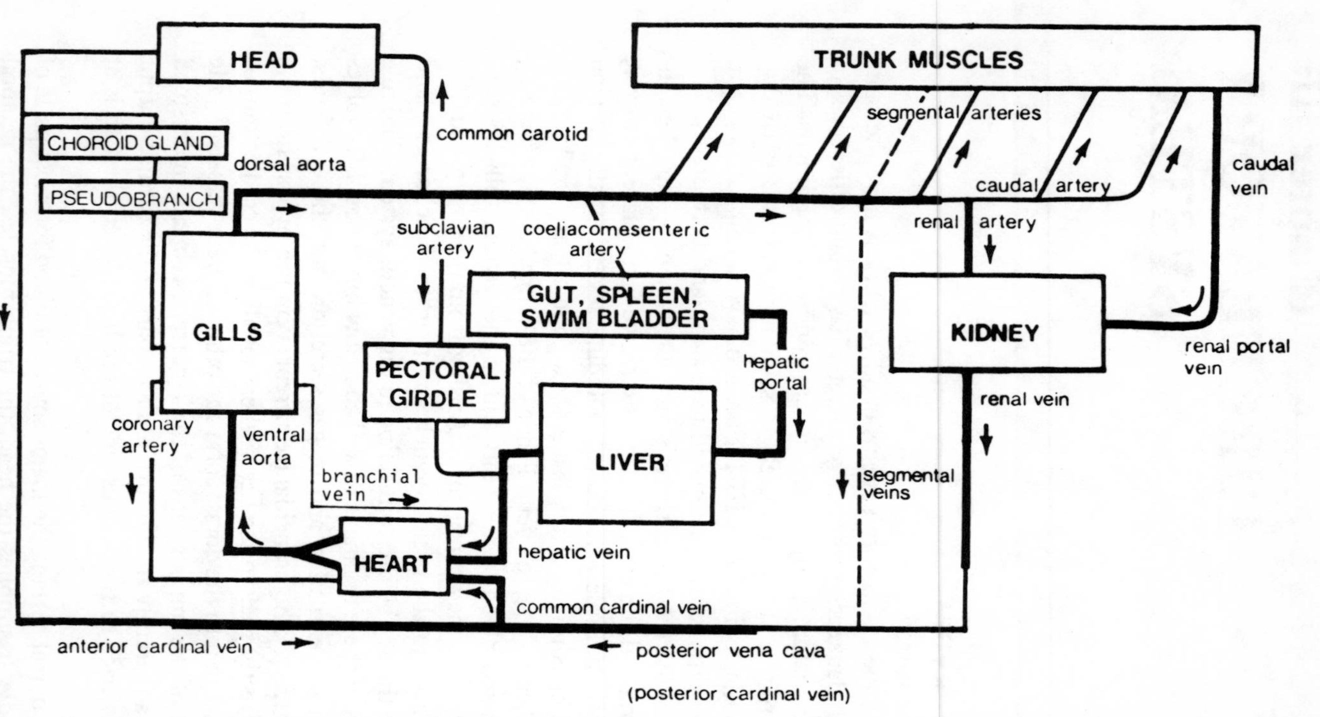

Fig. III-1. Block diagram of the circulatory system of salmonids. The sequence of routing of blood through various tissues can become important for interpreting the findings from analysis of blood samples taken from different locations. For example, oxygen content would be expected to differ dramatically between blood samples taken from the heart and from the dorsal aorta.

girdle, the kidney, and all of the visceral organs—all of the major capillary beds. Once through the main capillary beds, there are three major venous routes returning blood to the heart. The head drains to the heart through a pair of anterior cardinal veins which join a single common cardinal. The common cardinal is also joined by the posterior vena cava and a variety of small veins from the anterior body musculature. The posterior musculature, however, drains mostly into the caudal vein which leads into a capillary bed surrounding the kidney tubules. Since this is the second capillary bed in series after oxygenation of the blood, it is given the special designation of renal portal system. The veins draining the visceral organs (except the liver) lead to a similar portal system in the liver. Both portal systems seem to serve as major pools of blood into which metabolic products can diffuse with little increase in the concentration of these products. The renal portal system occurs only in fish and amphibians, although higher vertebrates have an hepatic portal system.

The lymphatic system has been little studied in fish, but exists in two alternative forms. In some fish, e.g. hake, the major lymphatic duct is axial and centrally located above the nerve cord inside the neural arch. In other fish, e.g. salmonids, the lymphatic system is peripheral with a major duct under each lateral line canal and along the dorsal midline. There are also short lymphatic ducts in each gill filament which collect filtered fluid from inside the gill filament and route it back to the heart via their connection to the recurrent vein at the base of each filament. In all cases, the lymph ducts serve to return tissue fluid to the circulatory system in tissues where not all of this fluid returns to the downstream portion of the capillaries. There do not seem to be any lymphatic glands or nodes in fish comparable to those in mammals.

2. General Pattern of Circulatory Function

The circulatory system serves many functions, but perhaps the most general is transport. Examples include transport of gasses between tissues and gills and lactate transport from muscle to gills and liver and then glucose back to the tissues. Foreign materials are transported to the kidney where soluble ones are excreted and cellular ones phagocytized. Eventually this results in antibodies being returned to the circulatory system later. Digestion products are transported from gut to liver and then to the rest of the body. Blood cells travel outward from their site of formation in the head kidney to all parts of the body. Clotting factors and thrombocytes combine at any site of damage to plug up major leaks in the circulatory plumbing without blocking the vessels themselves. In general, something of the functioning of every organ in the body can be seen in the blood sometimes.

3. Composition of Blood

Blood has two major components—cells and plasma. The functions of the two components are sometimes separate and sometimes shared by both components, clotting and antibody production, for example. These two functions will be discussed below in more detail. The remaining components of plasma (or serum, after the fibrinogen has been removed from plasma by allowing the blood to clot) include a limited number of inorganic ions and a wide variety of organic compounds relating to most metabolic functions. Cellular composition, on the other hand, consists of discrete cells having distinctive anatomy and discrete functions.

Mature blood cells can be identified by their shape and staining characteristics as seen under the light microscope (Figure III-2). These nucleated red cells predominate in numbers (about 1.1 million/cubic millimeter) and are consistent enough in size (9 x 13μm in salmon) to serve as a handy measuring stick for estimating the dimensions of other cells. In addition to measuring the external dimensions of cells, they can also be characterized by the ratio of the volume of the nucleus to the volume of the cell (Table III-1). These ratios combined with descriptions of staining characteristics are used in some reports instead of photographs.

The major types of blood cells in addition to red blood cells include the lymphocytes and thrombocytes. The lymphocytes are usually characterized as having relatively large nuclei and little cytoplasm, but can be grouped into size groups. Immature thrombocytes look like lymphocytes and may be derived from lymphocytes but change during maturation into oval cells which are smaller overall than lymphocytes and have smaller nuclei.

A. Four lymphocytes and six erythrocytes. Note the smooth periphery of the three clustered lymphocytes in contrast to the rough, pleomorphic margins of the cytoplasm of the single lymphocyte. The erythrocytes are typical, except that the granules in the edge of the nucleus may be heparin artifacts.

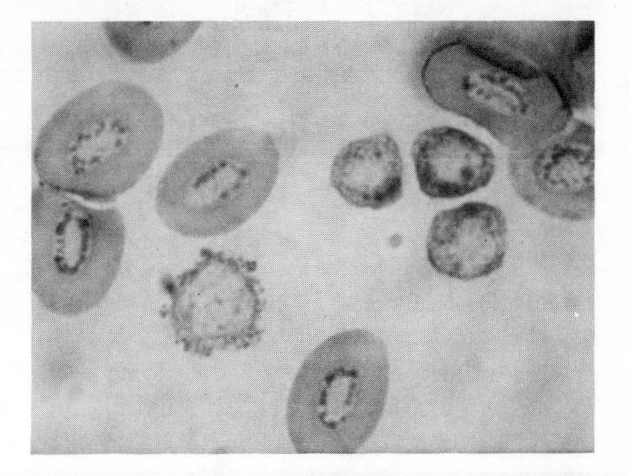

B. Two thrombocytes and one pleomorphic lymphocyte. A very thin or achromatic veil of cytoplasm is characteristic of thrombocytes. As these cells mature or become activated, their cytoplasm can become drawn out, forming polar spindles on either or both ends of the cell.

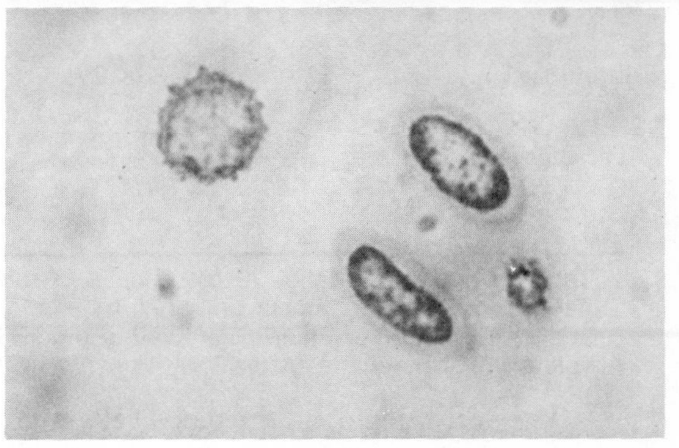

C. A mature, polymorphonuclear leucocyte is considered by many researchers as analogous to the mammalian neutrophil, although their function has not been clearly demonstrated in salmonid fishes. This cell type may be rarely seen in normal blood, but was common in a group of rainbow trout whose gills were infested with the copepod, *Salminicola*.

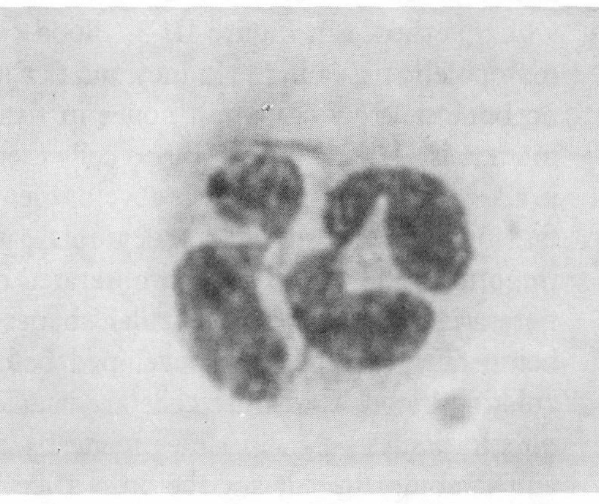

Fig. III-2. Cells from peripheral blood of rainbow trout. All photomicrographs taken at 1000X using heparinized blood stained with Leishman-Giemsa stain. Photographs courtesy of Don H. Bark, University of Washington Fisheries Research Institute.

Cell Type	Diameter (microns)	C:N Ratio	Shape
Hemocytoblast	10-20	1.5-1.6	Round, nucleus round and slightly eccentric
Large lymphoid hemoblast	12-17	1.5-2.0	Round to ovoid, nucleus round and central
Large lymphocyte	10-15	----	Slightly ovoid, nucleus often eccentric or irregular
Small lymphoid hemoblast	10-15	1.25-1.5	Round, nucleus central
Small lymphocyte	7-10	slightly more than 1	Round, almost no cytoplasm
Granulocytes	10-15	1.4-1.6	Granules in cytoplasm, nucleus eccentric and indented
Erythrocyte	12-16×7-9*	3.5-4.0	*Cell elliptical, nucleus oval and compact
Thrombocyte	----	----	Cell and nucleus oval, almost rectangular, can produce 1 or 2 pseudopods

Table III-1. Characteristics of blood cells of rainbow trout. Cells whose names end in "blast" occur primarily in hematopoietic tissues of kidney and spleen; the remainder in circulating blood. C:N ratio stands for cell volume:nuclear volume. (After G.W. Klontz, Univ. Idaho, unpublished.)

The derivation of all the blood cells is not entirely understood, although many workers follow the interpretations of Klontz, which is shown in Figure III-3. Blood cells are produced in hematopoietic tissue in the kidney and perhaps the spleen. There is no bone marrow or lymph nodes in fish comparable to that in mammals. The naming of blood cells mostly follows mammalian practices, however, because development proceeds in a similar fashion. There is a basic hemocytoblast which gives rise to all of the other cell types. These proliferated cells gradually differentiate and take on their particular shapes and functions, usually being fairly distinctively developed before they enter into circulating blood. Immature cells are usually seen only in hematopoietic tissues—e.g., in slides made by slicing the head kidney and daubing the cut surface on a microscope slide—unless the hematopoietic system is stimulated by stressful conditions. Thus the appearance of many immature cells in circulating blood can signify the presence of a disease or other pathological condition.

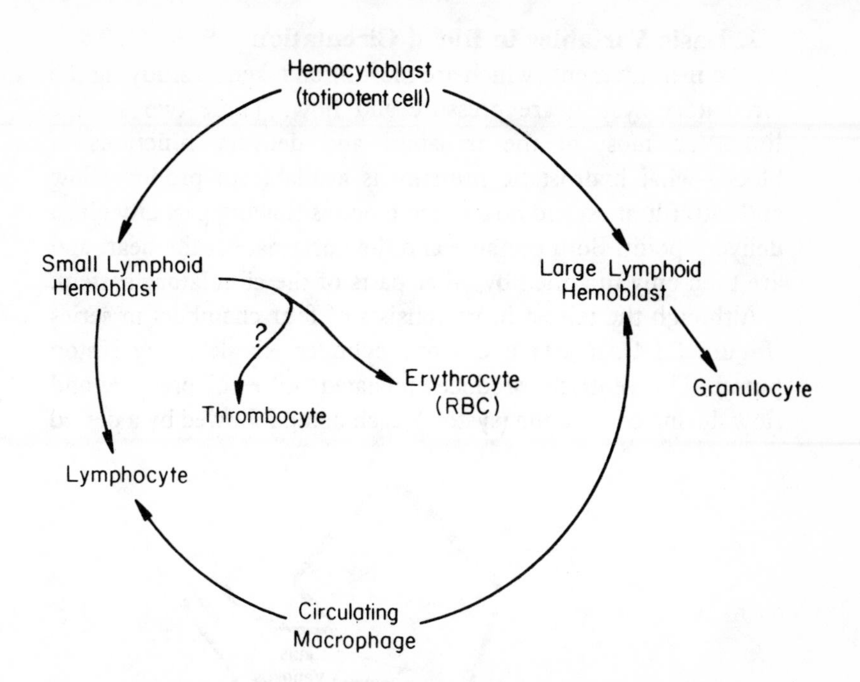

Fig. III-3. Theoretical sequence of development of blood cells in rainbow trout. The erythrocytes comprise the great majority of cells in the circulating blood with the remainder being less common. (After Klontz, reprinted from Wedemeyer et al., 1976, with permission of TFH Publications.)

B. HEMODYNAMICS

In most respects the circulatory system of fish (and most other vertebrates) is typical of any hydraulic system. It follows the same laws of physics as any other system involving a pump, pipes, valves and a viscous fluid. On the other hand, the flow is pulsatile rather than steady (comparable to a single-acting piston pump rather then a centrifugal or other rotary pump) and some of the pipes can stretch both lengthwise and radially. Further, the fluid being pumped is half full of flexible particles, which produces some flow characteristics which make blood sufficiently different from "normal" fluids that engineers describe blood as an anomalous fluid, i.e., it doesn't always follow the rules. Finally, where the pipes have thin walls, as in the capillaries, the system develops a controlled leakiness which serves to transport a variety of substances into and out of the blood.

1. Basic Variables in Blood Circulation

The measurements which are most often taken in studying the circulatory system are pressure and flow. These two set the limits on most of the transport and delivery functions of blood—what hydrostatic pressure is available to produce flow and ultrafiltration and how much blood is flowing past any given delivery point. Both pressure and flow originate at the heart and are then only modified by other parts of the circulatory system.

Although the teleost heart consists of four chambers in series (Figure III-4), it acts like a one cylinder, single-acting piston pump. The ventricle produces repeated pulses of pressure and flow during contraction (systole), each being followed by a period

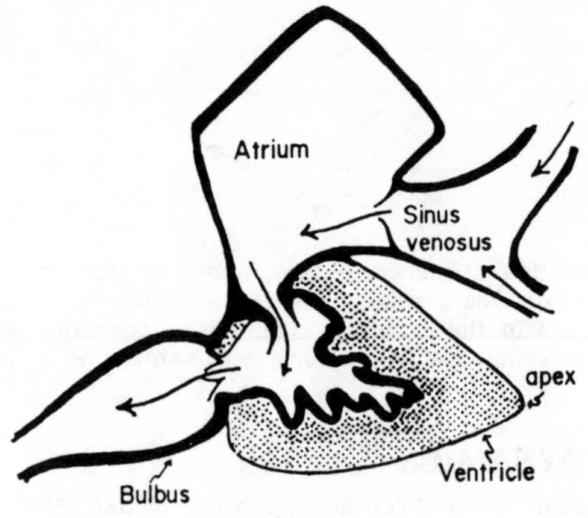

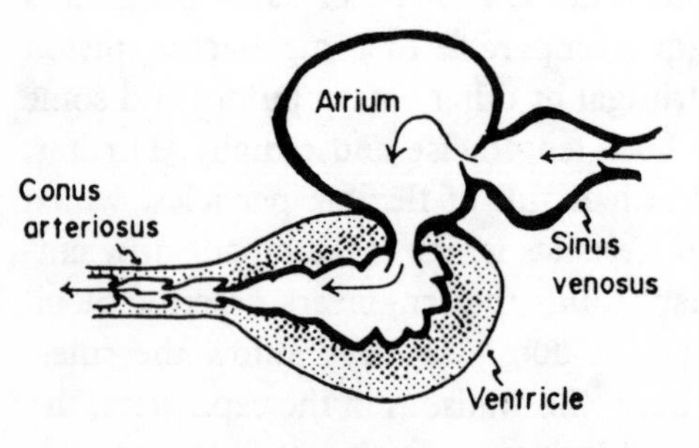

Fig. III-4. Diagrammatic cross sections of trout and shark hearts, showing the heart chambers, valves, and routes of blood flow. (From Randall, 1968, with permission of author and publisher.)

of relaxation and refilling (diastole). The sinus venosus and the atrium are both contractile but produce little pressure and serve primarily to fill the ventricle. The bulbus plays an important role of maintaining blood flow during ventricular relaxation. Under pressure its elastic walls expand considerably so that when the ventricle relaxes and one-way valves between the bulbus and ventricle close, the elastic rebound of the bulbus maintains a continuing pressure and flow into the aorta during ventricular relaxation. The expansion of the bulbus during systole also dampens the impact of the systolic pressure pulse which might otherwise rupture thin-walled blood vessels or at least cause excessive ultrafiltration. The relationship of these pressures as the wave of contraction starts at the sinus venosus and spreads over the heart is shown in Figure III-5.

Once having left the ventricle and bulbus, there is no further source of energy to maintain blood pressure, so the pressure falls in relationship to the friction encountered in the blood vessels. The smaller the blood vessel the greater the friction (in proportion to the fourth power of the radius of the vessel). Thus rather small changes in vessel diameter serve as a very effective means of controlling blood flow. For example, a change in internal diameter from 0.58 mm (.023″) to 0.76 mm (.030″) increase the ease of flow approximately 30-fold. This was calculated for rigid plastic tubes such as are used in cannulating blood vessels, but should be similar in blood vessels. The contractile site which regulates flow in blood vessels is the arteriole, just upstream from the capillaries. Small changes in diameter of arterioles can be produced by the sympathetic nervous system or by adrenalin and acetylcholine, which cause changes in resistance to flow.

Changes in resistance to flow cause predictable changes in blood pressure, depending on the location of the resistance. Increased resistance in the gills causes the major drop in blood pressure to occur in the gills—blood pressure in the ventral aorta is high, that of the dorsal aorta is relatively low. The pulse pressure (the differential between the maximum systolic and minimum diastolic pressure during any one heart cycle) is the reverse. Pulse pressure in the ventral aorta would be small because with minimal flow the bulbus stays expanded for most of

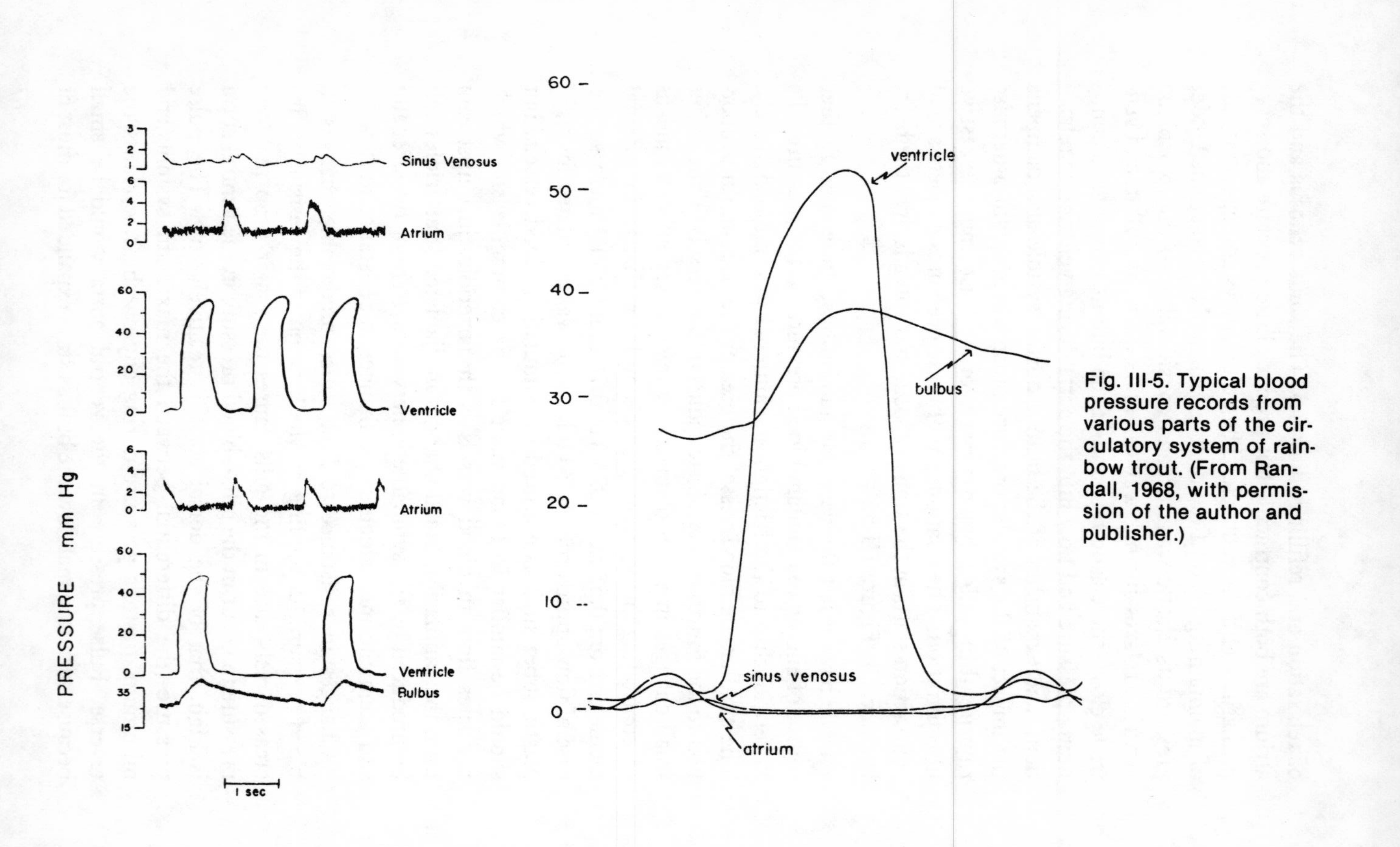

Fig. III-5. Typical blood pressure records from various parts of the circulatory system of rainbow trout. (From Randall, 1968, with permission of the author and publisher.)

the heart cycle. Pulse pressure in the dorsal aorta would be even smaller than in the ventral aorta, since there is a general tendency toward smoothing of flow throughout the entire circulatory system because of friction and somewhat elastic blood vessels everywhere. Conversely, if resistance in the gills is low, then average blood pressure would be low and pulse pressure high. The blood would flow out of the gills almost faster than the heart could supply blood and the relatively small volume supplied by the bulbus would hardly last through diastole. Thus systolic pressure would not be very high and diastolic pressure would fall quite low, but flows would be maximal. The pressure drop across the gills would be minimal because of minimal resistance to flow, so the blood pressure in the dorsal aorta would be closer than usual to that of the ventral aorta and the pulse pressure would be larger than usual. Thus, by taking records of blood pressure from several sites and analyzing them, one çan deduce changes in resistance to flow in most parts of the circulatory system.

By the time that blood has passed through the gills, then through a second set of capillaries, such as those in muscle tissue, and finally through a portal system, either the renal or the hepatic, little blood pressure and no pulse pressure remains. Swimming activity (muscle contractions compressing veins) contributed more to venous return to the heart in one experiment in cod than did residual blood pressure. Thus any kind of muscular activity directly affects the heart partly because increased venous return increases cardiac output.

Three other variables which can be measured during studies on circulatory function are heart rate, stroke volume and total blood volume. Heart rate is usually measured either by counting the pressure pulses in a blood pressure record or the electrical pulses in the electrocardiogram. Increasing the heart rate is an obvious way to increase cardiac output, although increased heart rate occurs in differing degrees in different species of fish (see below). The other variable in measuring heart performance is stroke volume. In a piston pump, stroke volume is the difference in the volume of the cylinder between the two extremes of travel of the piston and is fixed by the radius of the crankshaft. The

Animal	Temperature °C	Stroke Volume ml/stroke/kg	Cardiac Output ml/min/kg	Method
man			60-100	
dog			139	
cat			690	
frog	20	1.0 (approx.)	57	Ventricular displacement
octopus	7-9	.31-2.01*	5-32.2	Fick principle
Opsanus tau (toadfish)			10.1	Fick principle, minimum values.
Tetraodon maculatus (puffer)			15.5	As above
Stenotomus chrysops (scup)			13.7	As above
Gadus morhua (cod)		.31-doubles with increased venous return.	9.3	electromagnetic flowmeter
Myoxocephalus scorpius (sculpin)	15-18		27.75	Fick principle
Squalus acanthias (dogfish)	11-17	.25-.94 to 1,88 when ventricle perfused.	9.0-24.8†	Ligation and weighing of heart. Flow from cut ventral aorta.
Salmo gairdneri (rainbow trout)	12-18	.85-2.0	65-100	Fick principle
	4-8	.15 (resting) .7 (moderately active)	7.5 (resting) 38 (moderately active)	As above
Oncorhynchus nerka (sockeye salmon)		resting-fully active	resting-fully active	
	5	.67-1.39	20.74-65.07	Fick principle-
	15	.60-1.12	31.08-92.94	resting-VA
	20	.65-1.04	49.65-101.89	70%satd. active-VA
	22	.73-.99	62.65-103.7	0% satd.

† Calculated from a heart rate of 36 bts/min.
* Calculated from a heart rate of 16 bts/min.

Table III-2. Estimates of cardiac output for representative vertebrates and several species of fishes. (From Davis, 1966, with permission of author.)

heart's volume is not fixed, however, and varies in response to venous return, the sympathetic nervous system and adrenalin and acetylcholine in the blood. To increase its volume per beat, the ventricle relaxes more deeply—stretches longer and wider before contracting—and can be filled more completely and forcefully by the auricle. At high heart rates the stroke volume may decrease because the ventricle has insufficient time to fill. Cardiac output is thus determined by the combination of heart rate and stroke volume. Some comparative values for cardiac output and stroke volume are shown in Table III-2.

When blood volume and cardiac output are known, then circulation time can be estimated—i.e., how often an average round trip in the circulatory system takes. This can be expressed either as blood volume over heart rate x stroke volume or blood volume over cardiac output (depending on which variables were measured). More active animals usually have shorter circulation times, higher cardiac outputs and perhaps larger blood volumes than less active animals.

Some comparison of blood volumes in several species of teleost fish is given in Table III-3A and B. Blood volume determinations usually involve the measurements of the dilution of some substance added to either the red blood cells or to the plasma or both. The methods vary with different waiting times, different injected materials and different circumstances, so it is difficult to compare these data. Because of the scarcity of such data, however, two generalizations were attempted. One of these was that blood volumes have increased with continued evolution, i.e., primitive fish have smaller blood volumes than more recent fish. The other is that more active fish have larger blood volumes than sedentary fish. While these two statements may appear contradictory, both may be true depending on which species of fish one examines. For example, at a given level of activity, the primitive vs. recent comparison may be true, but the size or the effect of activity may obscure the evolutionary status as could be the case with salmonids. Salmonids have comparatively large blood volumes but are relatively active and relatively primitive.

A

Scientific Name	Blood volume (ml/100 g)	Method
Salmo gairdneri	3.5±0.9	Evans blue
Salmo gairdneri	3.3±0.9	I^{131} albumen
Salmo gairdneri	2.8±1.0	Cr^{51} RBC
Gadus morhua	2.4	Fluorescein
Gadus morhua	1.9	Exsanguination
Carassius carassius	2.5-3.0	Hb washout
Solea vulgaris	1.8-2.4	Hb washout
Belone vulgaris	4.0-5.9	Hb washout
Ophiodon elongatus	2.8	Evans blue
Several teleosts	1.8-3.8	Evans blue

Table III-3. Blood volumes of teleost fishes. A. Blood volumes of several fishes determined by several methods, values taken from the literature prior to 1966. (Opposite page): B. Blood volumes of three salmonids by several methods. The use of Evans Blue and repeated blood sampling via a cannula in the dorsal aorta gave larger estimates of blood volumes than any other method except possibly the hemoglobin-washout method in A above. (From Smith, 1966, with permission of the author and publisher.)

2. Changes in Blood Circulation With Activity

The circulatory system usually responds strongly to activity in most fish. The most crucial circulatory function during activity is oxygen transport because increased muscular activity without increased oxygen availability soon exhausts the anerobic energy system (see Chapter 5) and fatigue results. Fish with larger amounts of dark (aerobic) muscle are usually more active and have larger cardiac outputs (Table III-2) than less active fish. The basic response of the circulatory system to increased rates of activity thus is increased cardiac output, but the amount of increase and the mechanisms used to accomplish this increase vary with different species and lifestyles.

B

Fish	Test No.	Weight (g)	No. of fish	Hematocrit (% RBC)	Environment	Plasma (ml/100 g)	Blood (ml/100 g)	Method
Coho salmon								
(Oncorhynchus								
kisutch)	1	537 ± 108[a]	7	21 ± 8.1[a]	Sea water	4.8 ± 1.3[a]	6.2 ± 1.6[a]	Evans blue
	2	1049 ± 173	10	30 ± 8.3	Sea water	4.3 ± 0.7	6.1 ± 0.9	Evans blue
	3a	949 ± 23	3	33 ± 7.5	Sea water	3.0 ± 0.5	4.6 ± 0.3	Evans blue[b]
	3b	949 ± 23	3	33 ± 7.5	Sea water	4.7 ± 0.3	7.2 ± 0.1	Evans blue[c]
	4a	926 ± 112	7	25 ± 5.4	Sea water	4.6 ± 1.4	6.2 ± 2.0	Evans blue
	4b	930 ± 104	8	20 ± 7.3	Sea water	3.4 ± 1.0	4.5 ± 1.5[d]	Reinjected Evans blue
	5	1012 ± 120	6	—	Sea water	—	2.3 ± 0.4	Exsanguination
Sockeye salmon								
(O. nerka)	6	1814 ± 266	8	26 ± 3.8	Fresh water	4.0 ± 0.6	5.4 ± 0.8	Evans blue
Salmo gairdneri								
Steelhead								
trout	7	533 ± 127	13	21 ± 7.8	Sea water	5.6 ± 1.8	6.9 ± 1.8	Evans blue
Rainbow trout	8	548 ± 49	4	—	Fresh water	—	2.4 ± 0.4	Exsanguination

[a]Mean and standard error.

[b]Estimates made while the fish were under M.S. 222 anesthesia on an operating table (belly up, gills irrigated with oxygenated sea water) from blood samples taken at 20, 25, and 30 min after injection of dye.

[c]Estimates made on the same individuals as in 3a, but after they had been placed in aquaria and allowed to recover from the anesthesia. Blood samples were taken at 1, 2, and 3 hr after the first dye injection without an additional dye injection.

[d]Values obtained by use of reinjected Evans blue into the same individual fish as for test 4a, with random selection as to whether Evans blue or reinjected Evans blue was used first. The reinjected Evans blue consisted of plasma from donor coho salmon which had received large doses of Evans blue 5-6 hr before centrifuging and reinjection. The second injection was given 48 hr after the first.

The control of the heart is based on two mechanisms. The most obvious is the presence of inhibitory (cholinergic) nerve fibers from the vagus (cranial nerve X) nerve. There is usually some inhibitory activity by the nerve, so one way to accelerate the heart rate is to decrease the vagal inhibition. The heart also has receptors which respond to adrenalin in the blood which the heart is pumping. Increased amounts of adrenalin have been found in the blood in the heart during cardioacceleration, but the source (e.g.—sympathetic nerve, interrenal glands, etc.) of the adrenalin was not found. Then, in at least some fish, the heart is accelerated by increased temperature acting directly on the pacemaker (see "Electrocardiogram" section below).

The effect of increased swimming rates on the blood pressure and heart rate of sockeye salmon appears in Figure III-6. Heart rate increased steadily over most of the range of swimming velocity, tending to plateau toward the maximal swimming velocity. Measurements of blood flow during this type of stepwise increases in swimming velocity have shown that approximately a five-fold increase occurred. Since heart rate rarely even doubled, the increased flow must have been mostly the result of increased stroke volume.

Major increases in blood flow which occurred with little or no increase in blood pressure must have been accompanied by major decreases in resistance to flow, i.e., there was vasodilation at least in gills and muscle capillary beds to absorb the increased flow.

The pattern of change is typical of all salmonids examined so far except for jack sockeye salmon. (Jack salmon are precocious males which return to spawn a year before the majority of the males of that year class). Blood pressure was stable or rose slightly as before, but heart rate began fairly high and decreased with stepwise increases in swimming velocity. There was no explanation for this.

The recovery from strenuous activity (including fatigue and then reduced water velocity in this case) was a gradual return to normal blood pressure and heart rate (Figure III-7). A synchrony of heart beat with respiration was common during the resting period, a feature which was not seen during activity. This was somewhat surprising because there would seem to be obvious ad-

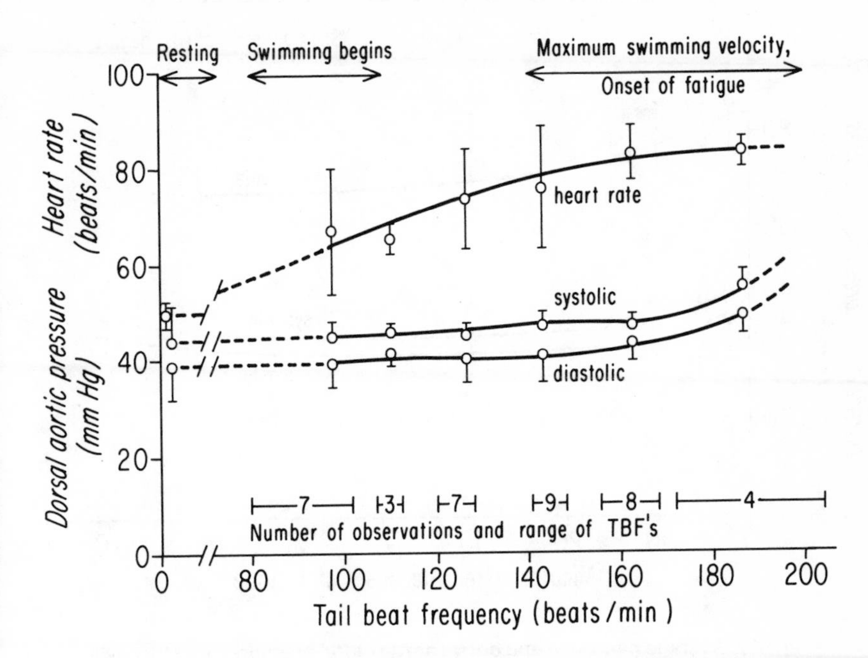

Fig. III-6. Average heart rate and dorsal aortic blood pressure (± 1 std. deviation) in 10 adult sockeye salmon at 15°C during stepwise increase in swimming velocity (expressed as tailbeat frequency). Tailbeat frequency was suggested as indicating the effort put forth in the swimming better than the actual swimming velocity. (From Smith et al., 1967, with permission of author and publisher.)

vantages of efficiency to couple the period of maximal blood velocity through the gills with the period of maximal water velocity over the gills.

Tench seem to respond to the need for increased cardiac output mostly by increasing the heart rate. Figure III-8 compares the effects of handling in tench and trout. The response had already taken place when the recording of the heart rate began, so only the recovery phase is shown. Swimming activity, presumably through increased venous return, has also been shown to produce greatly increased heart rates in goldfish and gray cod.

In both tench and trout, however, a cessation of gill ventilation produces a decrease in heart rate. During anesthesia, for exam-

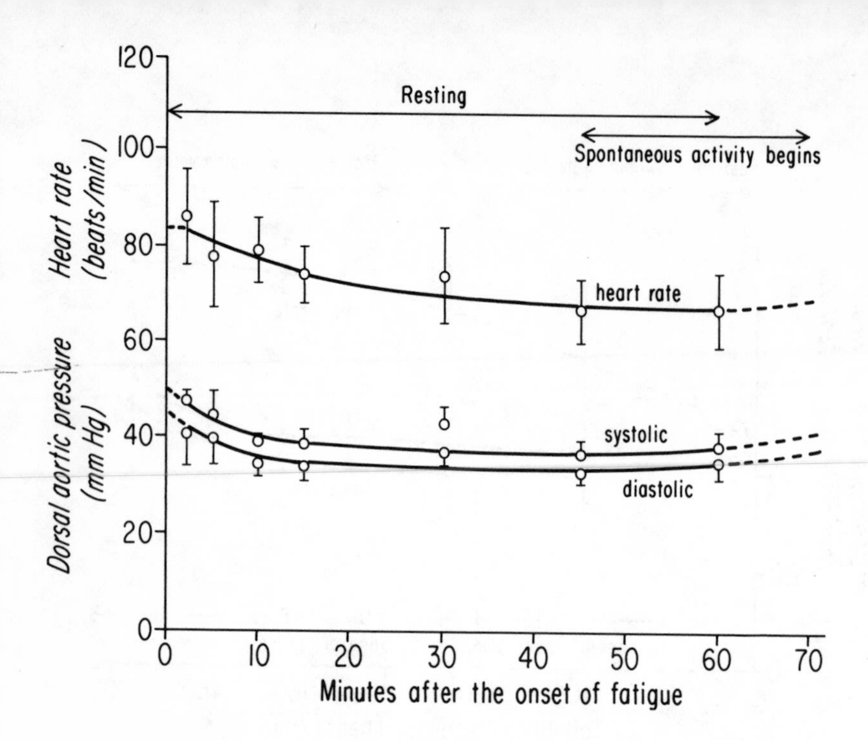

Fig. III-7. Average heart rate and dorsal aortic blood pressure (± 1 std. deviation) in 10 sockeye salmon after the onset of fatigue. This is a continuation of the data shown in Fig. III-6. Oscillations in both heart rate and blood pressure were smoothed because they were not conspicuous in individual fish. (From Smith et al., 1967, with permission of author and publisher.)

ple, if respiratory arrest occurred the heart rate in trout decreased to about ⅓ of its resting rate. Forced ventilation restored heart rate to nearly normal, while lack of forced ventilation was accompanied by further decreases in heart rate until eventual cardiac arrest a few minutes after respiratory arrest. (See Chapter XII for further discussion of hypoxia.)

3. Electrocardiogram

The electrocardiogram (ECG or also EKG, so as to minimize confusion with EEG or electroencephalogram) is a recording of the sequence of electrical events which occur during the heart contraction. A wave of electrical excitation (depolarization) begins at the sinatrial node (pacemaker) between the sinus and

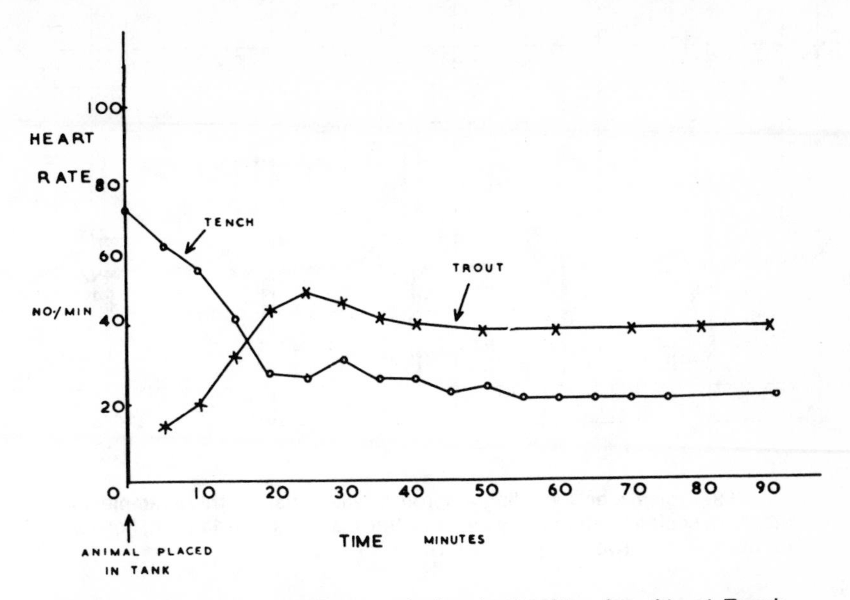

Fig. III-8. The effect of handling on the heart rate of tench and trout. Tench respond to stress with major increases in heart rate, shown here as initially elevated and then recovering from handling. Salmonids inhibit heart rate, especially during exposure to air, and then increase it during recovery. (From Randall and Smith, 1967, with permission of author and publisher.)

atrium and sweeps over the heart muscle. Giant fibers leading to the tip of the ventricle conduct the electrical impulse faster than the muscle, causing the ventricular contraction to start at the tip of the ventricle and meet the excitation wave coming from the atrium. This is supposed to provide a stronger contraction of the ventricle than if the wave of excitation spread from only a single point.

The appearance of the electrocardiogram in fish (Figure III-9) is comparable to that of other vertebrates, although the conduction velocities of the excitation appear to be slower than that in mammals. Typically there is a P wave, then a QRS complex and finally a T wave. In the eel, a V wave was recorded preceding the P wave which was associated with the contraction of the sinus venosus. In most fish the sinus venosus is more of a distinct chamber than in mammals, where it is largely incorporated into the wall of the atrium and any sinus venosus contraction would occur as part of the atrial contraction.

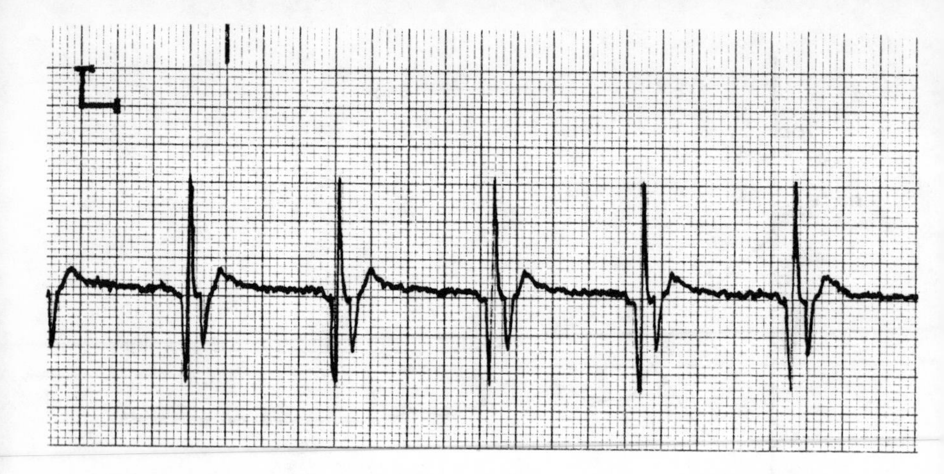

Fig. III-9. Typical electrocardiogram from a teleost fish, in this example, the staghorn sculpin, *Leptocottus armatus*. Vertical mark = 0.1 mVolt; horizontal mark = sec. (Courtesy of John Nichols.)

C. BLOOD CLOTTING

Fish blood clots similarly to blood of other vertebrates. There are three major components to the blood coagulation system: (1) a series of enzymes which lead to the production of fibrin fibers from fibrinogen, (2) blood cells (platelets in humans, thrombocytes in fish) which can become sticky and form into clumps or adhere to damaged surfaces and (3) a fibrinolytic enzyme system which dissolves clots. The fibrinolytic system has been directly demonstrated in fish (L. Smith, unpublished, 1977) and serves to dissolve clots. The coagulation system operates in a remarkable fashion to plug up leaks in blood vessels (some of which may be of considerable size) without blocking flow in the smaller blood vessels, by adjusting the balance between these opposing components.

Fish face two problems which make blood clotting a more difficult process for them than for terrestrial vertebrates. Being aquatic, any blood on the surface of the body has a difficult time clotting because the necessary enzymes and clotting components are diluted or are washed away before a clot can form. Thus there is an advantage for fish in having a clotting system which operates very rapidly so as to minimize the dilution problem. On the other hand, when blood stands still it can clot spontaneously, and fish have large amounts of white muscle in which blood moves very slowly or may even stop flowing, particularly during buildup of lactate in the muscle cells. This could occur after emergency (burst, non-sustainable) swimming or other stressful occurrences when lactate levels in muscle are high. Thus there is also need to minimize the blood clotting rate at some times and places. Viewed in the perspective of these contradictory needs, it is not too surprising to find that fish have a blood coagulation system with adjustable clotting times.

Coagulation time for whole blood in rainbow trout decreased in response to two minutes of stressful activity (Fig. III-10).

Fig. III-10. Effect of stress (severe muscular activity) and blood sampling on blood clotting times in hatchery trout (top) and wild trout (bottom). Solid line and dots, stressed group; dashed line and open dots, control group. Points represent mean ± S.E. Both groups responded similarly to stress by decreasing their blood clotting times (increasing the blood clotting rates), but the wild fish recovered more rapidly than the domesticated trout. The latter responded more strongly to blood sampling in the control group than the wild trout. (From Casillas and Smith, 1977, with permission from *J. Fish Biol.* 10: 481-491. Copyright by Academic Press, Inc. [London] Ltd.)

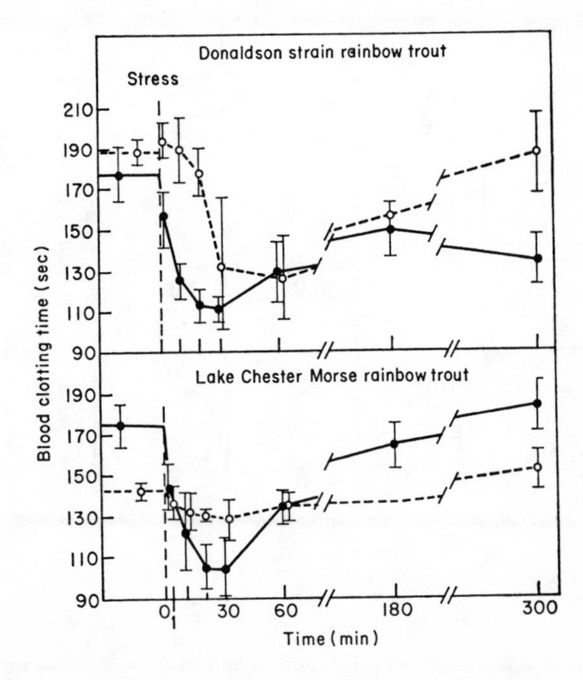

	Control	Chamber control	100 fsw +1 min	100 fsw +10 min	100 fsw +60 min	200 fsw +1 min
TPP (mg/ml)	21.6 ±3.57	21.27 ±2.96	21.28 ±3.05	23.81 ±4.32	29.08* ±8.1	22.9 4.08
n	*10*	*9*	*10*	*10*	*10*	*9*
Fibrinogen (mg/ml)	1.38 ±0.61	1.31 ±0.43	1.25 ±0.56	1.43 ±0.47	1.01 ±0.28	0.46‡ ±0.49
n	*10*	*10*	*10*	*10*	*9*	*9*
PT (s)	34.9 ±3.5	31.2* ±2.0	42.0† ±5.0	41.3† ±4.0	41.2† ±3.8	57.7† ±18.0
n	*10*	*9*	*10*	*10*	*6*	*5*
PTT (s)	146.7 ±58.4	176.5 ±102.3	238.7† ±115.8	217.4* ±96.6	316.7‡ ±200.3	377.8‡ ±189.3
n	*17*	*19*	*20*	*13*	*10*	*6*

All values are expressed as the mean ± SD. TPP = total plasma protein. PT = prothrombin time. PTT = partial thromboplastin time.

*Significantly different from controls ($P < .05$).
†Significantly different from controls ($P < .01$).
‡Significantly different from controls ($P < .001$).

Table III-4. Effects of rapid decompression on fingerling salmon. All values were compared using the Student's *t* test.

Starting from average resting times of three minutes, blood clotting times decreased to 1.8 minutes after 20 minutes and then returned to normal. Under the more stressful conditions of rapid decompression (bubbles in blood stimulate the clotting system), fingerling coho salmon first decreased their clotting times slightly and then increased them greatly (Table III-4). (PT and PTT times are indicative of two different parts of the enzyme cascade which leads to the production of fibrin fibers.) The later lengthening of clotting time appears to be caused by the consumption of the raw materials used in producing clots, in this case fibrinogen. There are two consequences of this depleted status. First, it can indicate that extensive blood clotting is occurring; in situations where there is no injury to provide a site for clotting, then there may be extensive clotting in peripheral blood vessels. The medical terms for this are consumptive coagulopathy and disseminated intravascular coagulation (DIC). DIC has been proposed as one possible cause for delayed mortality in fish after a stressful event such as capture and release. Second, the depletion of materials for making blood clots leaves the fish unable to cope for awhile with further injury involving blood loss.

Another aspect of how fish accelerate their clotting system is an increase in the numbers of thrombocytes in circulating blood (Fig. III-11). Thrombocytes were uncommon in blood from various undisturbed salmonids but rapidly increased in number during and after a stressful situation. It appeared that these additional thrombocytes were released from a storage site (perhaps the spleen), but it was unclear whether these cells were being consumed and eventually replaced by new cells or whether they returned to storage ready to be used again.

There are still unexplained areas as well. Anyone working with whole fish blood knows that it can clot in as little as 20-30 seconds (compared to 7-8 minutes for human blood). In comparison, blood taken through a cannula (i.e., taken without injury from undisturbed fish) has stood for an hour in clean glassware without clotting (L. Smith, unpublished). Whole blood clotting times accelerated proportionately much more than the increases in PT and PTT times or thrombocyte counts. A possible explanation for this is that fish tissues are unusually (by mammalian standards) rich in clotting stimulators which are released when cells are injured, but this idea has not been tested yet. There were greater changes in clotting rates and thrombocyte counts in wild rainbows (Chester Morse) than in a domesticated

Fig. III-11. Changes in numbers of circulating thrombocytes in response to stress and blood sampling in the same fish as shown in Fig. III-10. Solid lines and dots = stressed fish, dashed lines and open dots = control fish. Points represent mean ± S.E. Wild fish (lower graph) responded slightly more than domesticated fish (upper graph) and recovered more rapidly. (From Casillas and Smith, 1977, with permission from *J. Fish Biol.* 10: 481-491. Copyright by Academic Press, Inc. [London] Ltd.)

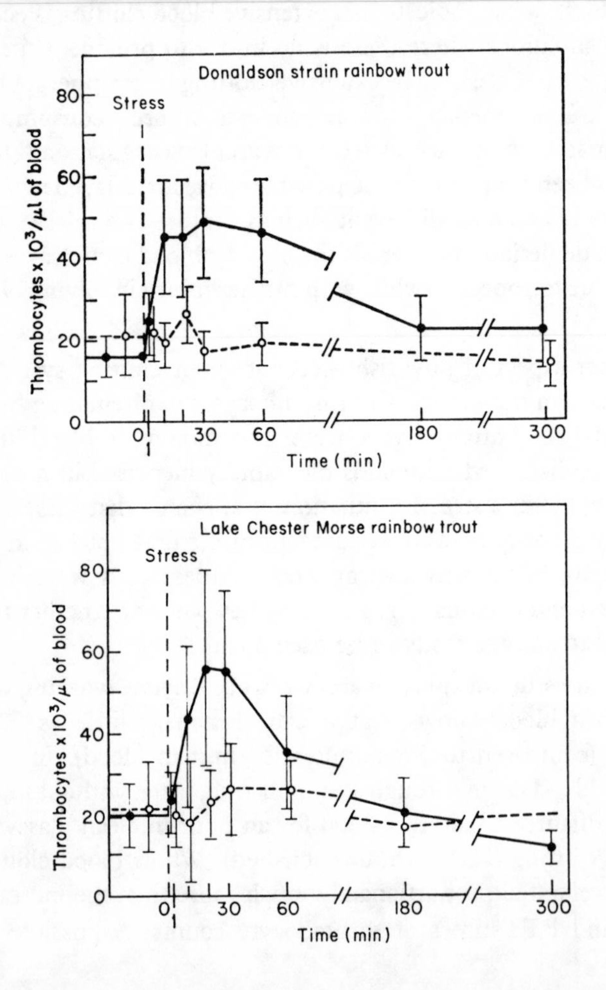

strain (Donaldson) as shown in Fig. III-10 and Fig. III-11. Whether such differences have survival value remains to be seen.

The general conclusions about blood clotting in fish are that the basic mechanisms are much the same as in mammals, but some of the details are different. For example, blood clotting times in man accelerate under stress about 30 percent and platelet counts increase similarly, but whole blood clotting times in fish can accelerate 3-5 fold. Also, the clotting cells of mammals (platelets) are non-nucleated, short-lived and expendable, while fish thrombocytes are nucleated and *may* be longer-lived and reusable, although both probably perform the same functions.

D. IMMUNE FUNCTIONS

It was first thought that teleost fish did not have an immune response, but an immune response to an antigen was later shown to occur in salmonids, though somewhat more slowly than in mammals and mostly in the middle and higher parts of their environmental temperature range. Hagfish barely show an immune response after a very long exposure to an antigen, lampreys show a somewhat stronger response and sharks form antibodies about as well as teleost fish. Thus there appears to have been a progression of the development of immune responses during vertebrate evolution, with the teleost fishes (mostly salmonids have been studied) somewhere in the middle between hagfish and mammals.

Production of an antibody is a process which occurs in lymphocytes in response to the presence of foreign protein material, including intact cells such as bacteria. The lymphocytes are sensitized by the foreign protein (antigen) and, if they survive long enough, release a new protein material (antibody) into the blood plasma which neutralizes the foreign protein. The foreign proteins or cells become covered with antibody molecules which neutralize their toxic capabilities and allow them to be safely engulfed by phagocytic cells. Antibodies usually have more than one reactive site and so can cause foreign particles and cells to stick together (agglutinate) when different cells stick to the same antibody molecule. Antibody molecules with one end attached to a bacterial cell can also activate a non-antibody protein called complement. Antibody and complement molecules together can cause lysis (rupturing) of bacterial cells. Complement is produced by the fish from its own globulin proteins in blood and is non-specific (works with any antibody). Antibody production occurs mostly in the spleen and head kidney, in the liver and peripheral blood to a small degree, and basically wherever lymphocytes congregate.

The production of antibodies (immune response) is usually a part of the process of inflammation. In man, external signs of inflammation include pain, swelling, redness and fever. In fish there is no way to measure pain, redness is usually obscured by the opaqueness of the skin and fever is impossible under water (although a change in temperature preference toward higher temperature has been described as a "behavioral fever"). This leaves swelling as the major sign of inflammation in fish. At the cellular level, however, fish and mammals are quite similar. Histamine and other substances released at the inflammation site cause vasoconstriction to block off blood flow in the area and prevent spread of pathogens. This is also what causes the swelling. A fibrin network also forms around the injury site. Although capillaries in the area may swell and would otherwise cause reddening, in fish skin inflammation also usually stimulates dispersion of the melanocytes and the skin darkening obscures any reddening. The mechanism for melanocyte concentration is disrupted by inflammation, but not the dispersion mechanism, so the affected area darkens. After the area is walled off phagocytic cells enter, ingest the foreign material, return to the kidney and spleen (mostly) during a latent period of up to two weeks and eventually an antibody is released into the blood.

A disease organism would not be passive during these inflammation processes, of course. Many bacteria produce toxins which kill lymphocytes and other cells, which then appear as pus. Bacteria may also produce lytic agents to dissolve the fibrin wall around the inflamed area and allow bacteria to gain entry into the blood stream or other body spaces. Upon gaining entry, disease organisms such as *Vibrio anguillarum* and IHN virus (infectious hematopoietic necrosis) can kill salmonids in as little as 5 and 9 days, respectively. This is less time than is required for these fish to produce antibodies. Thus an immune response is only useful to fish which survive their first bout with an infection.

Fish are not totally unprotected from the first attack of a given disease organism, however. First, scales and mucus act as mechanical and chemical barriers to entry. Mucus seems to be an excellent bactericide and fungicide in many species. Some commercially important species of fish store better if the slime layer is left intact when the fish are put on ice. Antibodies produced during an immune response may also sometimes appear in mucus. Some species of fish apparently have a natural resistance to diseases which readily infect closely related species. Most vertebrates have natural agglutinins which agglutinate a variety of antigens. Salmonids also have interferon—a general antiviral agent—in their blood. Entry through the intestinal tract is made difficult by the low pH of the stomach and the proteolytic activity of enzymes in both stomach and intestine. One response to stress which has been proposed, however, is gut stasis (no peristalsis), which allows anerobic fermentation of the gut contents and, at the same time, the intestinal enzymes may attack the gut wall. This would allow rapid proliferation of many pathogens as well as easy entry into the blood stream. Gut stasis is seen in mammals as a common route of entry of pathogens and has been proposed in adult salmon. In general, most fish appear to have a variety of more or less potentially pathogenic organisms on them, inside them and around them just waiting for one of the entry barriers to falter.

Passive immunization is the introduction of preformed (in other fish or in mammals) antibodies to give immediate protection (passive immunity). In active immunization the disease organism itself, treated to inactivate its disease producing capability, is given to stimulate a fish to produce its own active immunity. Finding routes for introduction of antigens is a considerable challenge which involves consideration of the form of the antigen, the physiology of the fish and practical considerations as to the size and numbers of fish to be vaccinated. Various inoculation media are used depending on whether the antigen is cellular, viral or soluble in saline or oil.

Several routes of introduction of vaccines have been tried with varying success. One of the more common routes is by injection.

Interperitoneal injection provides relatively slow uptake and prolonged exposure to the antigen. Subcutaneous injection gives similar uptake. Intramuscular injection is rarely used in fish because uptake is poor and an abscess often forms at the injection site—i.e., the inoculum is successfully walled off and completely rejected by the fish. For more rapid uptake, material is sometimes injected on the dorsal midline into the space between the epaxial muscles and the vertebral neural spines. Uptake here appears to occur through a dorsal lymphatic duct.

Injecting vaccines into a large number of fish or very small fish is laborious and thus costly, so other routes have been tried. Oral vaccination is logically possible by adding vaccine to the fish's food, but the results have been undependable. One of the problems appears to be that of getting the antigen past the digestive enzymes and into the fish's blood stream intact. An equally difficult problem is that by the time the fish grower becomes aware of having sick fish, the fish have stopped or at least greatly reduced their feeding. The fish in a population which most need the vaccine are least likely to eat it. Oral immunization becomes primarily a preventive measure rather than a cure. Giving antibiotics to fish in their food to cure a disease has the same limitation. Another way to give vaccines which is still experimental, but rather exciting, is hyperosmotic delivery. Fish are put into a strong (5.3 percent) salt and urea solution for two minutes and then into the vaccine (or other) solution. In the hyperosmotic solution they lose up to 6 percent of their body weight (presumably across the gills) and regain up to one third of the loss in the vaccine. The route of entry appears to be (in salmonids) through the lymphatic duct immediately beneath (and perhaps connected to) the lateral line canal. The method has also been used for uptake of things as large as bacteria and India ink, so the entry route will accept comparatively large particles. More recent vaccination methods simply dip fish in the vaccine without any hyperosmotic solution and achieve satisfactory uptake, presumably through the lateral lymphatic duct again.

Chapter IV: RESPIRATORY SYSTEM

A. OVERVIEW AND GENERAL PRINCIPLES

1. Availability of Oxygen in Water and Air

There is little oxygen in water compared to air, even under the best of conditions. Air is about 20% oxygen, with the remainder being mostly nitrogen with small amounts of other inert gases. In air, the volume and partial pressure of each of the gases is proportional to the quantity of each gas (number of molecules) and total atmospheric pressure. Thus a liter of air contains about 200 ml of O_2 (20% x 1000 ml) at a partial pressure (P-O_2) of about 150 mm Hg or 0.2 atmospheres (20% x 760 mm Hg or 20% x 1 atmosphere). In water which is in equilibrium with air—i.e., saturated—the percentage composition of the dissolved gases and their partial pressures are the same as in air. The quantities of gas dissolved in water differ drastically from those in air even when the partial pressures are identical to air. A liter of water at 15 °C contains only about 7 ml (= 10 mg) of oxygen and about twice that much inert gas because oxygen is not very soluble in water and nitrogen is even less soluble.

The low quantity of oxygen in water has several ramifications. First, fish (or other aquatic animals) must either pump large quantities of water over their respiratory surface to obtain reasonable amounts of oxygen or else be limited to relatively low metabolic rates (see Chapt. V for comparative aspects). Not only is the volume necessarily large, but water is also 800 times denser than air and therefore comparatively costly to pump. Second, as fish remove a relatively large proportion of the O_2 from the water, the partial pressure (P-O_2) decreases in proportion to the fraction of the total O_2 which was removed. In air, for example, removal of 5 ml of O_2 from a liter of air removes 5/200 of the oxygen, causing a change in P-O_2 from 150 to 146 mm Hg. In water, removal of 5 ml of O_2 removes 5/7 of the total gas, leaving a P-O_2 of only 43 mm Hg. Fish thus quickly lose the diffusion gradient needed to transfer O_2 into their blood as soon as they remove very much O_2 from the water. Therefore fish hemoglobins generally operate at lower partial pressures (have a higher O_2 affinity) than hemoglobins of air-breathing vertebrates. Finally, fish are prevented from having extra-large respiratory surfaces because of the osmoregulatory problems that would be created (Chapt. II).

The relatively low availability of O_2 in water is further lessened by both natural and man-made circumstances. The solubility of O_2 in water decreases as temperature rises and, of course, is zero at boiling. Solubility also decreases with salt content, so that normal seawater contains about 20% less O_2 than freshwater at the same temperature. Thus a tropical sea can be a difficult place to respire. Many kinds of man-made and natural pollution also consume oxygen, sometimes to the point of leaving no O_2 in the water. An obvious evolutionary alternative for survival in tropical swamps, where high temperatures and rapid decay of vegetation frequently produce an anoxic environment, is to breathe air—and a number of fish do so. A goldfish sucking at the surface of the water in a too-small fish bowl is attempting to do the same thing—escape the confines of low P-O_2—and gain a bit of respite by using the thin layer of air-saturated water at the surface or perhaps by actually breathing some air. The solubilities of O_2 at various temperatures in freshwater and seawater are given in Table IV-1.

2. Ventilation Requirements

Measuring the oxygen content of a fish's inhaled and exhaled water and then knowing the rate of oxygen consumption for that fish allows one to calculate its ventilation requirements. If a fish inhales air-saturated water at 15 °C and removes 30% of the oxygen, this means that the inhaled water contains about 7 ml O_2/liter and exhaled water about 5 ml O_2. If the fish consumes 70 ml O_2/hr., then the fish must pump 35 liters of water over the gills. The fish can change this only by changing its rate of oxygen consumption or the proportion of oxygen removed from the res-

TEMP. (°C)	FRESHWATER		SEAWATER	
	ml O_2/l	mg O_2/l*	ml O_2/l	mg O_2/l*
0	10.22	14.60	8.20	11.71
5	8.93	12.75	7.26	10.37
10	7.89	11.27	6.52	9.31
15	7.05	10.07	5.93	8.47
20	6.35	9.07	5.44	7.77
25	5.77	8.24	5.00	7.14
30	5.28	7.54	4.58	6.54

Freshwater data from Weiss, 1970
Sea water data from Sverdrup, Johnson, and Flemming, *The Oceans*, 1946
1 mg O_2 = ml O_2 (STP)
*ppm

Table IV-1. Oxygen content of freshwater and seawater at selected temperatures and in equilibrium with one atmosphere of air pressure. (Data selected from several tables in Weiss, 1970, with permission of publisher.)

piratory water. Sedentary fish such as flatfish, for example, tend to have low rates of oxygen consumption and remove up to 80% of the oxygen in the water. Combining both of these features produces a comparatively low ventilation rate.

Most fish fall into a category of either an oxygen regulator or an oxygen conformer, depending on the way they adjust their ventilation rates. Oxygen regulators maintain a relatively constant rate of oxygen consumption during decreasing P-O_2 by increasing both the ventilation volume and the proportion of O_2 extracted from the water. Salmonids regulate their O_2 consumption rate from an environmental P-O_2 of about 5 mg O_2/liter and higher (see Fig. IV-7 for example). Oxygen conformers, on the other hand, do little to change their ventilation volume but make large adjustments in their rate of oxygen consumption in direct proportion to the availability of O_2 in the environment. Sole and flounder are typical oxygen conformers, as are many other sedentary fish.

3. Carbon Dioxide Transport

In most respects, the chore of transporting gases which are dissolved in water is a strenuous one because of the low solubility of the gas and density of the water. Carbon dioxide, however, is an exception, being very soluble in water. At 0 °C, CO_2 is 35 times more soluble than O_2 and is still 25 times more soluble at 30 °C. This fact makes for several important differences between air-breathers and water-breathers. First, this large solubility of CO_2 minimizes the partial pressure of CO_2 in the water. If you exchange equal volumes of O_2 and CO_2 in a fixed volume of water, the P-O_2 decreases drastically, but the P-CO_2 increases only slightly. In air the two partial pressures would be equal. This relationship is illustrated in Fig. IV-1, and ventilation volumes for different oxygen removal rates are also included. Second, combining this high solubility with the fact that most natural water has little CO_2 content makes CO_2 levels in fish blood very low compared to air-breathers. Almost all of the CO_2 diffuses out of the venous blood when it passes through the gills. The P-CO_2 of venous blood rarely exceeds 10 mm Hg in fish and most often

Fig. IV-1. O_2 - CO_2 tension diagram showing the simultaneous O_2 and CO_2 tensions which occur when air is breathed, air R = 1 (equal amounts of O_2 consumed and CO_2 produced), and when water (20 °C) is breathed, water R = 1. The inspired O_2 and CO_2 tensions are the same in air and in water. The numbers following V_G (gill ventilation) and V_A (air ventilation) indicate the ventilation volumes (ml/min) in each case required to exchange 1 ml O_2 at the particular O_2 tensions (P-O_2). (From Rahn, 1966, with permission of author.)

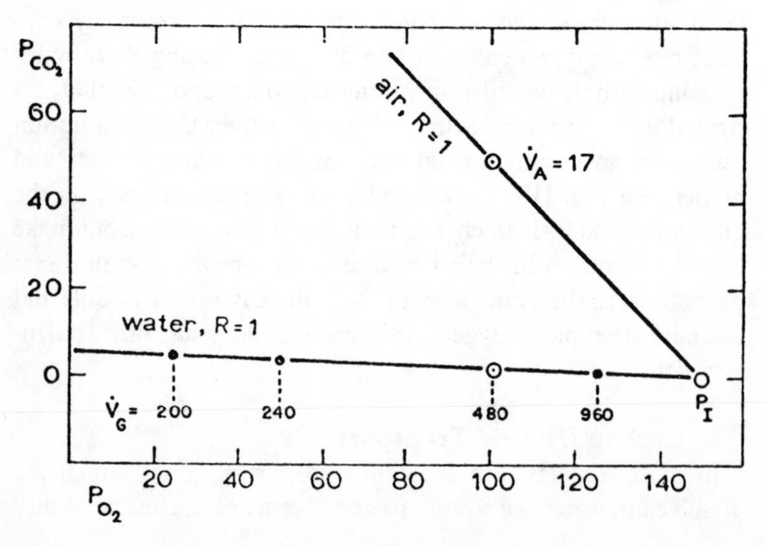

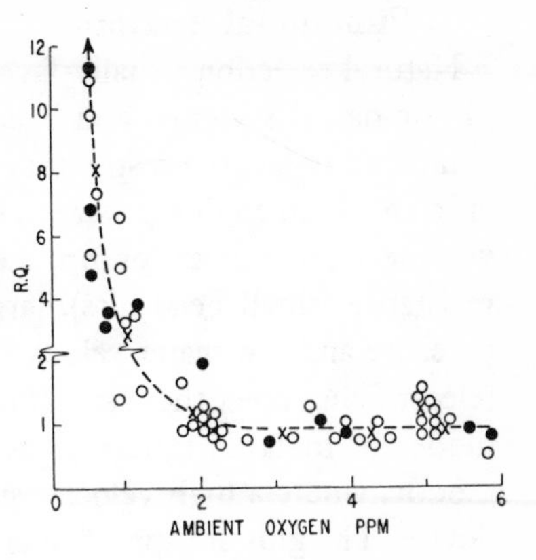

Fig. IV-2. The respiratory quotient ($R.Q. = \frac{CO_2 \text{ produced/unit time}}{O_2 \text{ consumed/unit time}}$) in relation to ambient oxygen concentration in *Tilapia mossambica* in freshwater at 30°C. The solid and open circles are data from two different methods for measure CO_2 and the X's are mean values for specific groups of fish. See text for further explanation. (From Kutty, 1971, with permission of author and publisher.)

is less than half of that, while air-breathers rarely have less than 40 mm Hg. This means that air-breathers and fish have differently behaving kinds of hemoglobin (see "Hemoglobin Functions" in this chapter). It also means that fish which live part-time in air have problems getting rid of CO_2 while in air and that fish cannot regulate their respiratory system by its changes in CO_2 levels as humans do—the changes are too small—but must sense O_2 changes instead (which humans cannot do).

Carbon dioxide can also be excreted as HCO_3^- as well as molecular CO_2 depending on the environmental pH and carbonate levels. If arterial P-CO_2 rises because of increased external CO_2, then venous P-CO_2 also rises a few mm Hg above the arterial P-CO_2—i.e., the diffusion gradient is still outwards. Varying the amounts of CO_2 excreted as CO_2 or HCO_3^- is a way of regulating blood pH.

One application of measurements of CO_2 production is in determination of the respiratory quotient (R.Q.). R.Q. is the amount of CO_2 produced per unit time over the O_2 consumption for the same time. Generally the R.Q. falls between 0.8 and 1.0. When less than 1.0, the interpretation is that the fish's metabolism is oxidizing some other substrate in addition to carbon—nitrogen (from proteins) or hydrogen (from fats). Under low environmental O_2, however, the fish turns increasingly to anerobic metabolism, oxygen consumption decreases and the R.Q. increases, as in the case of *Tilapia*. (Fig. IV-2).

B. THE RESPIRATORY PUMP

1. Functional Anatomy

Natural selection usually favors the conservation of energy in physiological systems, and because water is so heavy to pump compared to air, the respiratory pump of fishes should be highly efficient. If an engineer were to write the specifications for a high volume, low power pump, he would specify one with low resistance (small head loss), large diameter water passages, low pressure and low water velocity. The respiratory pumps of most teleost fishes meet these specifications quite well. Most fish have moderate to large mouths, excepting fish with small, tubular mouths where a high velocity suction stream serves as a feeding device. The gills are spread over considerable area so that there is little resistance to flow (little head loss) and the moveable parts of the pump (lower jaw and opercula) are large so that they don't have to move very far or very fast.

The resulting water flow is pulsatile. Further efficiency probably would be possible if the water could be made to flow smoothly. However, the respiratory pump of fish is functionally comparable to a double piston, single-acting pump (Fig. IV-3). Even by operating the two pistons (lower jaw, opercula) slightly out of phase—operculum acting slightly after the lower jaw—there is still no possibility for smoothly continuous flow. Being slightly out of phase decreases the importance of the buccal valve, however, by maintaining a negative pressure behind the gills (opercular cavity still expanding) after the lower jaw starts to close, thus minimizing reverse flow out of the front of the mouth. For example, in mature male salmon with strongly hooked snouts the buccal valve in the roof of the mouth cannot possibly meet the lower jaw because the hook prevents the jaws from closing that far, but respiratory pumping seems unaffected.

The respiratory pumps of teleosts and sharks differ in at least one major respect. In teleosts, muscles actively open as well as close the opercula. In sharks there is no operculum, but cartilage rods support the walls of the branchial cavity. The refilling of the branchial cavity at the end of the exhale cycle depends entirely on the elasticity of the cartilage. Sharks can speed up the emptying of the branchial cavity but not the refilling. Teleosts can

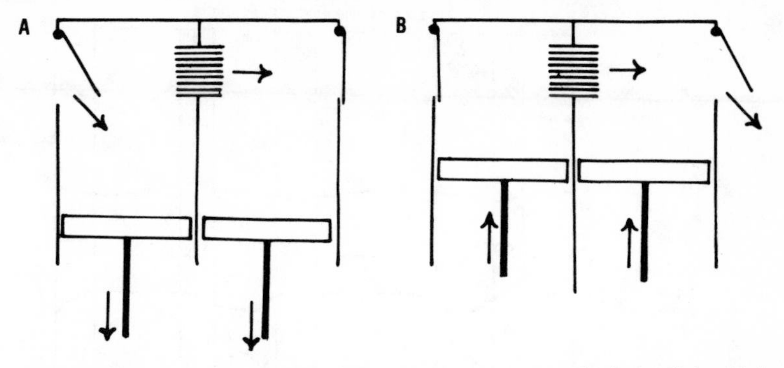

Fig. IV-3. A mechanical model of the respiratory pump of teleost fish which could be described as a two cylinder, single-acting pump. Diagram A shows inspiration into the buccal cavity, B shows expiration through the gills and opercular cavity. (From Alexander, 1967, with permission of author and publisher.)

forceably inhale as well as exhale because muscles operate both phases.

Salmon swimming at moderate and high velocities cease to use their respiratory pump and utilize ram ventilation instead. Water moved past the gills during forward motion with the mouth held partly open meets respiratory needs. This is not to say that ram ventilation is free, because holding the mouth open must produce more drag than keeping it closed. Another way of looking at ram ventilation is that the work of respiratory pumping has only been transferred from the buccal pump to the caudal fin. Some fish such as the menhaden and tunas routinely use ram ventilation most of the time.

2. Pumping Pressures and Volumes

The mechanical motions of the respiratory pump and the pressures produced by the pump in trout appear in Fig. IV-4. The predominant phases indicated across the top of the figure are the opercular suction pump in phase 1 and the buccal pressure pump in phase 3. Phases 2 and 4 are transitional between the other two. The valving is also indicated, with the buccal flap valve being either extended (C) or retracted (O) and a gradual opening and closing of the opercular valve. The pressures shown are relatively low and characteristic of a resting fish. Even in a trout which is vigorously pumping water, the

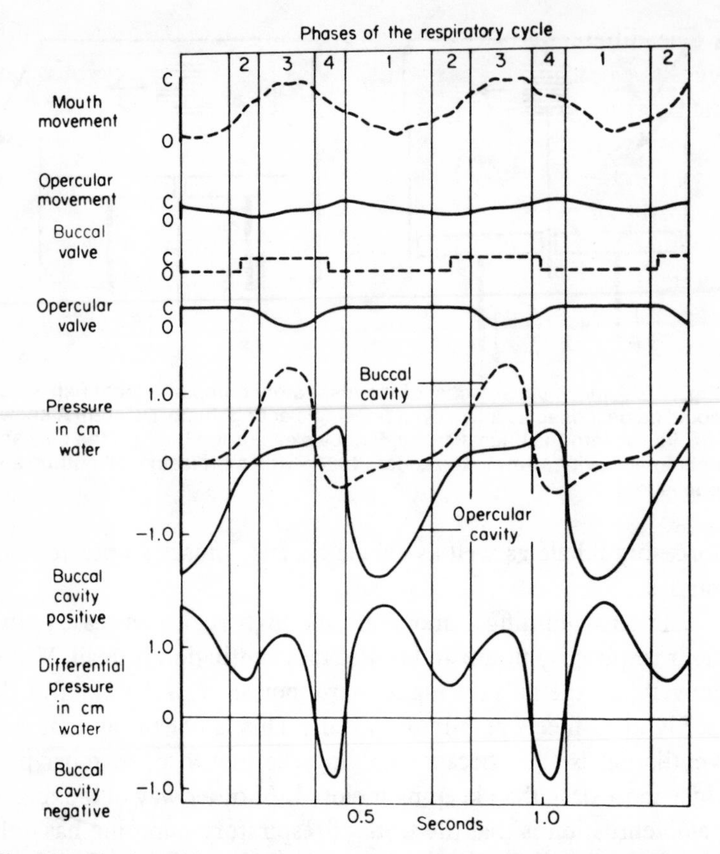

Fig. IV-4. The breathing movements of the mouth and operculum of a 70g trout together with the associated changes in the buccal operculum, and their associated valves. (From Shelton, 1970, after Hughes and Shelton, 1958, with permission of author and publisher.)

pressures would not go more than about 3 to 5 times higher. Note that the units in the figure are in cm of water rather than in mm of mercury (1 mm Hg = 1.3 cm H_2O). The use of water units probably relates back to earlier times when mercury-filled manometers were the standard pressure-measuring devices, but water-filled ones were substituted to gain greater sensitivity for low pressures and small changes. Since most recent measurements are made with electronic devices and the inertia of the mercury is not a problem, the use of mercury height units is more common. In either case the pressures shown fully meet the specifications of a low pressure pump.

The differential pressure shown at the bottom of Fig. IV-4 gives the best idea of the actual water flow because it indicates the pressure gradient between the buccal and opercular cavities. A positive differential indicates flow from buccal to opercular cavity and a negative differential the reverse. Thus in trout there is mostly a continuing and somewhat pulsatile flow over the gills, but there is also a brief reversal of flow. Most fish also have a "cough" reflex for clearing the gills of irritating matter with a more severe reversal of flow.

Other species have other patterns of differential respiratory pressures. Respiratory pumping pressures for carp are shown in Fig. IV-5. The respiratory cycle is much slower than in the trout in the previous figure and almost avoids having any reverse flow at rest (Part A). With more strenuous pumping (Part B) there is no reverse flow, although the flow is strongly pulsatile. There are still other variations on respiratory pumping patterns which depend on such things as the relative sizes of the buccal and opercular cavities, whether suction is used for obtaining food, whether the fish swims a little or swims a great deal and has the opportunity for ram ventilation, etc.

Respiratory volumes are difficult to measure directly because fish show irritation to most restraining procedures which have been devised to channel the water flow through a measuring device. Results from one restraining method (cuff of a rubber glove attached around the fish's mouth) are shown in Fig. IV-6 for rainbow trout. Resting ventilation volumes in the vicinity of 40 ml/min (175 ml/kg/min) increased to as much as 160 ml/min (700 ml/kg/min) while the fish was moderately active in the confinement chamber. During the increase in flow rate, the pressure differential between buccal and opercular cavities increased steadily, but the resistance to flow changed very little.

Water flow is also measured indirectly by measuring the fish's O_2 consumption and the difference in oxygen content of the water before and after passing over the gill lamellae and then calculating the water flow required to deliver the consumed oxygen. An example of data from the indirect measurement method is shown in Fig. IV-7 for adult coho salmon in seawater. The fish were swimming at a steady "cruising" speed of about one body

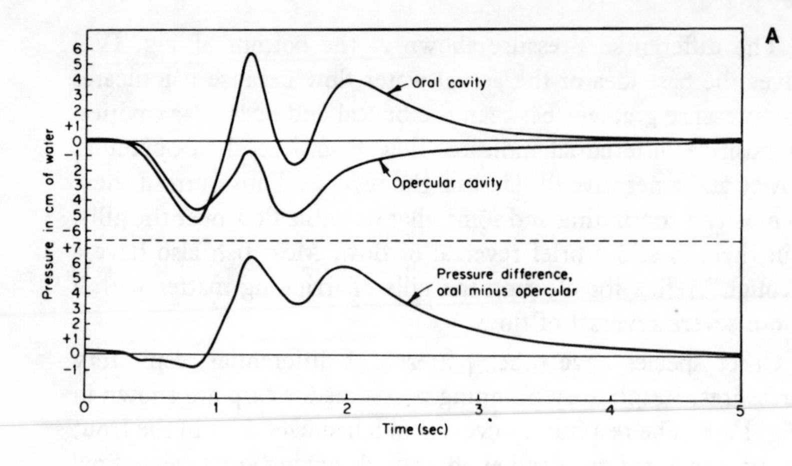

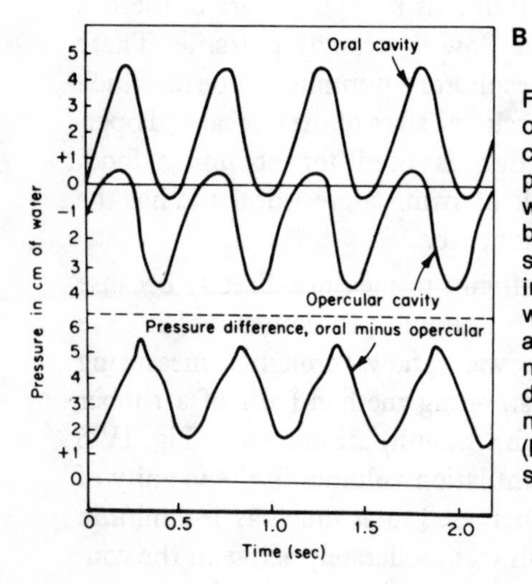

Fig. IV-5. Pressure changes in the oral and opercular cavities of carp, together with differential pressure curves. The records were taken from (A) quietly breathing fish and (B) fish stimulated to breathe strongly by increasing the CO_2 content of the water. Carp show only small amount of reverse flow (shown as negative pressure differential) during quiet breathing (A) and none during strong breathing. (From Shelton, 1970, by permission of the author and publisher.)

length/sec during both moderate and decreasing O_2 levels. Respiratory volumes considerably higher than those measured for trout were not unreasonable, considering the difference in activity levels. There was a further rise in ventilation volume during a period of decreasing O_2 level. The oxygen consumption rate decreased at the beginning of the reduction in environmental oxygen, but then recovered to nearly normal by the end of the observation period. This is an example of oxygen regulation.

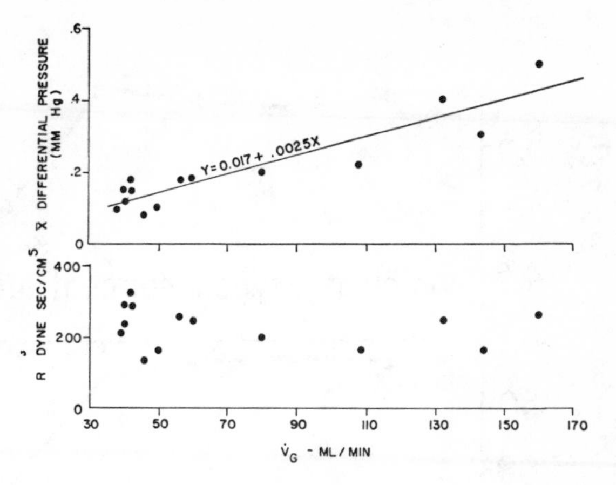

Fig. IV-6. The upper panel shows that the average pressure differential ($\overline{X}$ differential pressure) between buccal and opercular cavities increases steadily with increasing gill ventilation (V_G) but that the resistance to water flow (R) in the gills changed very little (lower panel). (From Davis and Randall, 1973, with permission of author and publisher.)

3. Efficiency of Oxygen Extraction and Energy Cost of Respiratory Pumping

Since respiratory pumps are relatively versatile in adjusting to changes in activity and the oxygen content of the water, it is difficult to make broad generalizations about their capabilities and limitations. One feature which is relatively common is that extraction efficiency (percentage of the available oxygen which is removed by the gills) decreases when ventilation volume increases. An example of this appears in the bottom graph of Fig. IV-7. The less active the fish and the lower the ventilation volume, apparently the greater the extraction efficiency can be. Benthic, sedentary fishes such as sole have been observed to extract up to 80% of the available O_2 when resting very quietly. One factor necessary to do this seems to be a tolerance for very low oxygen levels in the major tissues, which allows low venous $P\text{-}O_2$. The water leaving the gills can have a $P\text{-}O_2$ no lower than the venous blood; if the water had a lower $P\text{-}O_2$, oxygen would diffuse from blood to water.

The change of extraction efficiency (also called utilization) with activity varies with the species of fish. In salmonids, although extraction efficiency decreases with increased ventilation, the decrease is not great. In goldfish and carp where the resting extraction efficiency is high, the active extraction efficiency is quite low. Additional comparative data are presented in Fig. IV-8, which shows a wide range of species and changes.

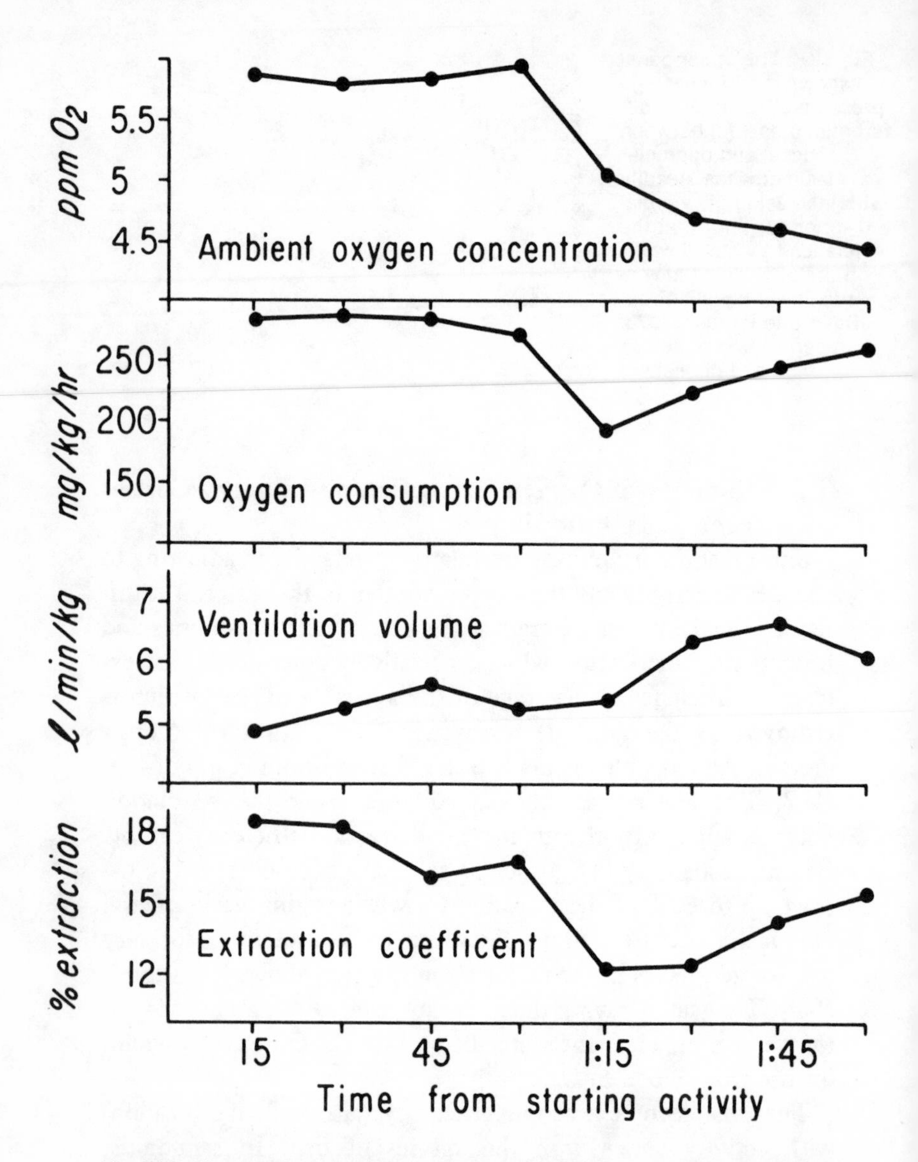

Fig. IV-7. Changes in respiration of adult coho salmon in seawater (20-28%) swimming steadily through water of decreasing oxygen content (top panel) at 12-15°C. Oxygen consumption decreased, but recovered to near normal. Ventilation volume increased, but the proportion of the available oxygen removed from the water passing over the gills (% extraction) decreased with some recovery toward the end of the observation period. (From Smith, et al., 1971.)

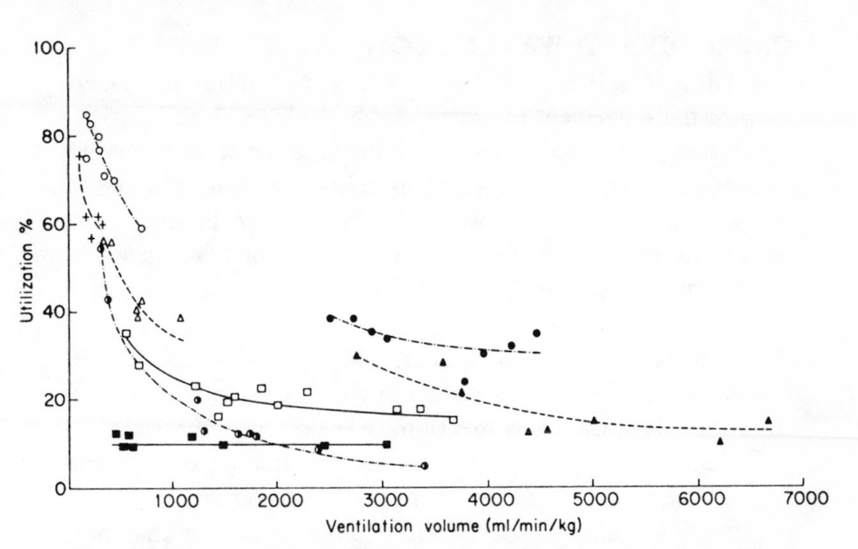

Fig. IV-8. Relationships between utilization (extraction) and ventilation volume under different circumstances in four species of fish. Fish included carp (circles), sucker (triangles), trout (squares), and dogfish (pluses). Conditions included hypoxia (open symbols), exercise (black symbols), recovery from exercise (pluses) and a combination of hypoxia and excess carbon dioxide (half-black circles). (From Shelton, 1970, with permission of author and publisher.)

With such a wide range of ventilation volumes and utilizations possible, different attempts to assess the energy involved have produced widely varying results. In salmonids, Alexander (1967) estimated that the energy cost was equivalent to 0.5% of the resting O_2 consumption and up to 15% of the active O_2 consumption. For an oxygen consumption increase of about 5-fold, this meant an actual increase in energy cost of about 150-fold. In tench, the energy cost for respiratory pumping ranged from 18-43% of the resting oxygen consumption to 44-69% at a ventilation volume three times greater than resting. By comparison with a resting energy cost in humans of about 2% of the oxygen consumption, it seems likely that fish actually do pay a considerable metabolic penalty for having to pump water instead of air across their gills. It is also instructive to note that air-breathing fish have as little as 20% of the gill surface area of their completely aquatic counterparts—aquatic respiration really does require relatively large surface areas to minimize the pumping costs.

C. BLOOD AND WATER FLOW PATTERNS IN GILLS

1. Characteristics of a Countercurrent Exchange System

Since the movement of oxygen from water to blood is strictly by diffusion—no active transport or other energy-consuming system for oxygen is suspected to exist—the many millions of years of natural selection would be expected to produce some highly effective means of enhancing the diffusion of oxygen. The countercurrent exchange system found in some fish is such a means.

The basic idea of a countercurrent exchange system is contrasted in Fig. IV-9 with a system in which flow in both parts of the system is in the same direction. In the co-current system, the diffusion gradient is large at first and the amount of O_2 transferred is also large, but then rapidly gets smaller and smaller as the two fluids approach equilibrium. There is no way that more than half of the dissolved oxygen could be transferred from flow A to flow B. By reversing the direction of one of the flows, however, a whole new set of conditions occurs. As the deoxygenated

Fig. IV-9. Diagram showing how a typical teleost gill achieves countercurrent flow of blood and water. (From Wedemeyer, et al., 1976, by permission of TFH Publications.)

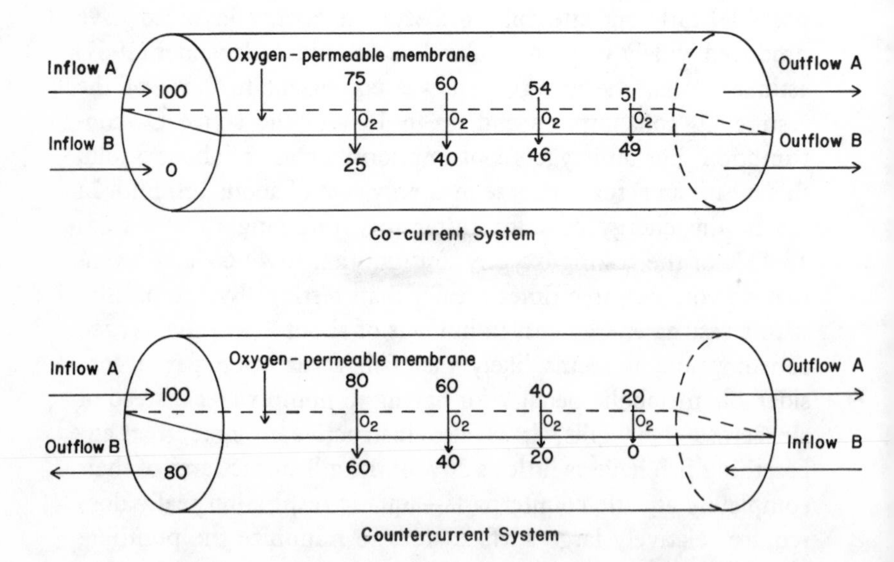

inflow (B) obtains some oxygen from the A side of the membrane, it then meets conditions where the oxygen content of A is higher and higher. Although not many fish achieve extraction values of 80% as shown in this hypothetical model, the possibility of getting more than half of the available oxygen is clearly workable because of the nearly constant diffusion gradient over the entire length of the countercurrent system.

The manner in which fish actually produce countercurrent flow is not as obvious as the two-tube system in the model above. Countercurrent flow occurs only at the finest level of subdivision of the gills, the lamellae, which are not obvious to the naked eye except in large fish. The lamellae project (Fig. IV-10) from the dorsal and the ventral surfaces of each gill filament. Lamellae from each gill filament may interdigitate with lamellae of adjacent gill filaments. Each pair of gill filaments is held by the gill bar in a V-shape so that water flows through the sides of the V rather than around it. Filaments from adjacent gill bars often touch at the tips so that there is no way for water to flow through the gills except past the lamellae. While gill filaments in most species are stiffened with a cartilaginous rod, even then the V-shape may be squeezed shut under high velocity water flow (as during rapid swimming), allowing significant amounts of water to bypass the lamellae without opportunity for gas exchange to occur. This is one reason for the loss of extraction efficiency noted above at higher ventilation rates. Some fast-swimming fish—tunas, for example—have the tips of the gill filaments joined to adjacent gill filaments to prevent the gill filament V's from collapsing and to maintain maximum gas exchange at high water velocities. (Muir and Kendall, 1968).

Cool

With water flowing consistently past the lamellae in one direction, countercurrent exchange is readily achieved by having the blood flowing in the opposite direction inside the lamellae. Thus, having the afferent (inflow) branchial arteries on the inside of the filament V's and the efferent (outflow) branchial arteries on the upstream side produces countercurrent flow in the lamellae. At this point the only thing that is special about the lamellar blood flow is the presence of the pillar cells inside the lamellae to prevent the lamellae from bulging outward under the relatively high

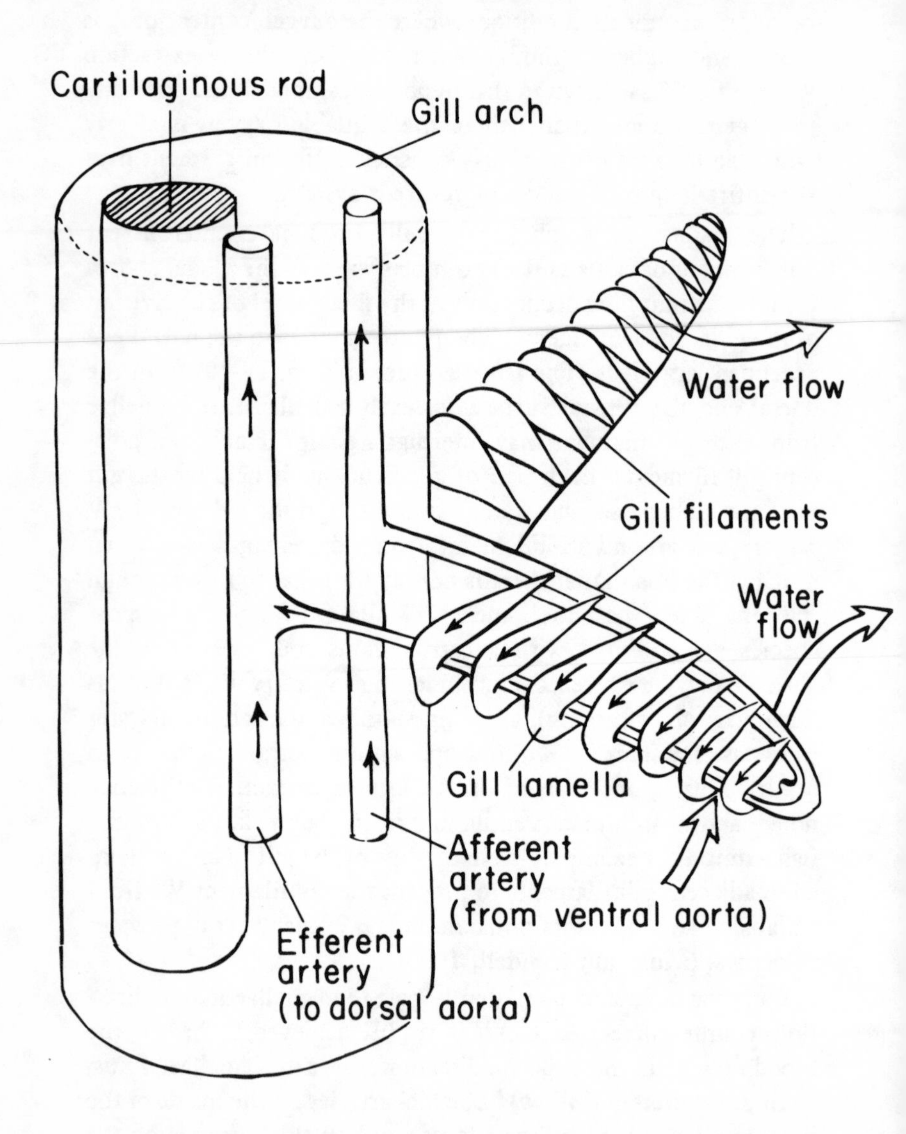

Fig. IV-10. A diagram to illustrate the pattern of blood flow through and water flow over fish gills so as to produce a countercurrent arrangement. (From Wedemeyer, et al., 1976, by permission of TFH Publications.)

blood pressure found in the gills. This keeps the blood in a thin layer and the diffusion distances short for maximum ease of oxygen uptake. The thickness of the lamellar tissue through which oxygen must travel between blood and water has been measured in various fish to be between 1 and 5 μ in thickness.

2. Alternate Pathways for Blood in the Gill Filaments

Since osmoregulatory problems can be avoided by having a minimal surface area for gas exchange, it should not be surprising to find that at least some fish have a means to adjust the effective surface area of their gills to suit their actual oxygen needs. Earlier workers described a blood bypass in which most of the blood from the afferent branchial artery was diverted into a central sinus in the gill filament under the influence of acetylcholine in resting fish rather than passing it through the lamellae. Adrenalin from excitement or stress caused most of the blood to flow through the lamellae. This was a simple and neat explanation of how a fish could regulate its respiratory and osmoregulatory exchange at the gills without having to physically change the size of the actual exchange surface.

However, recent work by Laurent and Dunel (1976) suggested a more complex system. Using scanning electron microscopy, they found that blood from the afferent branchial artery usually passed through the lamellae and then could either be diverted through a capillary bed into a central sinus or go to the dorsal aorta via the efferent branchial artery (Fig. IV-11). Regulation would still be through adrenalin or acetylcholine as before, but the point of regulation was further downstream than before. The additional complication was the presence of a new venous system, the branchial veins, which largely enveloped both the afferent and efferent branchial arteries and returned the bypassed blood directly back to the venous side of the heart. The central sinus also connected to the branchial vein at the base of each gill filament. The branchial vein should not be confused with the coronary artery, which supplies oxygenated blood from the gills to the heart muscle. The work of Laurent and Dunel (1976) also demonstrated that significant variations on the basic plan occur

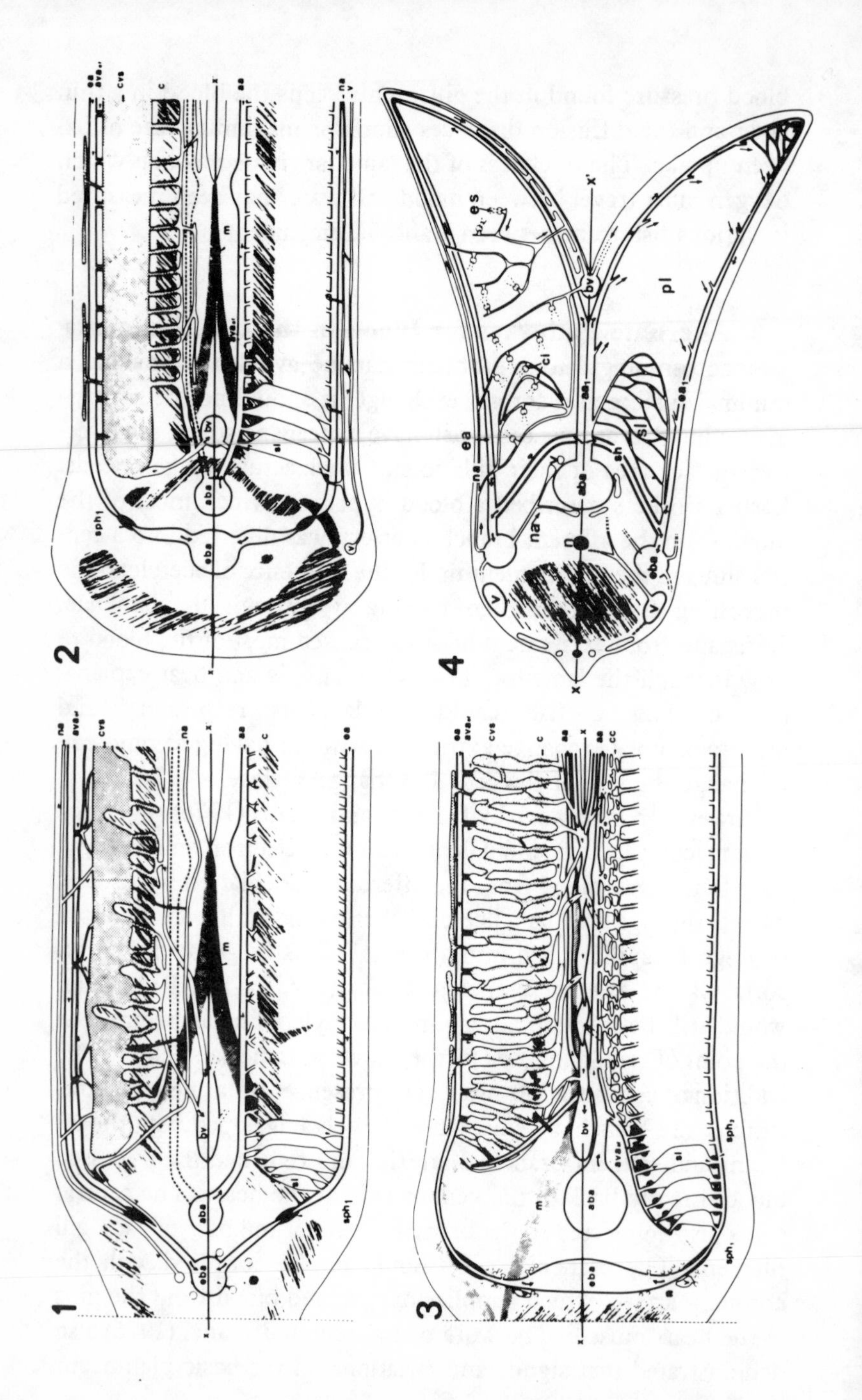
1
2
3
4
es
bv
pl
cl
sl
sh
ea
na
aba
eba
m
v
x
x'

in different species of fish, four of which appear in Fig. IV-11.

It is not clear yet how the branchial vein system might contribute to the minimizing of osmoregulatory problems at the gills. Since most of the blood goes through the gill lamellae most of the time, the osmoregulatory impact on the blood is maximal most of the time. By returning most of the blood directly to the heart, however, the osmoregulatory problem is mostly limited to a closed loop composed of the gill vessels, branchial vein, heart and ventral aorta. Whatever amount of blood goes into the dorsal aorta, however, would seem to be carrying its full share of osmoregulatory problems to the various organs and tissues although in relatively small volumes. The bioenergetic implications of the branchial vein system remain to be examined.

3. Ventilation-Perfusion Ratio, Control of Respiration

Provision of adequate gas exchange under varying conditions of activity and environmental oxygen levels should involve some coordinated regulation of the volume of water flow over and the volume of blood flow through the gills. If there were more water delivered to the gill lamellae than needed to saturate the blood, for example, there would have been a waste in terms of the effort expended for ventilation. Natural selection for a metabolically efficient system would be expected to lead to some kind of optimization of such factors.

Opposite page: Fig. IV-11. Schematics of the gill vasculature in 1) Rainbow trout *(Salmo gairdneri)*, 2) Eel *(Anguilla anguilla)*, 3) Sturgeon *(Acipenser baeri)*, 4) Lungfish *(Protopterus aethiopicus)*. aa - afferent artery; aba - afferent branchial artery; ava_{af} - arteriovenous anastomosis, afferent side; ava_{ef} - arteriovenous anastomosis, efferent side; bv - branchial vein; c - cartilage; ci - cisterna; cc - corpus cavernosum; cvs - central venous sinus; ea -efferent artery; eba - efferent branchial artery; es - extracellular space; m -muscle; n - nerve; na - nutrient artery; pl - primary lamella; sl - secondary lamella; sph_1 - efferent artery sphincter; sph_2 - pre- or post-lamellar sphincter; v - vein. Note that all four species have the same three major vessels -aba, eba, and bv—but that the smaller vessels, presence and placement of sphincters, and detailed flow patterns are different, i.e., the gills of all species are not alike even though they may appear similar externally. (By courtesy of Pierre Laurent from Suzanne Dunel and Pierre Laurent, "Functional organization of the gill vasculature in different classes of fish," in B. Lahlou [Ed.] Epithelial *Transport in Lower Vertebrates*. Cambridge University Press [in press]. See also Laurent and Dunel, 1976.)

In comparison to air-breathing vertebrates which typically have ventilation-perfusion ratios of about 1:1, fish have much larger ventilation volumes and rather variable ones. The ventilation-perfusion ratio in fish varies from 10:1 in carp and dogfish to as high as 80:1 in trout. This is a reflection of the relative scarcity of oxygen in water as compared to air.

There is also a tendency for the heart to beat in some degree of synchrony with the ventilation cycle so that maximum velocity of blood flow coincides with maximum velocity of water flow. Presumably this would produce more efficient gas exchange than having maximal flows of blood and water occurring in an uncoordinated fashion. There may be a 1:1, 1:2 or 1:3 ratio between heart and breathing rates, and it is common for the heart beat to occur during a particular phase of the respiratory cycle. In trout synchrony between heart and ventilation rates occurs mostly during periods of maximal rates of gas exchange, as during recovery from severe hypoxia. Other observations of ventilation-perfusion coordination include quietly resting fish in which there may be oscillations in the breathing rate which are followed by the heart rate. During deep anesthesia in which ventilation ceases, the heart also stops shortly afterward. There is little evidence at present to explain how ventilation and perfusion rates are coordinated, but the fact that such regulation occurs seems to be well accepted.

There are relatively large gaps in our knowledge of the control of respiration in fish, such as the lack of any sensors being clearly identified yet as providing information to the respiratory control center in the brain. It is clear that fish respond mostly to decreasing oxygen levels in the water rather than to increased CO_2 as in mammals. Artificial perfusion of the gills with water which was alternately oxygenated, deoxygenated and high in CO_2, all at the same temperature and flow rate, produced immediate increases in respiration rate with deoxygenated water and little change with CO_2 unless the CO_2 levels were increased to many times the normal physiological range. The speed of the response suggested that the sensor(s) were most likely in or on the gills rather than further downstream in the brain or dorsal aorta. It was considered least likely that the sensor(s) would be responding to venous $P\text{-}O_2$ because this would be the most slowly changing part of the system, even though it might be the most indicative of the actual oxygen demand from the tissues.

There is a respiratory center in the medulla of teleost fish which produces the rhythmic respiratory activity which is modified by the input from various presumed external receptors. The anatomy of the respiratory center includes a relatively large number of nerve cells which are distributed rather widely instead of being clustered together into a distinct body. Exactly how this system produces an appropriately oscillating output for the rather complex branchial musculature is unclear at the time of writing.

4. Respiratory Surface Area of Gills and Skin

The surface area of gills relates to ventilation to the extent that an increased surface for a given weight of fish might decrease the ventilation requirements, presuming that a greater area would produce greater extraction of O_2. In real life, of course, there would be added osmoregulatory problems that would accompany the increased gill area, increased drag and thus increased pumping costs. It is perhaps surprising, then, that fish have a fairly wide range of gill surface areas. Sedentary fish such as the Atlantic toadfish *(Opsanus)* have gill surface areas around 2 cm^2/g of body weight. More typical values are around 4 cm^2/g, and active fish such as mackerel *(Scomber)* and herring *(Clupea)* have gill areas up to 10 cm^2/g. Thus gill surface areas seem to correspond well with activity levels and oxygen consumption rates.

There are relatively few measurements of gill surface areas, perhaps because no one has found an easy way to do it. The present method involves measurement of the surface area of representative gill lamellae, counting the number of gill lamellae on representative gill filaments and then counting and measuring the length of all the gill filaments. Respiratory area of skin is determined by measuring the blood capillary area underneath representative portions of the skin. Either kind of measurement is a tedious chore.

The respiratory surface area in or under the skin has been measured even less often than the gill surface area and seems to vary widely. Capillary surface areas range around 0.5-1.5 cm^2/g with these areas representing from 10% to 25% of the total respiratory surface. Anguillid eels which were respiring in air caused the blood flow to the skin and fins to increase compared to water respiration so that up to 60% of their total oxygen uptake occured via the skin. This is particularly noteworthy when considering that the skin of eels is over 250 u (¼ mm) in thickness while typical gill epithelium is in the range of 1-5 u thick. A flounder using some degree of skin respiration had a skin thickness in the range of 31-38 u, which is somewhat more comparable to that of gills than eel skin. However, there did not seem to be much general correlation among species between the degree of skin vascularization and the habits of the fish or thickness of the skin.

Perhaps some of the vascularization of the relatively thick skins of adults is a left-over feature from larval stages where skin respiration might be important. One well known instance of larval skin (mostly fin) respiration is in the live-bearing sea perches (Embiotocidae). The larval fish develop large fins with elaborate capillary loops which serve to exchange gases with the ovarian fluid (Fig. XI-5). These loops are rapidly reabsorbed, however, immediately after birth.

D. HEMOGLOBIN FUNCTIONS

1. Loading and Unloading Curves

Hemoglobin enables blood to carry much more oxygen than would be carried solely in solution in plasma. The hemoglobin-oxygen association is a very special one since the bonding must be strong enough to take up large amounts of oxygen in the gills, yet weak enough to release most of the oxygen in the tissues—i.e., an easily reversible association.

This reversible association is usually described by a family of dissociation curves (Fig. IV-12). The basic function described by the curves is the proportion of oxyhemoglobin produced by a given $P\text{-}O_2$. A group of curves is required because of changes in the basic function in response to varying temperature, pH and $P\text{-}CO_2$. These changes serve to enhance the unloading of O_2 in the fish's tissues by moving the curves to the right and

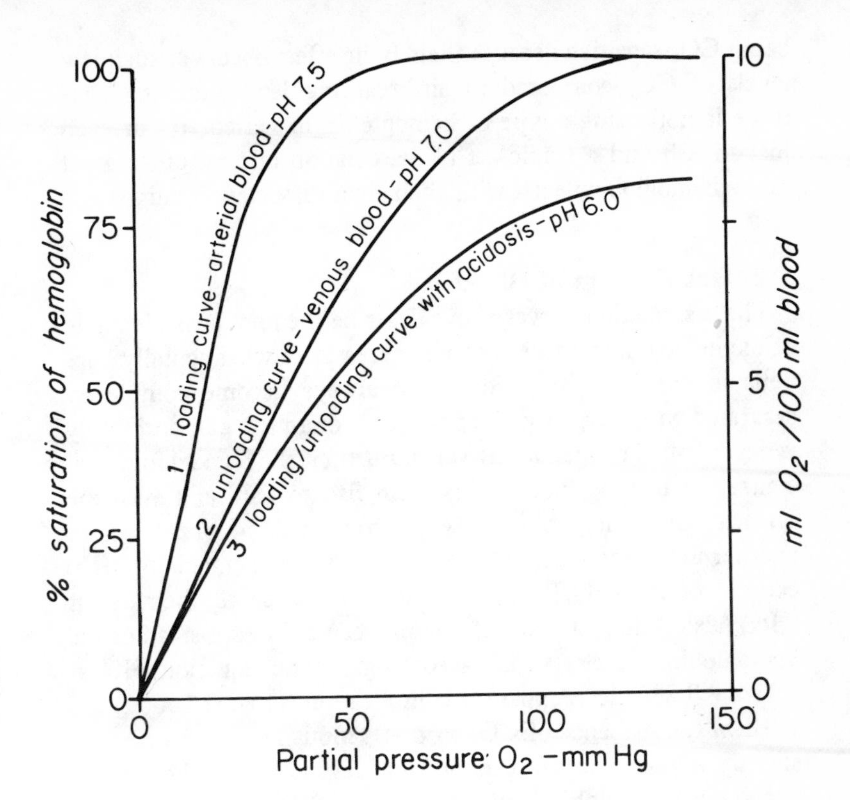

Fig. IV-12. Typical curves for the loading and unloading of oxygen from the hemoglobin of teleost fish. Carbon dioxide decreases oxygen capacity.

downward. The shift to the right is called the Bohr shift and the extension of that effect which produces a lowering of the curves is called the Root shift, both being named after the people who first described each phenomenon. Both the Bohr and Root shifts result from changes in P-CO_2 or pH, the two being related since CO_2 dissolving in water produces an acid. The Root shift is most pronounced in fish which have gas glands (in a swim bladder—see section F-3 of this chapter) and is absent or nearly absent in fish which breathe air. In air-breathers the internal P-CO_2 is much higher than in their aquatic relatives, so a pronounced Root effect would greatly depress the oxygen-carrying ability of the hemoglobin. In general, fish hemoglobins are described as being CO_2-sensitive because their Bohr effect occurs at such low levels of CO_2 compared to air-breathing land animals. Thus there is not a single type of hemoglobin in vertebrates or even just in fish and a single set of dissociation curves, but a great many hemoglobins each with their own dissociation curves.

2. Gas Content of Blood

The dissociation curves above describe the total possible range of oxyhemoglobin values but not necessarily what actually happens in the fish. Blood does not always become completely saturated with oxygen while passing through the gills and rarely is devoid of all oxygen in the veins. Further, the hemoglobin concentration in the blood differs from fish to fish, and even the quantity of hemoglobin can vary from one red blood cell to another in the same fish. Of these variables the hemoglobin (Hb) content of individual cells is least variable, so the hematocrit (Hct) (also called packed cell volume) can be used instead of the hemoglobin concentration, according to the equation Hb = 0.424 + 0.289 Hct. A possible complication is that red bood cells swell in the presence of CO_2 so that venous hematocrit may be slightly higher than in arterial blood even though there is no change in hemoglobin concentration or cell numbers.

In some arctic fishes there is no hemoglobin at all and oxygen is carried solely in solution in the plasma—a situation which is workable because of their low metabolic rate and the high solubility of oxygen at extremely cold temperatures (Holeton, 1974). In these fish the maximum blood oxygen content is predictable entirely from solubility data.

The way a set of dissociation curves operate in fish blood is shown in Fig. IV-13. Hemoglobin characteristics alternate back and forth between the two curves with the changeovers occurring at the gills (loading curve) and in the various tissue capillaries (unloading curves) where oxygen is being delivered. Little or no change in O_2 content occurs during transport in either arteries or veins (dots at ends of loading/unloading curves). The unloading curve never extends to 0% saturation. In salmonids venous blood usually remains more than 50% saturated, for example. Thus the full extent of the dissociation

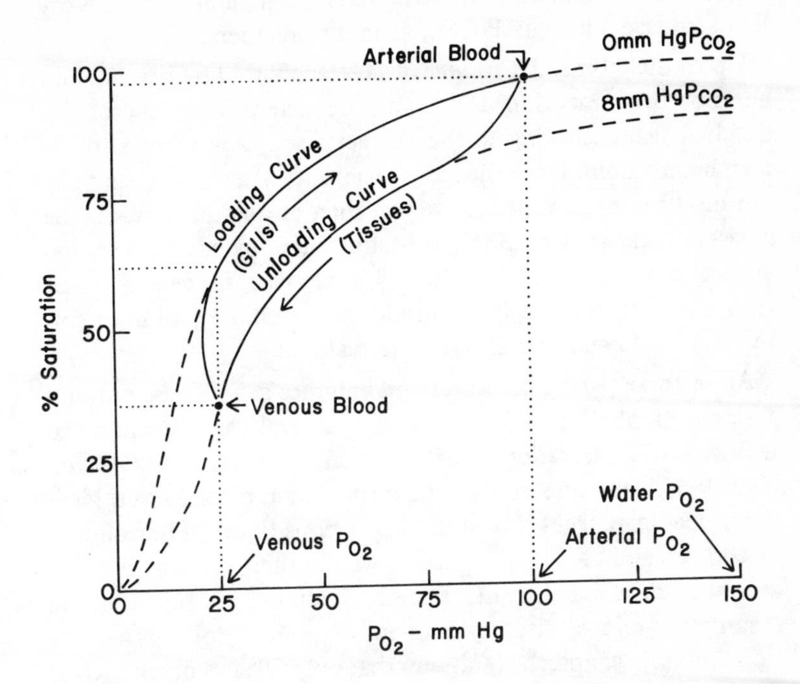

Fig. IV-13. The operational pattern of hemoglobin function following two of the curves from Fig. IV-12. Blood changes from its venous to its arterialized (oxygenated) condition in about the one-tenth of a second required to pass through the gill lamellae and then passes unchanged through the remainder of the arterial system. In the passage through tissue capillaries, the reverse reaction takes place, but then further changes may take place if the particular blood in question reaches either the renal or hepatic portal veins and passes through additional capillaries.

curves (dashed lines at ends of loading/unloading curves) are not used.

The value of having the Bohr effect is readily apparent in this hypothetical case (Fig. IV-13). Assuming that P-O_2 is constant at 25 mm Hg, with no Bohr effect (using only the 0 mm Hg P-CO_2 curve) the hemoglobin would unload no more than about 35% of its O_2. With the Bohr effect (using the 8 mm Hg P-CO_2 curve for unloading), the hemoglobin would unload about 65% of its O_2—nearly double the previous amount of O_2. The Root effect, on the other hand, is useful mostly in gas secretion into swim bladders (section F-3-c, this chapter) and could even be deleterious in situations where there is a comparatively high (by fish standards) venous P-CO_2, as in air-breathers.

Up to this point, the *volume* of O_2 transported by hemoglobin has not been mentioned. It must be determined in each individual fish according to the characteristics and amount of its own hemoglobin. Once the oxygen capacity of a certain hemoglobin has been determined, however, then one simply equates the maximum capacity to 100% saturation—i.e., by adding a second vertical scale to Figs. IV-12 and 13. After that, O_2 values can be read either in relative (% saturation) or absolute (volumes corrected to 0 °C and one atmosphere) values.

All of these partial pressures and volumes can now be put into a single graph (Fig. IV-14). Using mackerel as an example in which 100% saturation equals 12 vols % (12 ml O_2/100 ml blood), a single line can be drawn which represents the loading/unloading curves shown in the previous figure. The volumes of both O_2 and CO_2 can be read as well as their partial pressures at any point along the line. The CO_2 volumes are based on the assumption of a respiratory quotient (R or R.Q.) of 0.8 (see section A-4, this chapter), signifying that only 80% as much CO_2 is produced as O_2 consumed. A similar line can be drawn for the inspired and expired water. A similar graph could be drawn for any fish once the necessary measurements of partial pressures and gas volumes have been made. Since the measurements are relatively difficult to make, there are comparatively few such sets of data available. The data presented appear typical of active fish with high oxygen consumptions. Salmonids have somewhat less blood oxygen capacity—usually in the range of 8-10 ml O_2/100 ml blood. Even less active fish such as starry flounder have blood oxygen capacities in the range 3-6 ml O_2/100 ml blood. The arctic ice fish which have no hemoglobin should have blood oxygen capacities in the range of 0.8-1.0 ml O_2/100 ml blood at their normal environmental temperatures around freezing, because that is the solubility of O_2 at those temperatures.

The regulation of the oxygen content of blood is not understood at this time. The arterial and venous oxygen content in salmonids is quite stable over a wide range of environmental oxygen levels, and large adjustments in ventilation and perfusion

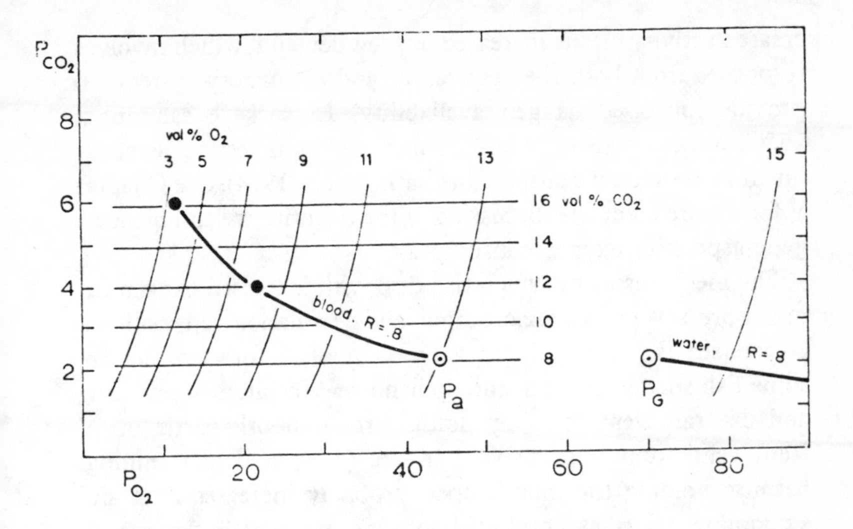

Fig. IV-14. Another way to show the gas exchange between blood and water of a mackerel, in this case. The line for water (R = 0.8) begins at 150mm Hg P-O_2 and O-mm P-CO_2, and ends at 70 and 2, respectively. The line for blood begins at 45 and 2, respectively, and shows two possible venous points. A grid behind this line shows the *volume* of O_2 or CO_2 represented by the present changes. (From Rahn, 1966, with permission of author and publisher.)

rates are required to produce this stability. The sensory system used to control the adjustments is not known but is probably complex. Presumably salmonids are typical of other oxygen regulators (section A-3, this chapter). Oxygen conformers, on the other hand, would be expected to decrease their blood oxygen content in proportion to decreases in environmental oxygen levels. In a starry flounder which I had first thought of as a likely oxygen conformer, it regulated its blood oxygen content quite effectively down to an environmental P-O_2 of 50 mm Hg (Watters and Smith, 1973).

E. RESPONSES OF THE RESPIRATORY SYSTEM TO EXTERNAL CHANGES

1. Increased Activity

Although no fish is ever completely inactive—heart beat, ventilation and postural movements never stop—there are many instances in which activity increases above the ordinary level. Increased activity means increased oxygen demand, which involves responses from both the respiratory and circulatory systems to provide increased oxygen availability. In sockeye salmon at 15 °C, oxygen consumption can increase from resting levels as much as 10-fold on a sustainable basis (Brett, 1964). See Chapter V for a more complete discussion of the quantitative and bioenergetic aspects of oxygen consumption.

The mechanisms by which fish deal with increased oxygen demand are several. Cardiac output and ventilation rate both increase (see Fig. III-6). If the activity involves forward motion, some fish such as salmon stop their normal respiratory pumping and use ram ventilation by holding their mouth partly open. Ram ventilation is probably not free (compared to pumping) because holding the mouth open probably increases drag and swimming effort as compared to swimming with the mouth closed. However, a number of fish (including tunas and menhaden) which swim constantly also use ram ventilation routinely, so it seems likely that there is a bioenergetic advantage to ram ventilation compared to pumping.

Some fairly subtle adjustments to activity occur which are not entirely explainable. Transfer factor, which is defined as the oxygen uptake divided by the oxygen gradient between water and blood at the gills, increases up to 5-fold during swimming. The mechanism for this change could include increased surface area of the gills, decreased diffusion distance between water and blood, and changes in the rate of chemical reactions for exchanging oxygen and carbon dioxide. The relative importance of these possibilities has yet to be evaluated.

2. Changes with Hypoxia

Changes associated with decreased environmental P-O_2 are similar but not identical to those occurring with increased activity. Heart rate decreases while stroke volume increases, maintaining the cardiac output at about a steady level in trout. Ventilation volume increases until some minimal P-O_2 is reached, and then both ventilation and oxygen uptake decrease because the compensating mechanisms can compensate no further. Transfer factor increases in part because venous P-O_2 decreases. Activity often increases as the fish appear to search for ways to leave the hypoxic area. Activity, ventilation and heart beat in salmon may all decrease rather suddenly at low P-O_2 levels in association with loss of equilibrium (turn belly up). In my opinion the fish may lose consciousness at this time and reduce all activity to reflex levels.

3. Changes with Increased Temperature

Increased temperature has two major effects. It dramatically increases the fish's metabolic rate and decreases the solubility of oxygen. Fish experiencing a drastic increase in temperature are thus squeezed between two converging problems—increased oxygen demand and decreased availability of oxygen. Brett (1964) interpreted his data on maximum sustainable swimming rates (which are limited by respiratory ability) to be temperature limited below 15 °C and decreased by oxygen availability above 15 °C. He demonstrated that maximum sustainable swimming rates could be increased considerably if more oxygen were made available as a supersaturated solution.

See also the discussions of bioenergetics in Chapter 5 and thermal stress in Chapter 12.

4. Respiratory Involvement in pH Regulation

In mammals, control of blood pH and CO_2 levels includes changes in ventilation, but this is not true in fish. There are at least two major reasons for this. First, neither aquatic environments nor fish blood has much CO_2, so a regulatory system based on CO_2 would either have to respond to very small changes in P-CO_2 or would respond only to those infrequent cases when there were relatively large changes in P-CO_2. Neither of these choices is particularly satisfactory.

That part of the pH regulation of fish that concerns respiration therefore involves bicarbonate ion (HCO_3^-) rather than CO_2. The two are still closely related as shown in the following equation:

◄More acid More alkaline►

$$CO_2 + H_2O \rightleftharpoons H_2CO_3 \rightleftharpoons H^+ + HCO_3^- \rightleftharpoons 2H^+ + CO_3^{--}$$

In words, this equation means that carbon dioxide reversibly reacts with water to produce carbonic acid (H_2CO_3), which then ionizes to produce hydrogen and bicarbonate ions. Bicarbonate ions predominate in either plasma or sea water (both slightly alkaline) so that little carbonate (CO_3^{--}) usually forms. Addition of CO_2 or removal of HCO_3^- makes the system more acid. These reactions are normally rather slow, but are greatly speeded up by the enzyme carbonic anhydrase, which is present in gill epithelium and in red blood cells. Carbon dioxide can be excreted across the gills of fish as either molecular CO_2 or as HCO_3^-. When excreted as HCO_3^-, it is often exchanged for Cl^- as a means of maintaining the electrical status of the gill membranes (see Chapt. II, part B-3).

Fish adjust this $CO_2/HCO_3^-/Cl^-$ system in a variety of ways. An increase in arterial P-CO_2, from increased muscular activity for example, results in an increased ventilation rate and increased excretion of CO_2. If the rise in arterial P-CO_2 is the result of increased environmental P-CO_2, then blood pH falls slightly because the arterial P-CO_2 remains elevated, but the pH is then corrected by elevating the plasma HCO_3^-. Also, blood becomes more alkaline as temperature decreases, partly because of the similar changes in the pH of water with temperature, and the increased alkalinity is accompanied by increased HCO_3^- while arterial P-CO_2 remains constant (Randall, 1974, 1975).

F. OTHER RESPIRATORY-RELATED ORGANS

1. The Pseudobranch

The pseudobranch is a gill-like structure (sometimes covered by a membrane or even reduced to a gland-like appearance) which occurs on the inside of the gill covers of many but not all teleost fish. Both its phylogenetic and physiological roles are continuing puzzles. It occurs irregularly among teleost species but can perhaps be said to occur predominately in marine species. Since many families of teleosts have both marine and freshwater species, this means that some closely related species differ in the presence or absence of a pseudobranch. However, most authors seem to agree that the name is correct—whatever the pseudobranch is, it is not a gill.

One of the reasons for saying that the pseudobranch is not a gill is that the blood supply to the pseudobranch has already been through the gills. An artery which branches from the first gill arch loops anteriorly and then posteriorly inside the operculum until it reaches the pseudobranch. Blood leaving the pseudobranch goes to the choroid gland (see F-2, below) and then into the veins draining the head.

The explanations of the functions of the pseudobranch are mostly speculative, although there are some fascinating fragments of information with which to speculate. Upon surgical removal of the pseudobranch, several species of fish darkened permanently, although they turned pale for a while after injection of a pseudobranch extract. The same darkening occurred from blocking the blood supply to the pseudobranch (Parry and Holliday, 1960). They also found that the choroid gland of the eye degenerated in the absence of the pseudobranch and vice versa. This is not so surprising because they are connected in series, so removal of either would block the blood supply to the other and cause degeneration. Presumably the oxygen supply to the retina also decreased, possibly producing blindness and explaining the darkening. Blood leaving the pseudobranch contained more carbonic anhydrase than blood entering, suggesting that the pseudobranch may be involved in pH regulation via ion exchange (section E-4, this chapter). On the other hand, elasmobranch pseudobranchs have no carbonic anhydrase and there is no choroid gland. Earlier work also showed that removal of the pseudobranch or addition of carbonic anhydrase inhibitors reduced the ability of the gas gland to secrete gas into the swim bladder in *Fundulus* but had no effect in *Perca*.

The study of the pseudobranch will undoubtedly continue into the future, although few investigators seem to have been working on it in recent years.

2. The Choroid Gland

The choroid gland is a horse-shoe shaped structure encircling the optic nerve on the back (medial) surface of the eyeball. It contains capillary loops similar to those in the gas secretion gland of physoclistous swim bladders and is supplied with blood from the pseudobranch as noted immediately above. Most fish which have choroid glands also have pseudobranchs. A few fish have choroid glands and no pseudobranchs or pseudobranchs and no choroid glands, and some fish have neither. Hagfish, lampreys, sharks and rays, ratfish and many primitive fish such as gars, coelacanths and sturgeons all lack choroid glands. The bowfin, *Amia*, appears unique among the primitive groups in having one. Taxonomic distribution of the choroid gland is of little help in understanding its function. There is some tendency for a choroid gland to occur in the marine members of a fish family rather than in the freshwater ones, similar to the occurrence of the pseudobranch (Wittenberg and Haedrich, 1974).

There are a variety of functions proposed for the choroid gland. There have been some measurements of P-O_2 in the liquid immediately in front of the retina which have been as high as 400 mm Hg (atmospheric P-O_2 is about 150 mm Hg). Such high P-O_2 levels could be produced by a countercurrent system such as is found in the choroid gland, but other authors point out the ease of contamination in making such P-O_2 measurements and are skeptical about the need for having such high P-O_2 levels (even though the retina has a high O_2 consumption rate). Wittenberg and Haedrich (1974) concluded that the choroid rete, working in combination with an exchange of HCO_3^- and Cl^- (mediated by carbonic anhydrase) at the pseudobranch, was a means of producing a large volume (not necessarily partial pressure) of oxygen at the retina without any increase of P-CO_2 at the same time. The question was by no means resolved to everyone's satisfaction at the time of writing.

3. The Swimbladder

Swimbladders are organs for adjusting the specific gravity of fish and for minimizing or preventing sinking. Without a swimbladder a typical fish is about 5% heavier than water. To achieve neutral buoyancy, a fish therefore requires an addition to its displacement (volume) of about 5% of some substance having little or no added weight. The secretion of various atmospheric gases into an internal cavity—the swimbladder—serves this purpose nicely, and the resulting neutral buoyancy has been computed to save from 5% (at rapid swimming speeds) to 60% (at slow speeds) of the swimming effort which otherwise would have been used to generate lift. Air is less buoyant with depth as it compresses, of course, until its specific gravity is about 0.7 at a depth of 7000 meters, where one swimbladder-bearing fish was caught. Even at a specific gravity of 0.7 the swimbladder would still be more buoyant than the other major kind of buoyant substances—fats, which have a specific gravity of about 0.9—so a swimbladder is still somewhat useful even at great depths.

Fish which lack swimbladders and seem to benefit from this lack include fast-swimming pelagic fish and fish which normally live *on* the bottom. Bottom fish such as sole, flounder, sculpins, etc., seem better able to remain camouflaged if they have enough weight so as to not be moved by ground swells or tidal currents. Tunas, on the other hand, rapidly go from surface waters to considerable depths to feed on squid, so volume control for a swimbladder would be difficult. A swimbladder also increases swimming drag because of increased cross-sectional area of the body, a factor which is most likely to be important at the rapid swimming velocities such as are found in tunas.

There are two types of swimbladders. In physostomous fish the swimbladder is thin, membranous and connects to the esophagous through the pneumatic duct. It was this kind of swimbladder in primitive lobe-finned fish which eventually gave rise to the simple amphibian lung and to the modern lungfish lung. In such fish most gas in the swimbladder enters and leaves (rapidly, if necessary) through the pneumatic duct. Salmonids are typical physostomous fishes.

In physoclistous fish the swimbladder is thick-walled and gas enters and leaves rather slowly via the blood stream at specialized secretory and reabsorptive gas glands. Typical physoclistous fish include rockfish, hake and the cod-type fishes. There are some fish, including carp, catfish and anguillid eels, which have both a gas gland and a pneumatic duct, although the duct may be atrophied and non-functional in the adults of some species.

The combination of physical and chemical processes which enables a secretory rete (gas gland) to transport gas against large

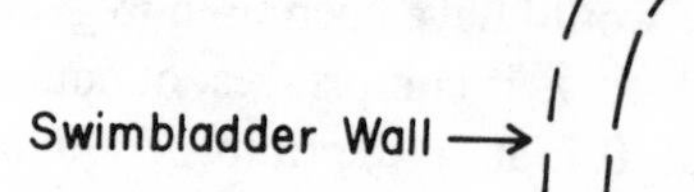

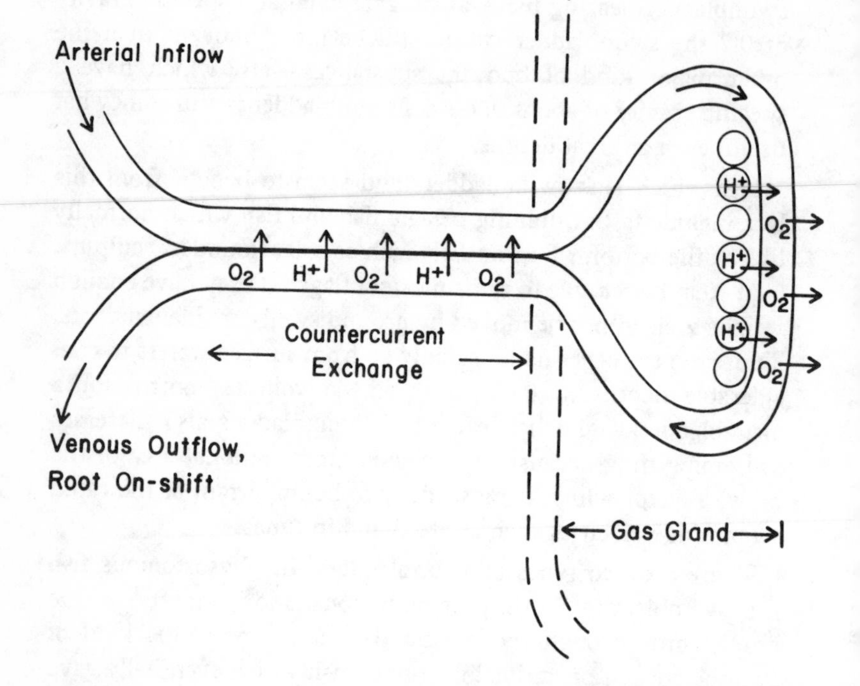

Fig. IV-15. Diagram of the capillary loops in the secretory gland (rete mirable) which acts as a countercurrent multiplier system to build up sufficient oxygen tensions in the blood so that oxygen will move into the swim bladder.

pressure gradients has been understood only in recent years (Fig. IV-15). First, the gas gland capillaries form hairpin or countercurrent loops. The longer the countercurrent exchange surfaces, the greater are the gas pressure gradients which can be maintained. Second, secretion of lactic acid into the blood in the loop portion of the capillary causes hemoglobin to unload oxygen into the plasma. This additional oxygen makes the plasma supersaturated. Excess oxygen diffuses from the venous to the arterial side of the countercurrent system, producing even greater supersaturation of the plasma when this incoming blood is subsequently acidified. The increased level of supersaturation causes greater diffusion into incoming arterial blood until the P-O_2 builds up to a level exceeding the hydrostatic pressure. Once sufficient countercurrent multiplication occurs, then oxygen diffuses into the swimbladder and often forms bubbles on the surface of the gas gland.

Even when forming bubbles, gas secretion is relatively slow—refilling swimbladders which had been experimentally emptied required a minimum of several hours and sometimes as much as several days before the fish achieved neutral buoyancy again. Fish which employ neutral buoyancy as part of their life-style cannot make major, rapid changes in depth, especially upward. Alexander (1967) calculated that any time the positive buoyancy of the swimbladder exceeded approximately 1% of the body weight (approximately 20% overexpansion of the swimbladder) the fish would float upward uncontrollably, faster than it could swim downward.

The countercurrent mechanism in the secretory rete of the gas gland was recognized early (1930-1950) in the study of physoclistous swimbladders, but the details of its operation were not clarified until recently. The acidification of the blood which causes hemoglobin to unload some of its oxygen (see Root effect, Fig. IV-13), called the Root off-shift, has a half-time of about 500 milliseconds. The Root on-shift, which occurs as lactic acid diffuses across the countercurrent exchange surfaces along with the oxygen, has a half-time of from 20 to 30 seconds. If the on-shift were as rapid as the off-shift, most of the multiplying effect would be lost. With the slow on-shift and countercurrent loops up to 1 cm long, swimbladders have been calculated to transport oxygen up to a P-O_2 of 3000 atmospheres from an environmental P-O_2 of 0.2 atmospheres. No fish actually does so, because it would have to live at a depth of 30,000 meters and there isn't any ocean that deep.

There are a number of peculiarities about the secretory rete. First, lactic acid is normally produced by most cells when they lack oxygen, but the epidermal cells of the gas gland produce lactic acid in the presence of high P-O_2 levels. Second, the gas in the swimbladder usually is mostly oxygen when the swimbladder is first filled, but later may contain quite a variety of other gases. Nitrogen can be secreted similarly to oxygen because the increased lactic acid reduces its solubility in the plasma and produces supersaturation, a process described as salting out. However, the possible P-N_2 which can be produced is limited. Thus the gas content in newly-filled swimbladders resembles air in fish living near the surface but has increasing amounts of oxygen in fish at greater depths. After remaining several weeks at a depth, the swimbladder of whitefish contained almost pure nitrogen, suggesting reabsorption of oxygen from the swimbladder. Carbon dioxide and argon have been found in other swimbladders in significant amounts.

Compared to gas secretion, reabsorption of gas from swimbladders is a simple process. Since the diffusion gradient is always outward, any non-acidified blood exposed to the gas will carry it away. Thus the reabsorptive (also called oval) gland has more of a problem to prevent unwanted loss of gas than to facilitate gas transport. There are a variety of configurations of reabsorptive glands, but most of them involve a movable mechanical barrier to prevent gas from reaching the reabsorptive site. Some, such as in hake, have a contractile membrane which curtains off the reabsorptive gland. Rockfish have their reabsorptive gland in a pocket at one end of the swimbladder with a closable pore connecting them. In the anguillid eels the reabsorptive gland is in the pneumatic duct some distance away from the swimbladder. The architecture of physoclist swimbladders is highly variable, particularly that of their reabsorptive glands.

G. AIR-BREATHING IN FISH

The reason for fish to breathe the air seems obvious—when oxygen becomes scarce in water or when even the water becomes scarce, then some degree of ability to obtain oxygen from air has obvious survival value. The idea is sufficiently useful that development of air breathing has occurred many times, although most often in tropical climates. It occurs in many different families of fish using a variety of structural modifications to breathe air. Opting to breathe air also brings changes and problems, and thus many fish breathe air only intermittently to avoid or at least minimize aerial problems such as drying and difficulties from internal build-up of CO_2. At the other extreme are cases such as the Australian lungfish which breathes air, but rarely seems to encounter adverse environmental conditions for which air-breathing would be beneficial. Thus there are probably many reasons for fish to breathe air in addition to the obvious one.

The structures for breathing air differ as widely as the reasons. First, most fish gills collapse in air and function poorly for gas exchange. A few fish (e.g., anguillid eels) have additional supporting structures which can support the gills in the absence of the buoyancy of water, but most air-breathing fish have some structure for gas exchange in air other than gills and have reduced or even completely lost the typical teleost branchial apparatus. A well known such structure is the physostomous swimbladder, which was presumably a simple lung in the primitive lobe-finned fish before they gave rise to the amphibians and before it was used as a buoyancy organ. The modern lungfish (all three species) also use the swimbladder as a lung. In the electric eel of the Amazon, the mouth lining is greatly folded and expanded to a point of effectiveness where the gills are almost vestigial. There are a great variety of expanded surfaces in the back of the mouth and under the gill covers, including several kinds of pouchings and various tree-like or coral-like proliferations of respiratory surface area. Many fish with these kinds of "air-gills," including the bettas of tropical fish fanciers, are obligate air-breathers which suffocate in water if denied access to air.

A few fish use their stomach or intestine for gas exchange by swallowing air. Such a great diversity of structural adaptations for breathing air is usually interpreted as indicating multiple evolutionary origins of the idea.

The physiological problems encountered by air-breathing fish are more subtle than just providing a new respiratory surface and preventing desiccation. The main problem is accumulation of CO_2. In water, CO_2 passes through the gills in solution at very low P-CO_2 levels. In air, CO_2 removal depends on diffusion alone, and considerable P-CO_2 (25-40 mm Hg) builds up in the fish's blood. This would not be a problem as long as the fish's hemoglobin has little or no Bohr effect and no Root effect (see Fig. IV-13). If these effects were present, the oxygen-carrying capacity of hemoglobin would be so depressed as to be almost useless. The hemoglobins of most air-breathing fish investigated so far show little or no Bohr or Root effects. P-CO_2 levels rise considerably and CO_2 excretion is thought to occur through the skin because of the high levels of carbonic anhydrase found there in air-breathing fish. Also, one of the reasons some air-breathing fish return to water at frequent intervals is to remove accumulated CO_2.

Other problems include adapting the pattern of blood flow so that oxygen gets delivered appropriately to the tissues. In lungfish, for example, the oxygenated blood from the swimbladder goes to the heart, then mostly through the first gill arch (which has no gas exchange capability) and then to the rest of the body. If oxygenated blood passed through a normal gill when the P-O_2 of the water was less than that of the blood, the blood would lose oxygen to the water. There are also problems of trying to reduce the mixing of oxygenated venous blood from the new (air-breathing) gas exchange organ with venous blood from other parts of the body. A few air-breathing fish have almost separated the oxygenated and non-oxygenated venous blood somewhat like mammals do. On the other hand, the control of aquatic and aerial breathing seems to be combined yet, sometimes leading to inappropriate responses of one of the two systems.

The lore about air-breathng fish is too large and diverse for this book. Readers desiring further information are referred to reviews by Johansen (1968, 1970).

Chapter V: BIOENERGETICS AND METABOLISM

A. OVERVIEW

Bioenergetics is the study of energy usage by living organisms. For animals, this includes analysis of energy sources, methods of obtaining energy, its distribution pathways inside the animal, rates of consumption under various conditions and the final energy status of products leaving the animal. The units of energy are most often calories (cal) or kilocalories (Cal or kcal) on a weight-specific basis. An energy consumption rate is usually given as kcal/kg (of body weight)/hr or kcal/kg/24 hrs. Making direct measurements of energy consumption in terms of heat production is difficult to do in small cold-blooded animals such as fish, so energy consumption is almost always measured indirectly in terms of oxygen consumption. Oxygen consumption is sometimes given as ml O_2/kg/hr with the O_2 volume corrected to a stated temperature and pressure (e.g., O °C and 760 mm Hg), but it is more conveniently stated as a weight of O_2—mg O_2/kg/hr—because this is independent of temperature. Volume and weight are readily interconvertible since 1 mg O_2 =0.70 ml O_2. The conversion from oxygen to kcal is 1 mg O_2/kg/hr = 0.00337 kcal/kg/hr or 0.081 kcal/kg/day. Also 1 kcal/kg/hr = 297 mg O_2/kg/hr (Brett, 1972).

Bioenergetics thus involves several organ systems at several levels of function. For animals the basic source of energy is food, but the energy in food is not available until the food is ingested, digested and assimilated by the digestive system. The energy is released from food mostly by oxidation, so most of the quantitative aspects of oxygen consumption appear in this chapter because bioenergetics most often is studied in terms of oxygen consumption. The molecular aspects of energy flow generally come under the heading of metabolism, which is mostly a combination of liver and muscle function, but also involves osmoregulation and excretion of the waste products from the metabolic machinery. Finally, an amount of energy and raw materials not needed for maintenance and routine demands of daily living is available for storage, growth or production of gametes. Thus efficiency of energy usage becomes an important part of the concept of ecological or evolutionary success.

Alexander (1967) expressed the general idea of energy distribution and the importance of efficient energy usage in an equation,

$$uF = g(G + H) + R + S,$$

in which F is food intake, G is growth (production of new tissue), H is gametes, R is basal metabolism and S is swimming or other activity. The factor μ reduces the gross food intake to include only that food which is actually assimilated and has values typically around 0.8. The remaining 20% of the food becomes feces, urine, or ammonia (excreted by the gills) without ever really entering the energy-yielding pathways of the fish's metabolism. The 0.8 factor is approximately correct for salmonids eating normal food but would be much lower if they were eating large quantities of raw starch for example, which is only about 10% digestible. The growth and gamete factor, g, is approximately 2 because it takes about twice as much food to produce new tissue as to maintain existing tissue. This is because the raw materials—amino acids, fatty acids, simple sugars—and the chemical energy needed to assemble them are about double that for maintaining existing tissue. Some typical percentages for the usage of food in Alexander's equation are:

$$0.8 \times 100 = 2(5 + 1) + 34 + 34.$$

There are times when such numbers are totally incorrect, such as in migrating or spawning fish which are not eating. In another example, flounders fed minimal diets continued to grow slowly, but produced no gametes that year—H equaled zero. Or activity levels (S) and basal metabolism (R) could both change with temperature and season. Nevertheless, these typical numbers are useful as a starting point, both conceptually and to use as approximate numbers when actual data from a particular fish species are unavailable.

The significance of the equation is to demonstrate the rather large effect on reproduction (H) of rather small changes in any of the other factors. For example, if food intake increases from 100 to 101, the gamete production increases from 1 to 1.4, a 40% increase. Or if a fish was able to obtain its normal food supply with 2% less activity (S = 32), then H would become 2, a 100% increase. Natural selection can thus result from relatively large changes in growth and reproduction which, in turn, are influenced by relatively subtle changes in bioenergetic efficiency. Growth and reproduction seem to be the summation and integration of a fish's past history in combination with its capability to cope with its environment.

B. PERFORMANCE ENVELOPES FOR FISH

One way to describe the bioenergetic capabilities of a fish is with a four-sided polygon which I call a performance envelope (Fig. V-1). The polygon's right and left sides are the upper and lower lethal temperature limits for that particular species, while the top and bottom are the minimal (basal standard) and maximal sustainable (active) rates of oxygen consumption. The upper boundary is usually in two parts. The portion to the left of 15 °C is thought to be temperature limited, while the right hand portion seems limited by the availability of oxygen. Most, but not all, fish show a similar discontinuity in the maximal oxygen consumption rate (see below and Fig. V-2).

Fish can perform outside the limits of the performance polygons shown in these figures, but only temporarily. Either the upper or the lower lethal limit can be exceeded for a few minutes to a few hours, depending on how far it is exceeded. The tolerable time of exposure outside the performance envelopes gets shorter and shorter as the number of degrees beyond either lethal limit is increased (see Chapter XII-E for further discussion). Fish exceed the upper limit routinely during "burst" swimming by expend-

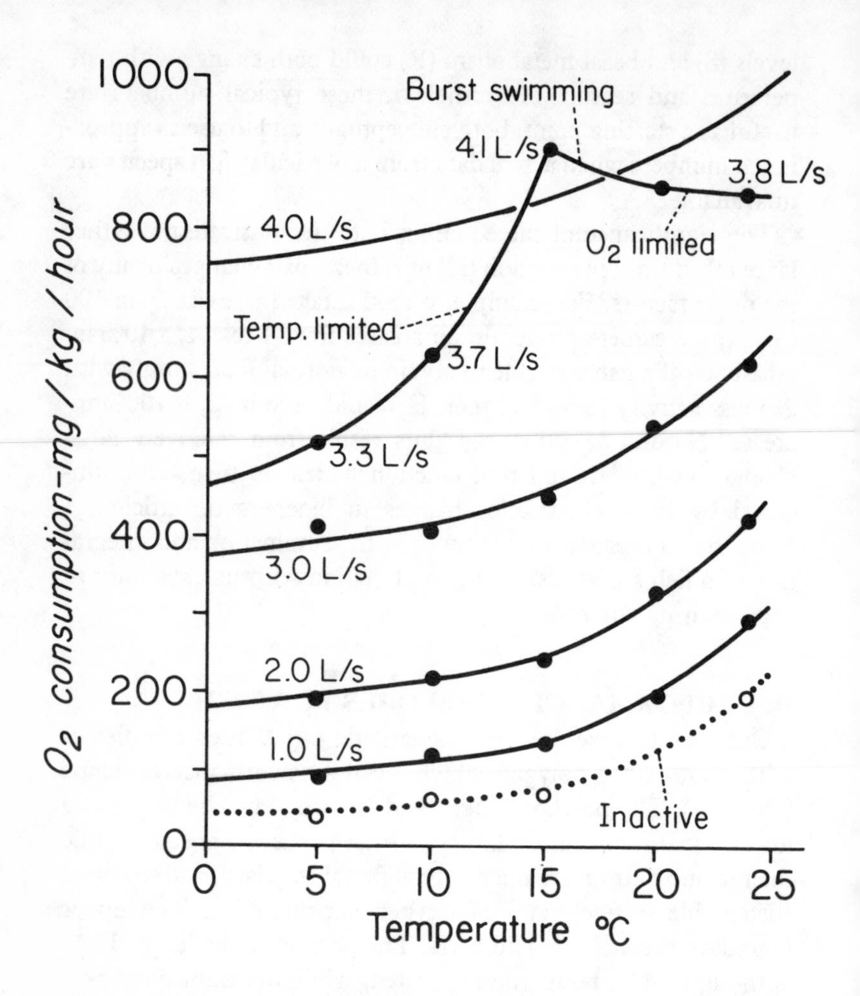

Fig. V-1. Activity and temperature "performance envelope" for sockeye salmon *(Oncorhynchus nerka).* Maximum sustainable velocities (those which can be sustained for at least one hour) are temperature-limited below 15°C and oxygen-limited above 15°C. Burst swimming includes any velocity exceeding the sustainable limit with maximal velocities of up to 10 body lengths per second (L/S) being achieved for only a few seconds. (From Wedemeyer, et al., 1976, after Brett, 1964, with permission of both authors and publishers.)

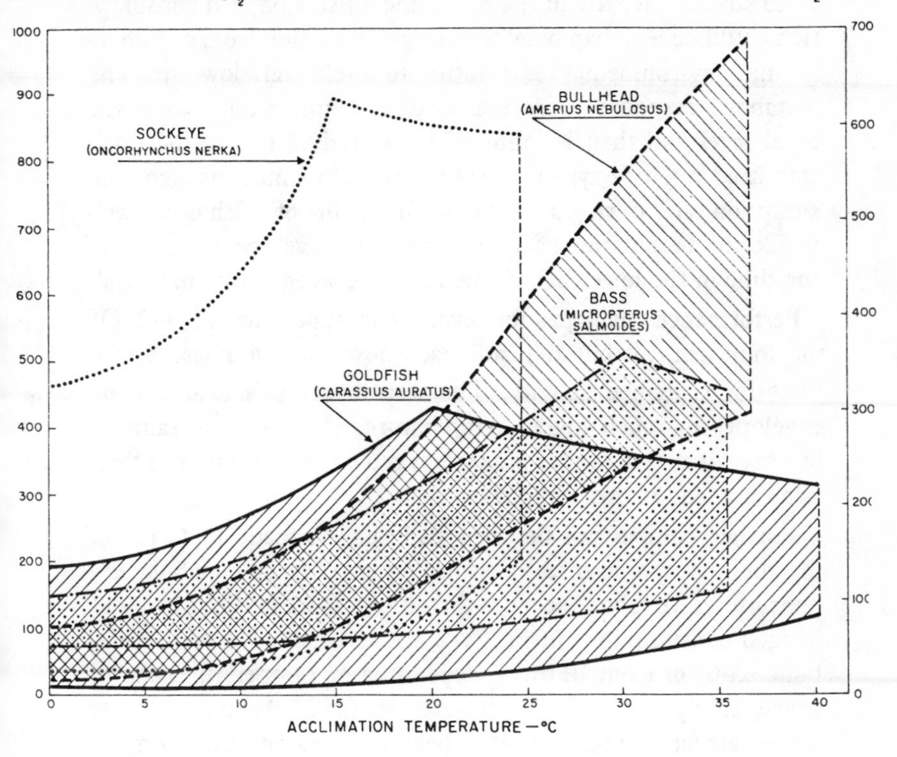

Fig. V-2. Comparison of rates of oxygen consumption in relation to temperature for four species of fish. Active and basal rates are indicated by the upper and lower lines for each species, terminated by a vertical line at the upper lethal temperature. (From Brett, 1972, with permission of author and publisher.)

ing energy faster than their oxygen consumption (using anerobic energy stores) and developing an oxygen "debt." Then they must rest at some later time during which their oxygen consumption rate exceeds that needed for their actual activity level. Like the limitations on the temperature tolerance outside of the performance polygon, the energy available for burst swimming appears to be a fixed, finite amount which can be used up in a few seconds at maximal burst swimming rates or used up over many minutes at swimming rates only slightly exceeding the maximal sustainable levels. The lower side of the performance polygon is less commonly passed because it marks the minimal energy needed to sustain life. About the only time a fish's oxygen consumption could be less than basal is during acclimation from a relatively high environmental temperature to a relatively low one. The metabolic response to decreased temperature undershoots the basal level and then is followed by a gradual rise to the final, stabilized rate of oxygen consumption. Thus most oxygen consumption rates during a typical day in the life of a fish occur well inside the boundaries of its performance envelope and most of the time in the lower half of the range between active and basal.

Performance envelopes for several fish appear in Fig. V-2. Of the four fish, all but the bullhead show the two-phase upper limit for sustained activity. The warmwater fish extend their envelopes to much higher temperatures than sockeye salmon, but may or may not have higher oxygen consumption rates than salmon at the higher temperatures.

Further comparison between fish can be made in terms of metabolic scope, also called scope for activity or metabolic expansibility. Metabolic scope is the ratio of basal to active levels of oxygen consumption. In Fig. V-2, sockeye salmon have a metabolic scope of from 10 to 15, depending on temperature, and of nearly 20 if you use the lowest basal rate (at 5 °C) and the highest active rate (at 15 °C). Goldfish appear to have a much lower metabolic scope, but it is similar to that for salmon because the basal level is so low. Of the four fish in the figure, bass have the lowest metabolic scope—from 2 to 4, depending on temperature.

Comparisons between fish and land vertebrates are also interesting. The performance ranges of fish are similar to those of amphibians and reptiles, but those of birds and mammals are much higher than fish. The active levels of fish barely overlap the lower ranges of basal metabolism for the larger mammals (larger animals have lower weight-specific metabolic rates than smaller animals) and fall short of the basal rates of birds by about half an order of magnitude. Brett's (1972) interpretation of such comparisions was that neither air-breathing (as in amphibians and some fish) nor warm body temperature (as in tunas and hot-climate reptiles) alone can produce the high levels of metabolism found in mammals and birds. High metabolic rates require the coupling of aerial respiration and homeothermy.

C. SWIMMING SPEEDS

When any object moves through water, it produces wake waves in proportion to its length. The velocity of these wake waves is proportional to the cube root of the wave length. The maximum velocity achievable by a "hull" in or under the water (not planing on the surface) approximately equals the velocity of its wake wave. At maximum displacement speed the hull length equals the wave length (crest to crest) of the wake wave. Thus the absolute speed possible for large objects moving through the water is greater than for small objects. Rowboats will never go as fast as yachts, porpoises will always lag behind whales, and adult salmon will always outswim salmon fry, all other things being equal.

Figure V-3 shows the maximum velocities which sockeye salmon of various sizes can sustain for 60 minutes (Brett, 1972). The increased swimming capabilities of larger salmon, when measuring absolute velocity, is readily seen. When watching various sizes of fish swimming at maximum sustainable rates, you get a clear impression that the little fish are working harder than the big ones—more tailbeats per second. Measurements of oxygen consumption (discussed below) would confirm this. Thus, very early in the studies of swimming performance of fish, the conversion of absolute to relative (body lengths (L)/sec) swimming velocity was adopted as a more realistic measure of the energy involved than absolute velocity. Thus in Fig. V-1, the size of the fish is not mentioned and activity levels are indicated only as swimming velocities in L/sec. I have added to Brett's data in Fig. V-3 some indications of relative swimming velocity which start at nearly 7 L/sec for small fish and decrease to less than 2 L/sec for large adult fish at lower temperatures. The curves of both the absolute and relative velocity lines emphasize the same 15 °C temperature optimum as shown in Fig. V-1 as the peak of oxygen consumption.

Comparable data for species other than salmonids are scarce. Preliminary studies in my laboratory (unpublished) showed that sablefish *(Anoplopoma)* might be able to sustain 2-2.5 L/sec and that catfish *(Ictalurus)* would be unlikely to sustain even 1 L/sec.

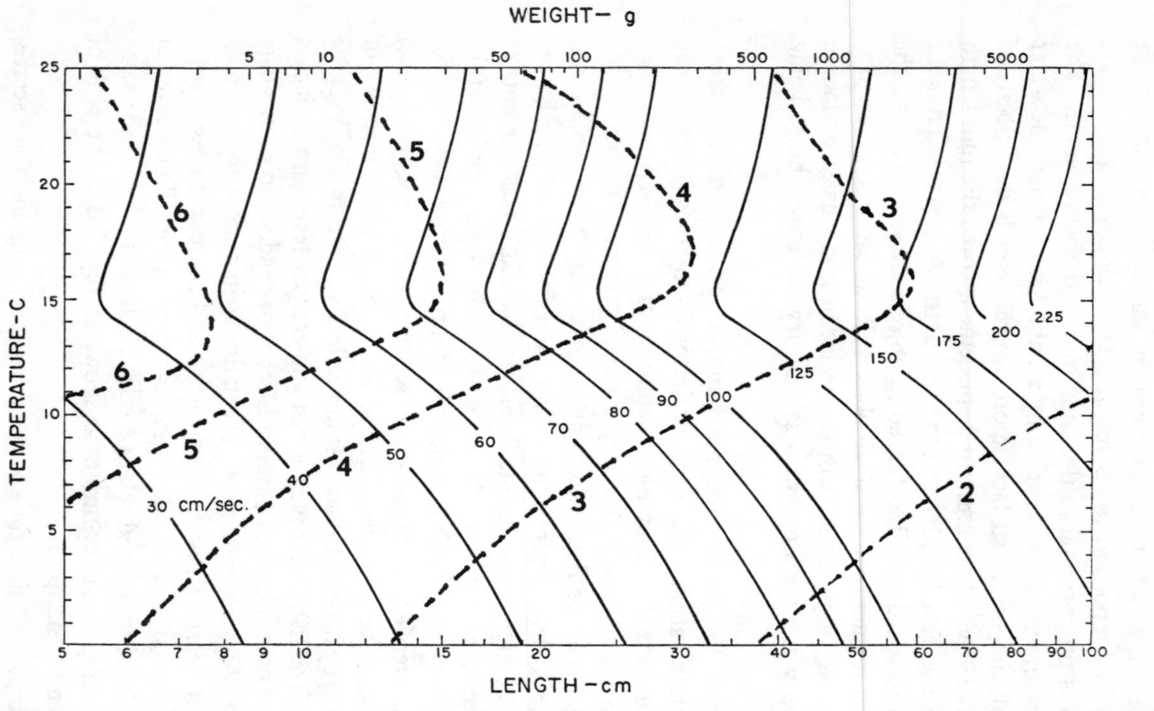

Fig. V-3. Critical swimming speed (maximum speed sustained for 60 min.) of sockeye salmon in relation to temperature, body size, and wet weight. The critical swimming speed is shown as absolute velocity in cm/sec. (solid line) and relative velocity in L/sec. (broken line). Both lines show that small fish work very hard and reach sustained velocities nearer their maximal burst speeds than larger fish. (Adapted from Brett and Glass, 1973, with permission of author and publisher.)

In my opinion the sablefish is well streamlined but doesn't have enough muscle power to go any faster, while the catfishes have neither the streamlining nor the power. On the other hand, observations of striped mullet swimming parallel to the beach during their spawning migration gave velocities of 4.9-8.0 L/sec for individual fish or schools of fewer than eight fish (average 5.6 L/sec) and average velocities of 7.6 L/sec for large schools. This suggests a hydrodynamic advantage of schooling (riding on someone else's wake saves energy, as in migrating flocks of waterfowl) and that salmon are by no means first-rank performers in swimming.

Sustained swimming speeds are an important measure of a fish's metabolic capabilities but do not necessarily represent the fish's everyday experience. Adult salmon on their high seas migration probably swim 15-25 miles/day or about 1 L/sec. Their upstream migration probably averages about 2.0-2.5 L/sec, but they are so skillful at finding back eddies and avoiding fast water that exact measurements are difficult. Salmon photographed while jumping at a falls were calculated to leave the water at a velocity of 14 mph or about 12.5 L/sec for a 50-cm salmon. The high seas cruising speed of tuna has been estimated at 2.0-2.25 L/sec, although they can probably sustain much greater speeds.

While discussion of high speed swimming may excite one's imagination, the reader should also remember that many fish occupy behavioral niches having minimal needs for high speed swimming or even for any swimming at all. Many bottom-dwelling fish use stealth or subterfuge to obtain food and avoid predators, or they use spines or armor-like scales for defense from predators. Such fish frequently lack dark muscle since they do not indulge in sustained swimming and generally live in a much slower and less vigorous lifestyle than the high performance swimmers.

D. EFFECTS OF BIOTIC AND ABIOTIC CHANGES ON OXYGEN CONSUMPTION

A list of the many factors which influence metabolic rate and therefore oxygen consumption was drawn up by Brett (1970) and is shown in Fig. V-4. While the effects of a number of these factors are relatively unknown and therefore only estimated, it is clear that temperature, oxygen and activity exert the greatest effects on metabolism.

Oxygen availability can be limiting for metabolism, even when water is saturated with air. The upper limit to the right of 15°C of the performance envelope in Fig. V-1 is interpreted as being oxygen-limited because sockeye salmon swam in sustained fashion at speeds considerably in excess of those shown when they were given oxygen-supersaturated water. At temperatures less than 15°C, temperature is thought to be limiting—i.e., the molecular reactions which produce energy cannot go fast enough to use even the available oxygen. Temperature and oxygen levels presumably set the upper performance limits for most other fish (e.g., Fig. V-2), but the relationship has not been tested as thoroughly for other species as for sockeye salmon.

Activity is by far the largest single factor affecting oxygen consumption. The relationship between oxygen consumption and

Fig. V-4. Various factors which may affect metabolic rate, with estimates of the extent of possible influence. Symbols denote increase (+), decrease (−), times (X), approximate range (∼), potential influence (⟶), dissociation involved (⇌), active rate (A), and standard rate (S). The range or circumstance considered for each factor is considered in parentheses. (From Brett, 1970, with permission of author and publisher.)

ENVIRONMENTAL FACTORS
& METABOLIC RATE
[STANDARD ———— ACTIVE]

	ABIOTIC	
1	TEMPERATURE (+10°C)	+ ·5∼3X
2	SALINITY (0 → 30‰)	− 10–15%
3	OXYGEN (10 − 1 8ppm)	→ −100%
4	CARBON DIOXIDE (2ppm)	10X = −20%
5	AMMONIA (0 3ppm NH_3)	⇌ toxic
6	PH (·5 − ·9)	?
7	PHOTOPERIOD	30%?
8	SEASON	30%
9	PRESSURE	?

	BIOTIC	
10	ACTIVITY (Max)	+ 5∼15X
11	WEIGHT ($M = aW^b$)	b = ·7 ∼ 1·0
12	SEX (♀ gonad = 12%-W)	?
13	AGE (vs Size)	?
14	GROUP (Schooling)	−20% ?
15	O_2 DEBT (Normal Load)	±20% ?
16	CONDITION (Exercised)	A +30% ?
17	STARVING (10 Days)	S −20% ?
18	DIET (Fat Ratio)	± 6% ?

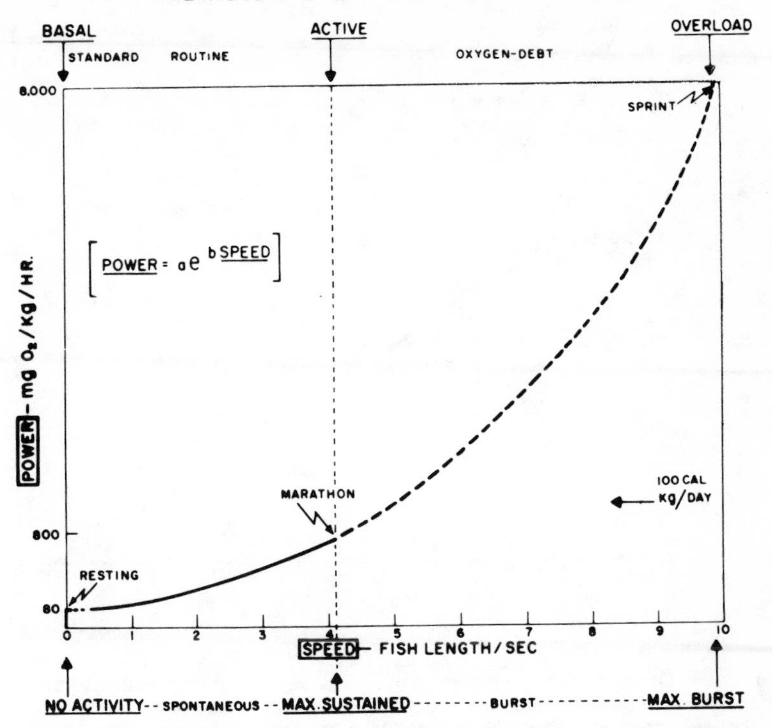

Fig. V-5. The relation between power (metabolic rate) and speed showing the energy demand associated with typical levels of performance such as "marathon" (sustained maximum) and "spring" (burst maximum). The basic curve is extrapolated to indicate the equivalent oxygen requirements for burst speeds which produce an oxygen debt. (From Brett, 1970, with permission of author and publisher.)

swimming velocity appears in Fig. V-5. The relationship between swimming velocity and oxygen consumption is a simple exponential one comparable in shape to the power-performance curve for any self-propelled device, either biological or mechanical. The various parts of the curve—basal (standard), active (maximum sustainable, marathon), burst (sprint)—are comparable to the same labels in the performance envelope of Fig. V-1. The oxygen consumption rates for power levels greater than "active" are not measurable, however, but extrapolated from the left-hand portion of the curve. Actually, an oxygen debt occurs because the oxygen delivery system cannot operate at such high rates. The

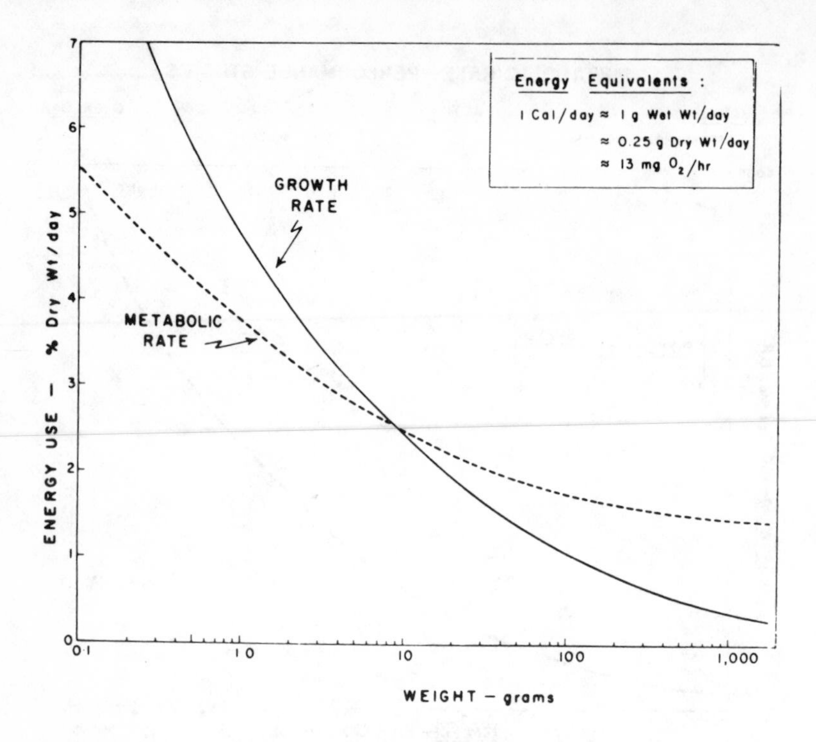

Fig. V-6. General relation between the energy deposited in growth and that expended in total metabolism, according to the size of the fish. Data are for sockeye salmon. (From Brett, 1970, with permission of author and publisher.)

slope of this power curve decreases at higher temperatures for sockeye salmon (Brett, 1970) and must also differ for different species of fish.

Changes in body weight make much smaller but highly predictable changes in the rate of oxygen consumption. While the total oxygen consumption increases with increased size, the oxygen consumption per unit weight decreases, as does the growth rate (Fig. V-6). Both of these decreases are characteristic of a wide range of animals, homeothermic or poikilothermic, aquatic or terrestrial.

E. MEASUREMENT OF ENERGY USAGE

The use of oxygen consumption rates as indicators of energy usage involves certain assumptions which need to be stated and the exceptions to these assumptions identified. The first assumption says that all energy production must be aerobic—i.e., that there are no incompletely oxidized molecules excreted and that there is no anerobic metabolism occurring during the oxygen consumption measurements. The first part of this assumption has been found to hold for several species of fish, except that in one of Brett's long-term swimming studies the oxygen consumption of unfed sockeye salmon smolts accounted for only 80% of their lipid loss during the swimming test of several weeks in duration. Thus the aerobic assumption usually, but not always, holds true. The second part of this aerobic assumption can be tested by measuring whether or not there is any increase in blood or muscle lactate (see below, section J and Chapter VII-B), which is the common end product of short-term, anerobic consumption of glycogen. This is of concern only in short-term experiments because fish cannot remain anerobic for long periods.

The second assumption concerns the caloric value associated with a given oxygen consumption. The caloric value used by Brett was that consumption of 1 mg O_2/kg/hr is equivalent to 0.00337 kcal/kg/hr or 0.081 kcal/kg/day. Also 1 kcal/kg/hr requires consumption of 297 mg O_2/kg/hr. (See also section K below for caloric values of food constituents.)

There are also certain conditions necessary for taking measurements of oxygen consumption, such as that the fish are healthy, are fasting and have recovered from anesthesia, handling and any "excitement" associated with the testing. In general, exact measurement of oxygen consumption requires careful control of many variables which is possible only under laboratory conditions. Meaningful measurements of the bioenergetic dynamics of fish in nature are very difficult.

F. ENERGY USAGE

The major steps in energy expenditure are shown in Fig. V-7. Brett (1970) developed this chart to show energy flow through a typical salmonid in which 100 calories were ingested and 80 of them actually became available to the fish. If 40 calories were used for basal metabolism (maintenance), then the remaining 40 calories could be used in various combinations of growth and activity. As shown, the entire 40 calories were expended as

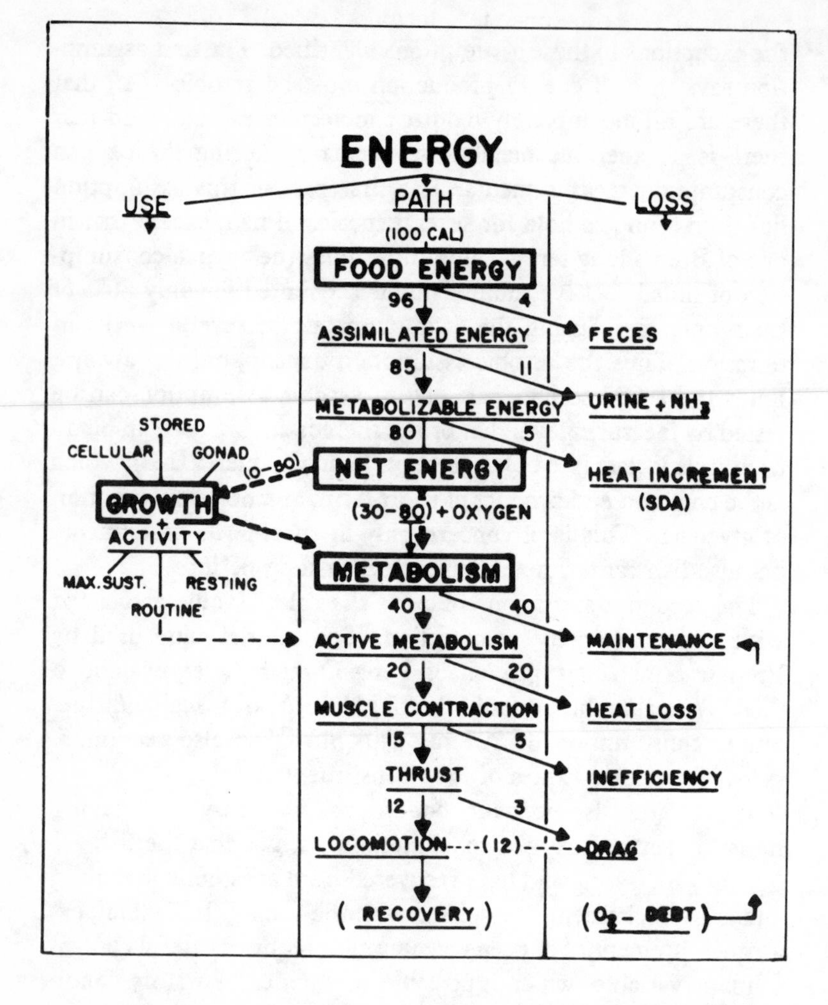

Fig. V-7. Flow chart illustrating the major steps in energy expenditure, with alternative routes and their accompanying losses and uses. The numbers provide an approximate estimate of the amount of energy used at each stage starting with 100 calories or 100% of the ingested food. Anywhere from 30 to 80 calories (%) of Net Energy may be expended for metabolism depending on how much is devoted to growth. The loss of 20 calories (%) as feces, urine, and heat is approximate, assuming an average state of activity. The heat increment labelled SDA stands for Specific Dynamic Action which is described in Chapter VI. (From Brett, 1970, with permission of author and publisher.)

muscular contractions for locomotion. All of the energy pathways, except that for storage, eventually end up as heat.

G. GROWTH EFFICIENCY

Any time that animals are grown on artificial food, it is of obvious importance to measure the relationship between food intake and growth. If growth is poor, the food may be inappropriate for the animal. The objective in commercial aquaculture is to maximize growth at minimal cost.

Most aquaculture operations regulate their feeding quantities as some percentage of the body weight of the fish. Samples of live fish are weighed periodically, often by counting the number of fish required to increase the weight of a preweighed container of water by one pound. Then by estimating the number of fish in the raceway or pond, the total biomass can be estimated and the growth compared to the amount of food fed. One such comparison is the conversion ratio:

$$\frac{\text{Food Fed During Interval}}{\text{Weight Increase During Interval}}$$

To understand any particular conversion ratio, one must know the kind of information which went into it, otherwise some ridiculous interpretations result. In many cases the weight of the food is simply its bulk weight as delivered by the manufacturer and the weight increase is in terms of the wet weight of the fish. Using dry food, it is possible to get conversion ratios of less than one, for example, which does not mean that the fish magically grew more than it was fed. The fish had added about three or four volumes of water to every volume of dry food assimilated and incorporated into new tissue. Conversion ratios are useful for empirical comparisons of growth efficiencies in circumstances where it is impractical to kill and analyze samples of the fish, but they shed little light on the actual bioenergetic efficiencies which are involved.

One answer to problems of using conversion ratios is to use dry weights for both the food and the weight gain. Then it is possible to compute a realistic growth efficiency as:

$$\text{Gross Growth Efficiency} = \frac{\text{Gain in Dry Weight}}{\text{Ration (dry weight)}} \times 100$$

For young sockeye salmon, Brett (1971) found a maximum gross growth efficiency from as little as 10% using marine zooplankton (frozen) to as high as 50% on specially balanced artificial diets. Since different metabolic demands, particularly the specific dynamic effect (see below), could be produced by different foods, another useful piece of information is the net growth efficiency in which the amount of food required for no growth (maintenance ration) is subtracted from the food input:

$$\text{Net Growth Efficiency} = \frac{\text{Gain in Dry Weight}}{\text{Total Ration—Maintenance Ration}} \times 100$$

This adjustment to the efficiency determination raised the efficiency levels noted above to about 20% for the zooplankton and to nearly 75% for the special artificial diet. Determination of the maintenance ration is difficult unless the fish are fed at several ration levels and the exact maintenance level determined by interpolation or extrapolation.

H. EXTERNAL FACTORS AFFECTING GROWTH

Growth and eventually, reproduction often seem to be indicators of how well a fish has coped with all of the other demands of living—from basal metabolism to the energy required to defend territory in its environment. While growth does not necessarily get last priority position during energy distribution—hormonal controls may dictate otherwise—in many cases growth and reproduction seem to get the surplus, if any, of energy which remains after most other functions, stresses and any other immediate responses have been taken care of. Thus growth and reproduction are, on the one hand, good indicators of how successfully a fish has met its environmental problems. On the other hand, both are influenced by virtually every factor imaginable.

In the routine daily life of a fish, two external factors have major effects on growth and growth efficiency—temperature and ration level. Brett et al. (1969) combined temperature, ration and gross conversion rate into a single graph for young sockeye

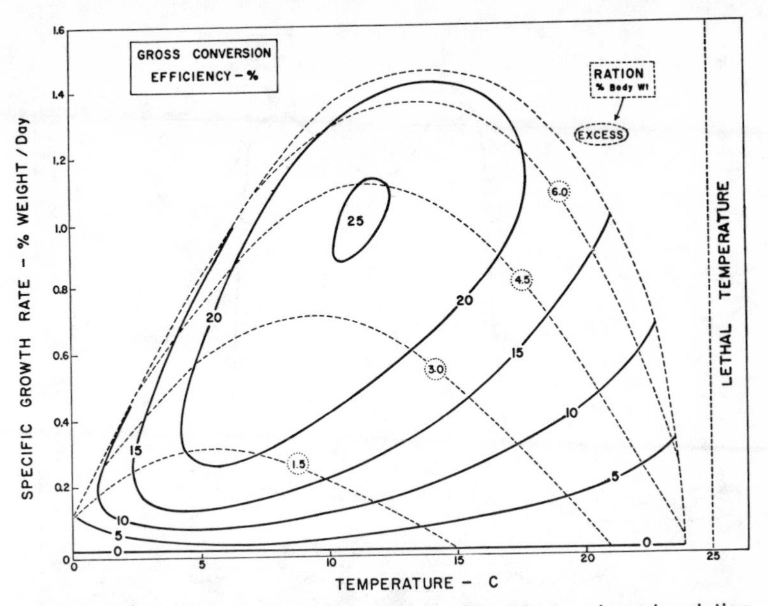

Fig. V-8. Gross efficiency of food conversion in sockeye salmon in relation to temperature and ration, drawn as isopleths (solid lines) overlying the growth curves (broken lines). Note that the points of maximum growth and efficiency shift to colder temperatures at lower ration levels. This corresponds nicely with the cold, oligotrophic nursery lakes used by sockeye salmon in British Columbia and Alaska. (From Brett, et al., 1969, with permission of author and publisher.)

salmon (Fig. V-8), which included several important findings. The combination of temperature and ration which produced the maximal gross efficiency centered at 11.5 °C and 4% (of body weight) ration per day. If one views these conversion efficiency isopleths as a contour map, then the 20% contour shows the efficiency peak as a relatively flat plateau which extends over much of the fish's temperature and ration range—i.e., sockeye grow relatively well over a wide range of temperature regimes. Even more important is the fact that the maximal growth rates at lower temperatures shift to the left—i.e., sockeye make maximally efficient use of the little food available in cold northern lakes. Using the contour map analogy again, the conversion efficiency plateau has a ridge which extends from the peak toward the lower left of the graph. Further, maximum growth efficiency is not at the maximum ration level.

Although not shown in this graph, the 1.5% and 3.0% ration

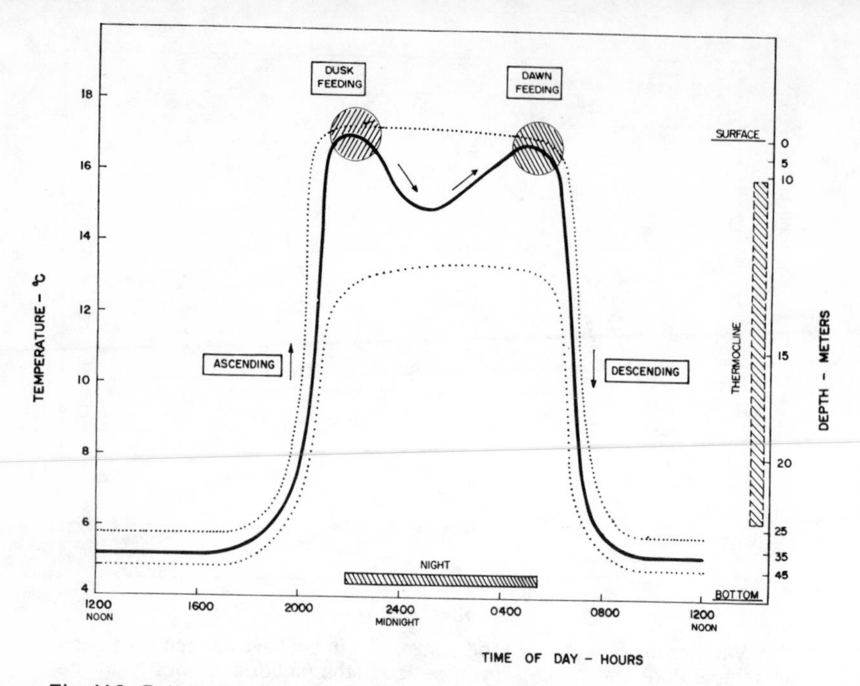

Fig. V-9. Pattern of diurnal migration of sockeye salmon in mid-summer with corresponding temperatures. Dotted lines indicate the general limits of the thermal experience. This behavior occurs only in the presence of a strong thermocline in the lake. For further discussion, see Chapter V, section H. (From Brett, 1971, with permission of author and publisher.)

lines continue to the right below the zero growth rate. At temperatures over 15 °C, sockeye can eat moderately well and still lose weight because of high basal metabolic rates. At temperatures close to either lethal limit, feeding usually ceases so growth obviously ceases, too. With no food intake at all, fish lose weight at all temperatures, although the loss is quite slow at 5 °C or less.

While data such as that shown in Fig. V-8 can be derived only from laboratory experiments, the resulting concepts have application to field situations. One such application is shown in Fig. V-9. In surveying the fish populations of Babine Lake (western British Columbia, Canada), sockeye salmon were readily captured at dusk and dawn at the surface and a few were catchable during the night, but none could be found during the day. Using sonar, fisheries biologists finally identified a diurnal vertical migration, and Brett (1971) suggested a bioenergetic explanation for the behavior. In oligotrophic northern lakes such as Babine, there is a strong thermal stratification for a month or two during the summer, and the sockeye's zooplankton food supply is above the thermocline. By feeding at dawn and descending into the colder water below the thermocline, the fish should minimize their maintenance (basal metabolism) costs and maximize their conversion efficiency at ration levels in the 1-2% range. Another preliminary hypothesis was that the fish were avoiding predators during daylight hours, but there were few salmon predators in the lake so this idea was discarded.

If the bioenergetic hypothesis explained the fish's behavior, then why did the fish not return to colder water after their evening feeding? Brett had data showing that gastric emptying times at the colder temperatures would be about 16-18 hours and surmised that the fish would not have been ready to eat again at dawn unless they stayed in the warmer temperatures. Finally, the migration occurred only while there was a marked temperature differential between surface and bottom waters.

I. EFFECT OF AGE ON GROWTH

The absolute rate of growth accelerates during the juvenile period and produces a somewhat S-shaped curve for total body weight over the lifespan of the individual fish as growth later slows down again. On a weight-specific basis, growth rate and basal metabolism have been shown to decline over the entire life span (Fig. V-6). A further amplification of this idea is shown in Fig. V-10, in which there were distinct changes in growth rates. Brett (1969) called each of these different slopes a growth stanza and said that perhaps changes in levels of growth hormone were responsible. The changes seemed to be independent of temperature, since they happened at several temperatures at the same time. Also, the changes did not correlate with any changes in photoperiod or season. Thus age appears to be primarily an internal factor for controlling growth, in contrast to the external ones discussed above.

J. METABOLIC PATHWAYS

Although the emphasis in this book is on functions of organs

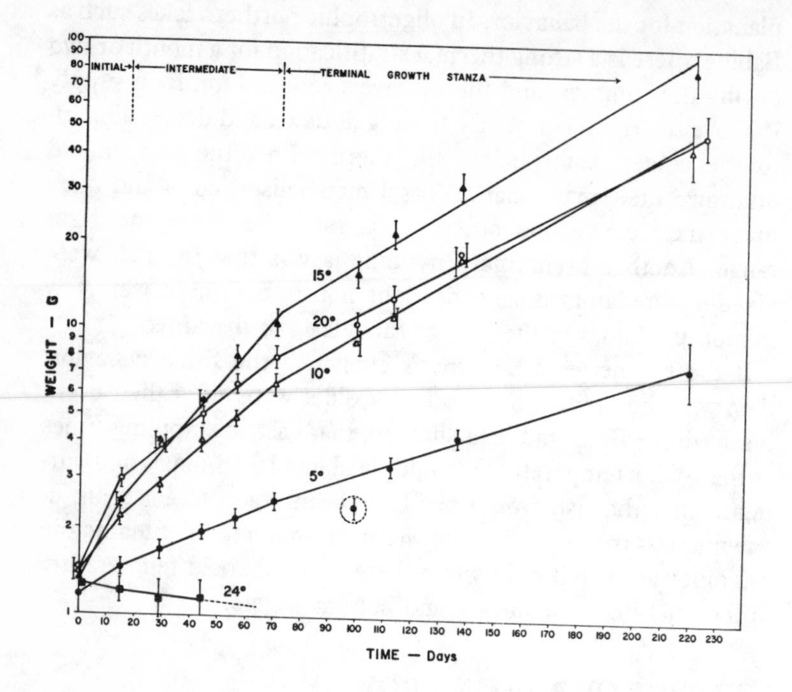

Fig. V-10. Mean weights of groups of 20 sockeye salmon held at five temperatures on excess ration starting on June 2. Limits for each point represent ±2 S.E. Three stanzas of growth are indicated according to inflection of the slopes. One point (circled) was discarded as being outside the 95% confidence limit of mean growth rate. (From Brett, et al., 1969, with permission of author and publisher.)

and whole organisms, most of the bioenergetic processes occur at the molecular level. Fig. V-11 is a greatly simplified chart which identifies the pathways followed by each of the three major food components—carbohydrates, lipids and proteins—and shows their locations at various stages of processing by a fish's molecular machinery. Food components start in some integrated form, usually as part or all of another organism which has been eaten, are separated by different processes of digestion and assimilation, but eventually funnel into a final common pathway in which their carbon and hydrogen atoms are combined with respiratory oxygen. Discussion of some of the highlights of each of these pathways follows. (See also Section K, this chapter.)

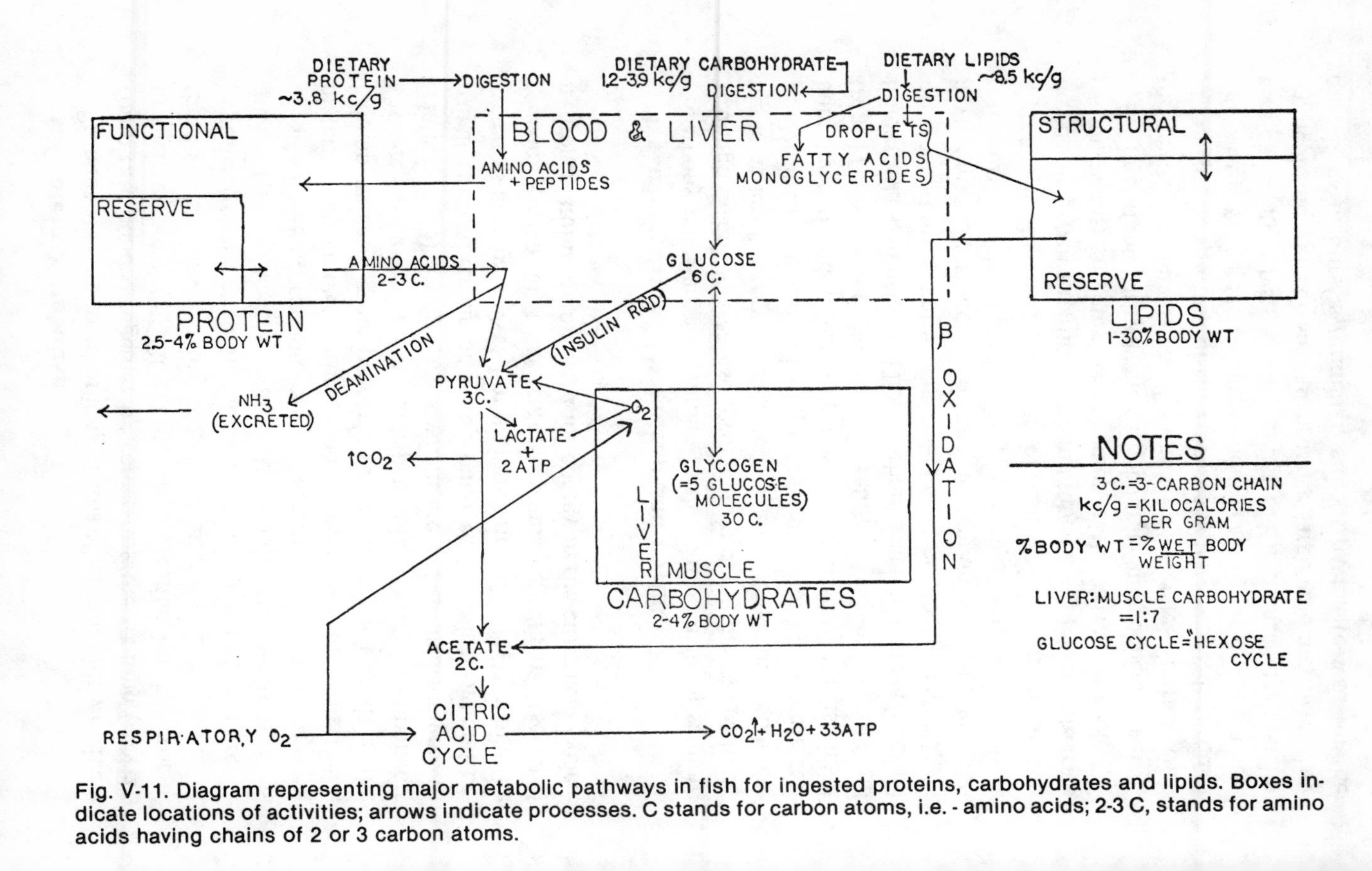

Fig. V-11. Diagram representing major metabolic pathways in fish for ingested proteins, carbohydrates and lipids. Boxes indicate locations of activities; arrows indicate processes. C stands for carbon atoms, i.e. - amino acids; 2-3 C, stands for amino acids having chains of 2 or 3 carbon atoms.

1. Protein Metabolism

Proteins are composed of long chains of amino acids which are complexly bonded together by connecting their amino ($-NH_2$) to their carboxyl ($-COOH$) groups. Digestion breaks most of these connections, leaving individual amino acid molecules or peptides (relatively short chains of amino acids). There are 21 different amino acids, of which 10 are indispensable in the diets of salmon—i.e., salmon cannot synthesize them from new materials or convert other food components to form them. Salmon will not grow properly or sometimes not at all in the absence of essential or indispensable dietary components (Halver, 1973). Amino acids which are not needed for growth usually have their amino group removed (deamination) and excreted as ammonia (NH_3) or ammonium ion (NH_4^+), while the remaining two or three carbon atoms and their associated hydrogen atoms are oxidized for energy in the citric acid cycle.

2. Carbohydrate Metabolism and Lactate

Carbohydrates are uncommon in the natural diets of salmon, although carbohydrates also form an important internal pathway for short-term energy management. Natural foods of salmon are predominately composed of proteins and lipids, so most carbohydrates found in fish are probably produced by the fish itself from two-carbon or three-carbon remnants of amino acids and lipids (discussed below). Those carbohydrates which are ingested usually end up as simple sugars—glucose or other related sugars having six-carbon atoms—during absorption from the gut and transport in the blood. They are then either stored as glycogen—a five-unit polymer of glucose—or further broken down into pyruvate (three carbon atoms) and then acetate (two carbon atoms) for complete oxidation in the citric acid cycle.

An important pathway in carbohydrate metabolism involves the reversible conversion of pyruvate to lactate. Starting with glycogen, the splitting of stored carbohydrate proceeds through glucose to pyruvate to lactate without needing any oxygen, a process called glycolysis. Thus, under the influence of adrenalin (epinephrine), glycolysis can produce large amounts of energy quickly as in burst swimming, without having to wait for the respiratory system to deliver additional oxygen. The anerobic reactions end at lactate, so lactate accumulates (mostly in muscles) until oxygen eventually becomes available. Lactate thus can be thought of as an oxygen "debt" which will eventually be paid back by oxidation of about one-fifth of the lactate to carbon dioxide and water to provide the energy to convert the remaining four-fifths back into glycogen.

Tolerable lactate levels have been observed to rise to 900 mg/100g of white muscle or up to about 150 mg/100 ml of blood in adult salmon. Levels exceeding these were often associated with delayed mortality following handling or other stress (Black, et al. 1962). In juvenile salmon which survived scale loss experiments, Mearns and Smith (unpublished) found lactate levels in excess of 500 mg/100 ml in blood. Mearns (1971) also showed that adult salmon had stable, low levels (5-20 mg/100 ml) of lactate present in their blood even when they were rested and inactive. Lactate rose only slightly during swimming until the respiratory system began to fall behind on meeting the oxygen demand of the muscles. A related observation (in Black's laboratory) was made by E.D. Stevens (personal communication) who noticed that blood lactate never increased as much as muscle lactate and investigated what would happen if it did. He injected enough sodium lactate into the blood of a large rainbow trout to produce about 600 mg lactate/100 ml blood and found that the fish developed rigor mortis which started at the tail and progressed anteriorly, killing the fish by suffocation when it stopped the branchial pump. Similar results were obtained in my laboratory by injecting lactate. We also have seen premortal rigor in fish which have been subjected to severe stress but did not measure the lactate levels in those fish.

The liver has a major role in lactate metabolism. The oxidation of lactate to pyruvate takes place in the liver, although there is now one report that the reaction also occurs in fish muscle. In addition to converting lactate to glucose, the liver stores about one-eighth of the total glycogen. Liver glycogen seems relatively unaffected during exercise but becomes depleted during starvation. Liver glycogen therefore probably is not part of the quick energy system in the same way that muscle glycogen is. Thus most muscle lactate travels in the blood from muscle to liver for oxidation and then returns to the muscle as pyruvate or glucose for storage inside the cell again as glycogen. As described above, death occurs when blood lactate gets too high, but it is not clear whether the lactate comes out of the muscle too fast or the liver processes it too slowly. Recent experiments suggested that gill tissues also can oxidize lactate.

Theoretically, an alternative way to deal with high levels of blood lactate instead of oxidizing it would be to excrete it. This would mean that the fish would be throwing away over 90% of the energy contained in the original glucose molecules, but that could be better than not surviving the high blood lactate. Mearns (1971) showed that lactate increased markedly in salmon urine during periods of high blood lactate but that the quantity of lactate was small—from 1% to 27% of the total blood lactate—mostly because the urine volume was small. Loss of lactate across the gills also seems possible but has not been investigated.

3. Lipid Metabolism

Lipid metabolism is the routine energy source for daily living in salmonids. Whereas white muscle operates from anerobic metabolism of glycogen stores and readily becomes fatigued, dark muscle obtains its energy primarily from oxidation of lipids and is virtually non-fatigable. In salmonids, dark muscle is situated laterally and is only around 8% of the total muscle mass but does most of the work during sustained swimming. Fatty acids, the primary kind of lipid which is oxidized by dark muscle, are chains of 16-22 carbon atoms with a carboxyl (acid) group on one end. These long chains are broken into two-carbon units (acetate) by a sequence of reactions called beta oxidation. The acetate is then fully oxidized by the citric acid cycle enzymes which occur in the mitochondria of all cells.

Lipids are a diverse group of substances and are therefore handled in a variety of ways for various purposes. Lipids are absorbed from the intestine either as droplets (chylomicra) of emulsified (chemically unchanged) lipid or as the component parts of triglyceride fats—glycerol and fatty acids. Triglycerides are then reconstituted by the fish using a wide variety of combinations of fatty acids, some of which must come from the diet (essential) and most of which can be converted from other lipids. Some lipids are simply stored (depot fat), often as triglycerides, for later use in energy production. Some triglycerides are converted into phospholipids by removing one of the three fatty acids from the glycerol and substituting a phosphate group. Phospholipids are important structural components of cell membranes and therefore are essential in producing new tissue. Still other lipids include steroids such as cholesterol and some of the hormones. Lipids not normally encountered by fish, such as hard (saturated) mammalian fats, can be digested and assimilated but usually cannot be utilized either for growth or for energy and just accumulate in muscle and as visceral fat.

4. Final Common Pathway

The citric acid cycle (also called Kreb's or TCA cycle) sooner or later is the eventual processor for most components of ingested food. It is also the major source of chemical energy with which fish (or any other organism) operate their molecular machinery. Starting from one unit of glucose, the citric acid cycle produces 33 units of high energy phosphate (ATP = adenosine triphosphate), which is a biochemical way to store energy until needed in other chemical reactions. Anerobic change of pyruvate to lactate produces only two ATP. Excreting a unit of lactate, therefore, would be a loss of 31 ATP. Converting lactate back into pyruvate requires input of two ATP. Lipids fit into the final common pathway at the acetate level, while amino acids, which may have either two or three carbon atoms, can fit into the final common pathway at either the pyruvate or acetate levels.

5. Nitrogen Metabolism and Other Pathways

There are at least three other routes for metabolites to leave a fish's body besides the final common pathway. Before oxidation of an amino acid, the amino group (NH_2^-) is removed and excreted across the gills as either ammonia (NH_3) or as ammonium ion (NH_4^+). In sharks, rays and ratfish (also reptiles and birds), two ammonia molecules combine with one carbon dioxide molecule to form urea, which is used in hyperosmotic osmoregulation (Chapt II-A). Components of DNA and RNA from ingested food and from the turnover of the fish's own DNA and RNA are excreted by the kidney as purines and pyrimidines. Finally, the bile from the gall bladder contains bile salts, which are a detergent-like product from the decomposition of cholesterol or other steroids, and bilirubin, which is a decomposed product of hemoglobin. In the land vertebrates bile promotes the emulsification of fats in the intestine and may be reabsorbed and recycled several times before finally being excreted in the feces. Although little studied in fish, bile probably functions similarly there.

6. Changes in Metabolism During Acclimation; Other Adjustments

Acclimation is the sum total of the adjustments which fish make to long-term changes in their environment. The changes are most frequently thought of in terms of seasonal or other temperature changes, but also can occur in response to changes in oxygen levels, salinity or other environmental factors. The changes are complex mixtures of adjustments in hormones, metabolic pathways, enzymes and behavior which occur at all functional levels, from the molecular and cellular to the whole organism and population. The subject is too large to discuss in any detail here (see reviews by Fry, 1971; Hochachka and Somero, 1971); only some major highlights will be given.

The typical pattern of changes in oxygen consumption during acclimation to a new temperature is first a change which is proportional to the temperature, but which is then modified. For example, a salmon transferred from 15 °C to 5 °C will immediately show approximately a 50% reduction in oxygen consumption, followed by a slow rise during the next three weeks or so. Similarly, a salmon moved from 10 °C to 20 °C will approximately double its rate of oxygen consumption and then decrease it slowly to reach a new stable level after about a week. With either increased or decreased temperatures, acclimation processes reduce the impact of the temperature change. One of the biochemical changes shown to be associated with adaptation to cold temperatures is a change in emphasis from hexose (glucose) metabolism to pentose (five-carbon sugars) metabolism (not shown in Fig. V-11).

After acclimation is completed, one of the results is increased tolerance. A salmon acclimated to 5 °C has lethal temperature limits of about 0 °C and 24 °C. The same fish acclimated to 22 °C has lethal limits of about 1 °C and 26 °C—i.e., warm-acclimated fish tolerate heat slightly better than cold-acclimated fish and vice versa. Fish also acclimate somewhat to low oxygen and perhaps to some toxicants.

There are several other metabolic adjustments which resemble acclimation. Lungfish decrease their metabolic rate during estivation in their mud cocoons, even though the temperature probably increases. Fever in mammals is a kind of metabolic adjustment used to combat disease. An equivalent change in fish was suggested by evidence that infected fish showed a higher temperature preference than healthy fish.

K. NUTRITION OF FISH

Nutrition is the study of the food requirements of fish to identify those substances which are necessary for normal growth and reproduction. Laboratory experiments have been performed to identify the individual amino acids, sugars, fatty acids and vitamins which are required by salmonids. The early work on diet development for salmonids involved the empirical testing of commercial feeds. The development of prepared feeds for new species now usually involves both laboratory and empirical tests. Commercial feeds are made from ingredients such as fish and meat meals, vegetable and fish oils and commodity grain products rather than chemically-pure compounds. Prepared foods for tropical fish sold in pet stores have been developed similarly from natural sources—brine shrimp, freeze-dried plankton, shrimp flakes, clams, tubifex worms, fish meal, oils, starches, etc. In most cases the suitability of various diets has been determined for a fish species by trial and error (empirical) feeding experiments, often using the fish's natural food as a starting point for the selection of commercially-available food components to be tested. The two main efforts in fish nutrition have concentrated on salmonids (by state and federal laboratories) and on tropical fish by the feed manufacturers and the hobbyists themselves. Feeds for cultured catfish began as an offshoot of salmonid foods (less protein and more carbohydrate than salmonid feeds) and more recently have become a research area in their own right. However, nutrition of salmonids is still the best known.

1. Nutritional Requirements of Salmonids

Whatever a fish eats, regardless of the composition and nature of the food, it should contain sufficient energy to sustain the basic metabolic processes and provide for growth and reproduction. These energy requirements vary widely with the size and species of fish and with the environmental temperature and the fish's level of activity. Halver et al. (1973) indicated that a 100 g growing rainbow trout needs about 3.1 kcal/day at 15 °C. A standard trout ration contains about 2200 kcal of metabolizable energy per kilogram so that the 3.1 kcal would be provided by 1.4 g/day of the feed. Fish having other than a 100-gram body weight would be fed at the rate of 1.4% of body weight per day. Fish at temperatures above or below 15 °C would have their feeding rate adjusted upward or downward, respectively. For comparison note that in Fig. V-8, Brett showed that sockeye salmon reach a maximal growth efficiency at ration levels of about 4.5% of body weight and that anything over about 6% of body weight was wasted.

Different food components provide different amounts of energy according to their inherent energy content and their digestibility. Fats contain the most energy—9.45 kcal/gram—and trout, salmon and catfish can obtain about 8.0 kcal/gram of that after digestion and assimilation. Proteins have an energy value of 4.65 kcal/gram, of which about 3.9 kcal/gram is available. Carbohydrates contain about 4.1 kcal/gram, of which highly variable amounts are available to fish depending on the digestibility of the particular carbohydrate. Glucose and other simple sugars provide up to 4.0 kcal/gram, while starches provided only about 1.6 kcal/gram and cellulose provides almost zero energy. These figures reflect the predatory habits of salmonids and that most of their natural food is of animal rather than plant origin—their metabolic systems deal quite effectively with proteins and lipids but poorly with large molecules like starch or cellulose.

There have been many feeding experiments to determine the maximal amounts of carbohydrates which salmonids can tolerate in their diet. Up to a point, carbohydrates have a sparing action on dietary proteins—i.e., carbohydrates are oxidized in the citric acid cycle (see Fig. V-11) rather than amino acids, so the amino acids are spared and available for growth. Further amounts of carbohydrate probably enhance the storage of glycogen in liver and muscle. Still further amounts of carbohydrate are then converted into lipid, which can appear as muscle oils, visceral fat or even result in fatty degeneration of the liver. In the latter case the liver enlarges, becomes yellow or whitish in splotches and presumably has reduced functional levels. A major reason for experiments with such potentially harmful results is that carbohydrates—usually in the form of cereal grains—are much cheaper than either proteins or lipids and maximal use of carbohydrates would reduce the cost of rearing salmonids.

A practical upper limit to carbohydrate levels in salmonid diets is about 20% (dry weight). While some trout diets may contain up to 50% raw starch, the low digestibility (20-30%) of raw starch reduces this to an effective level of only about 10% dietary carbohydrate. Dried skim milk, a common component of commercial diets, contains much lactose (milk sugar), which is about 60% digestible; and therefore skim milk should not exceed about 30% of the diet.

Gross protein requirements vary with age and temperature. Salmonid fry require about 50% protein in their diet to sustain their normal rapid growth rate. This decreases to about 35% protein by the time they are a year old. Chinook salmon at 7 °C required 40% protein for maximum growth while they needed 50% at 15 °C. Because of their need for large amounts of protein, salmonids will never be inexpensive fish to culture.

Requirements for specific amino acids include arginine, histidine, isoleucine, leucine, lysine, methionine, phenylalanine, threonine, tryptophan and valine. Salmonids show no growth or even suffer weight loss when fed proteins which lack one or more of these amino acids. Some of the dispensable amino acids also

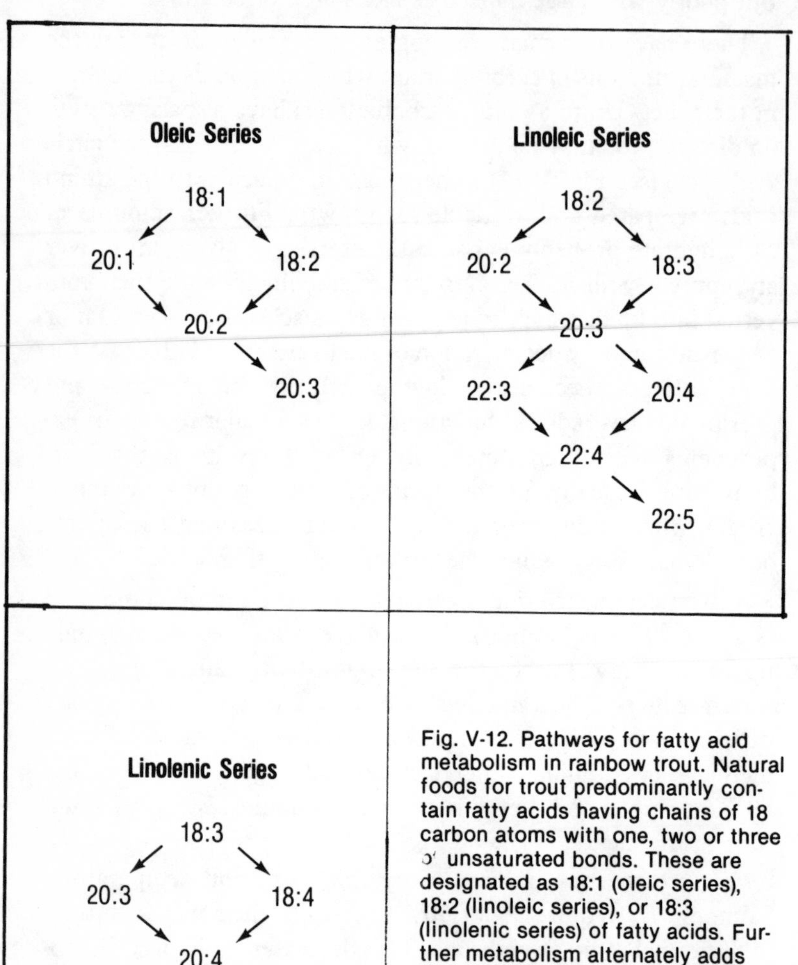

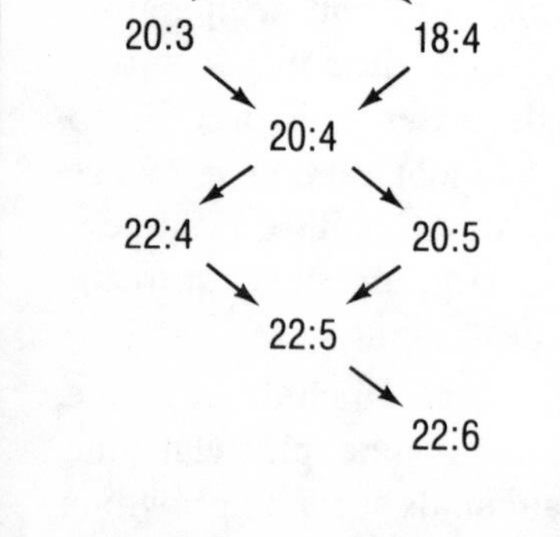

Fig. V-12. Pathways for fatty acid metabolism in rainbow trout. Natural foods for trout predominantly contain fatty acids having chains of 18 carbon atoms with one, two or three of unsaturated bonds. These are designated as 18:1 (oleic series), 18:2 (linoleic series), or 18:3 (linolenic series) of fatty acids. Further metabolism alternately adds two carbon atoms to lengthen the molecule and removes two hydrogen atoms to produce an additional unsaturated bond. The predominate fatty acids in rainbow trout have 20 or 22 carbon atoms and 5 or 6 unsaturated bonds. (adapted from Halver et al., 1973.)

have sparing effects on the required ones. Cystine, for example, replaces up to about half of the requirement for methione, and tyrosine replaces part of the required phenylalanine.

Lipid requirements are complex and quite different from those of most land, and particularly warm-blooded, animals. The most conspicuous difference is in the melting point of the lipids. Mammalian fats are solid at temperatures at which most fish lipids are still liquid. The chemical basis for this difference in melting points is in the length of the carbon chains in the fatty acid molecules and in the number and location of unsaturated bonds in the carbon chain. The shorter carbon chains and chains with more unsaturated bonds have lower melting points. Most lipids in natural fish diets come originally from plants and contain fatty acids having chains of at least 18 carbon atoms. These chains have from one to three unsaturated bonds and are designated as 18:1 (oleic acid), 18:2 (linoleic acid) and 18:3 (linolenic acid). These carbon chains can be further built up by zooplankton and fish to produce chains as long as 20:3, 22:5 and 22:6 by addition of units of two carbon atoms and removal of pairs of hydrogen atoms to create additional unsaturated bonds (Fig. V-12).

The 18:3 series of fatty acids is essential for normal growth in rainbow trout, while there appears to be little need for the 18:2 series. There should be at least 1% of 18:3 series in the diet and less than 2% of the 18:2 series fatty acids. The 18:1 series appears even less important than the 18:2 series. Salmon and catfish probably have similar requirements. The total amount of lipid essential for normal growth appears to vary widely. Rainbow trout will grow very slowly on a fat-free diet. In nature dietary lipid ranges from 3 to 15% (9-40% dry weight). In several commercially available trout diets dry pellets contained from 6 to 14% lipid, while moist pellets (which have to be stored frozen) contained 16-20% lipid (dry weights). Thus neither commercial diet equalled the maximal levels found in natural diets.

A major problem with producing artificial diets with high lipid contents is that the most effective fatty acids for producing maximal growth—e.g., 22:6—oxidize at the unsaturated bonds very rapidly in air and become rancid. Rancid fats produce poor growth and can even be toxic. The lipid levels in present commercial diets are about the maximum for which oxidation can be controlled by the use of antioxidant additives such as butylated hydroxytoluene (which are also used in many pastries for the same purpose). Unsaturated plant oils, which are more stable than fish oils, don't produce good growth or may even be toxic. Thus, even when kept frozen to slow down oxidation (some occurs even at −10 °C), commercial fish feeds have a limited storage life. Improved technology for handling highly unsaturated lipids is needed.

2. Nutritional Requirements of Other Species

The nutritional requirements of channel catfish appear to be similar to those of salmonids. Other catfish species appear to require slightly less protein than salmonids and also tolerate more carbohydrate.

The nutritional requirements for most cultured species have been worked out only at the practical level—i.e., they are known from trial and error. Particularly in southeastern Asia, where a considerable number of fish species have been cultured for several hundred years, much of the aquacultural art consists of knowing how to fertilize ponds or otherwise enhance the natural food supply. Much of the feeding of tropical fish by aquarium hobbyists is at a similarly primitive level, even though they may use freeze-dried preparations of natural foods which are high-technology products. The suitability of the diet is still determined on a trial-and-error basis. This is not to say that the lore developed by hobbyists and commercial aquaculturists is not of value to fish nutritionists. There is a wealth of information available there for someone to organize and use as a basis for designing scientific experiments. Some of this information is summarized by Halver et al. (1973).

Chapter VI: DIGESTION

A. OVERVIEW

Digestion is a combination of the mechanical and chemical processes by which ingested food is broken down and sorted into components which are either absorbed into the blood or remain in the gut until voided as feces. The general processes of digestion in fish have not been systematically investigated, but the fragmentary information available so far suggests that fish are similar to other vertebrates. The review by Kapoor, et al. (1975) covers many details of digestion in a wide range of species, mostly non-salmonids. Fish show many variations on the basic vertebrate plan which generally correlate with the food eaten. Some adaptations in fish are perhaps unknown in terrestrial vertebrates because some of the food sources available to fish are unique to the aquatic environment—e.g., the eating of coral by parrotfish or jellyfish by ocean sunfish. However, most fish are not highly specialized in their feeding habits and, even when they normally eat only a single kind of food, can switch to something quite different when the one food becomes unavailable. Becoming too specialized can be hazardous to the survival of the species.

While feeding and digestive structures of fish are exceedingly diverse, a few broad generalizations are possible. Most fish can be categorized as either herbivorous, omnivorous or carnivorous. Most herbivores eat a fairly limited range of plants and often have special chewing structures to get maximum nutritional value from them by ultra-thorough grinding. Eating detritus (a mixture of sediment, decaying organic matter and bacteria) can be fitted into the herbivorous category for lack of a better place. Omnivorous fish have a mixed diet and minimally specialized structures. Often they consume small invertebrates. Carnivorous fish eat larger invertebrates and other fish and may specialize on

Species	Feeding Habits	Relative Gut Length
	Characidae (Tetras)	
Pyrrhulina filamentosa	Insectivorous	1.0
Hydrocyon forskalii	Mainly carnivorous	0.8
Xenocharax spilurus	Plants & invertebrates	2.0
Distichodus niloticus	Plants & detritus	2.8
Citharinus congicus	Microscopic organisms	4.0
C. citharus	Microscopic organisms	up to 7.5
	Cyprinidae (Carps)	
Elopichthys bambusa	Carnivorous	0.6
Gobio gobio	Invertebrates	0.8
Barbus ticto	Invertebrates, plants	1.6
Ctenopharyngodon idella	Plants	2.5
Garra dembensis	Algae, invertebrates	4.5
Varicorhinus heratensis	Algae, detritus	up to 7.5
Hypophthalmichthys molatrix	Phytoplankton	13
Labeo niloticus	Algae, detritus	16.9

Table VI-1. Relative gut length (gut length/body length) of selected species of fish from families in relation to their feeding habits. (Adapted from Kapoor, et al., 1975.)

some particular type—insects or bivalves for example; these preferences may change with seasonal availability of particular foods. Some fish also fast during winter and during spawning. Herbivorous, omnivorous and carnivorous species of fish can be found in the same family, so it appears that feeding structures are highly adaptable and readily changed, at least in an evolutionary time frame.

Another generalization is that gut length correlates with diet—herbivores have longer digestive tracts than carnivores. Gut length seems to relate more to the amount of indigestible material in the food rather than to its plant or animal origin. Fish which ingest large amounts of mud also have gut lengths similar to those of herbivores. Data illustrating typical gut length in selected species of two families are shown in Table VI-1.

Other factors besides diet also influence gut length. Intestines with folds or other mechanisms for increasing surface area usually are shorter than those in related species with a simple tubular intestine. Gut length may change with the age of the individual fish, although this age difference may also include a change in diet. Gut length may be longer in individual fish grown on a sparse diet than on a plentiful diet. This may even change seasonally in some fish in response to changes in availability and kind of food. Thus the gut responds sensitively to many factors.

B. ANATOMY OF THE DIGESTIVE TRACT

The macroscopic and microscopic anatomy of the digestive tract will be dealt with here in somewhat more detail than any of the other organ systems of fish because a significant portion of the information on digestive function is inferred from its anatomy.

1. The Mouth and Teeth

Two extreme mouth shapes are readily identifiable, with large numbers of variations and intergrades between the extremes. One type of mouth has a large opening which extends along the sides of the head. Alexander (1967) describes this grinning type of mouth as being characteristic of predators, enabling them to grip prey crossways in the mouth. Typical examples would include barracuda, pike, salmon, bass, etc. The other extreme is a small, tubular mouth which maximizes sucking ability. For a given volume of water which can be "inhaled," the smaller the mouth opening the greater the velocity of the water. Maximizing intake velocity enhances feeding ability on zooplankton up to the point where the mouth opening is so small that it hinders entry of the food organism. Extreme examples of tubular mouths include the seahorse and pipefish, which are related types in the family Syngnathidae. Herring have a proportionately somewhat larger mouth opening than the seahorse, but they have special bone linkages which cause their lips to move forward while the mouth is opening and remain in a forward position during closing. Thus herring can reach forward and suck at the same time when capturing zooplankton.

Fish have a much greater variety of teeth and associated structures than any other group of vertebrates, thus only a brief overview can be given here. In the buccal cavity of the mouth, teeth (if any) are anchored to bones of the jaws and head. They are solidly fixed (a few rare cases are hinged) and named according to the bone to which they attach. Thus maxillary teeth are on the maxillary bones around the front edge of the mouth, palatine and vomerine teeth project from bones of the same name in the hard roof of the mouth, and lingual teeth project from the tongue, which is usually bony and relatively immobile in most teleost fish. Most of these teeth serve in capturing and holding prey and are commonly small. Large teeth get in the way or require the mouth to open wider. Most fish predators swallow their prey whole, although the piranha and the parrotfish are obvious exceptions which are very well equipped to bite flesh and coral, respectively.

In the posterior part of the mouth (pharynx and branchial cavity) are rather different types of teeth than in the front of the mouth. These teeth often attach to the upper or lower ends of the gill bars. The upper teeth often take the form of a pair of hardened discs in the soft palate. These pharyngeal pads, also called palatal organs, can be moved back and forth independently to "ratchet" prey into the stomach in fishes such as hake. In other fishes the pharyngeal pads may serve as a grinding surface for masticating food and may also produce mucus to lubricate it. The teeth associated with the lower ends of the gill bars usually serve a grinding function, often of plant material. These can be elaborately shaped pairs of pharyngeal teeth as in carp, which are largely herbivorous, or pile perch where they are used to crush and grind mussels. Pharyngeal teeth may chew against the pharyngeal pads in some cases.

Accessory organs abound—valves, membranes, taste organs, mucus-producing organs, accessory respiratory structures, gill rakers and others too numerous to describe here. These structures usually relate to particular feeding behavior, although the functions of a few—the epibranchial organs for example—are still questionable. Refer to Kapoor, et al. (1975) for details.

2. The Esophagus

The esophagus in most fish is a short, broad connection between the mouth and stomach. In stomachless fish it connects directly to the intestine or intestinal pouches (sometimes called gizzards) which serve somewhat the same function as a stomach. The functions of the esophagus are minimal outside of transport. Taste buds have been identified histologically (see Chapt. VIII—D), and mucus-producing cells may be abundant. There are several layers of muscles which may be more predominant in freshwater than in marine fish. This difference has been attributed to the "need" of freshwater fish to minimize water inflow during swallowing because of osmoregulatory disadvantages of additional water inflow as compared to marine fish which routinely drink seawater and should have minimal problems swallowing excess water with food.

3. The Stomach

The size of the stomach can usually be related to a typical interval between meals and to the size of the food particles. Fish which eat large prey at infrequent intervals have large stomachs, and those which eat small food items (i.e., are microphagous) more or less constantly have small or no stomachs. Taxonomic groups in which the ancestral species were stomachless but in which later species again became macrophagous may have intestinal pouches which look like stomachs but do not produce pepsin or acid. In stomachless fish the content of the entire digestive tract is alkaline, whether there are intestinal pouches or not.

There are several hypotheses to explain the evolutionary basis for development of stomachless species. In freshwater fish, the scarcity of Cl^- in ambient water might make the production of the Cl^- for acidification of the stomach difficult. If the food material contained shells with large amounts of $CaCO_3$, it would require large amounts of HCl to acidify it and then large amounts of alkali in the intestine to neutralize it, perhaps an excessive metabolic burden. In marine fish the drinking of seawater (which is alkaline) for osmoregulatory purposes might similarly make acidification of the stomach difficult. In microphagous fish eating large quantities of indigestible material, there may be little value in detaining food in a stomach. All of these hypotheses are

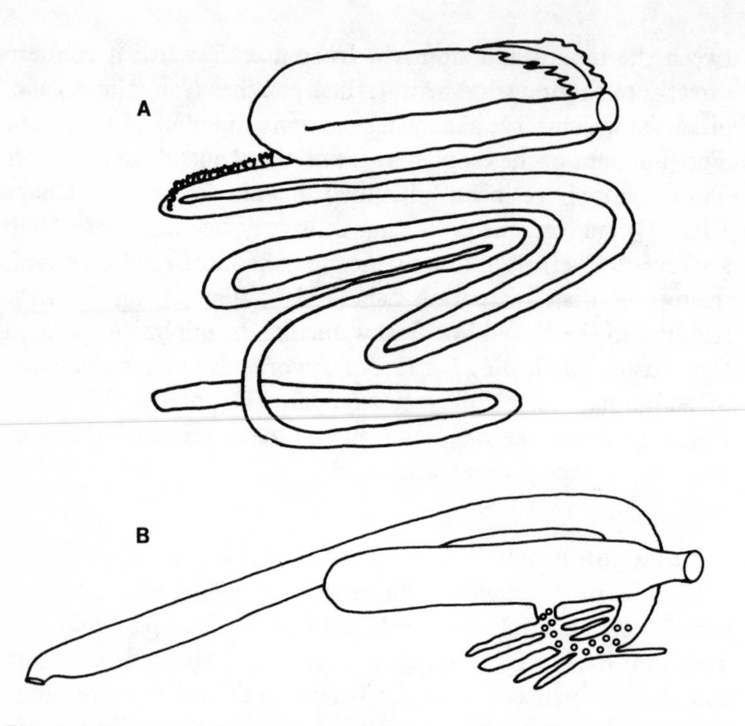

Fig. VI-1. Diagrammatic representation of the digestive tract of two fishes: A. A characin, *Distichodus niloticus,* having a gut length which is 2.8 X body length and which eats plants and debris (from Kapoor et al., 1975) and B. a salmonid, *Oncorhynchus kisutch,* having a gut length which is 0.6 X body length and eats smaller fish (adapted from Smith and Bell, 1976).

probably relevant for certain species, although none of them has been proved conclusively.

The digestive tracts of two fish shown in Fig. VI-1 illustrate two rather different stomach configurations. The gut in the characin is a continous tube and is perhaps the simplest situation, in which the stomach is simply an enlargement at the anterior end of the gut. In salmonids the intestine attaches to the side of the stomach rather than to the end. This salmonid plan appears to have possibly two functional adaptations. First, the stomach can extend posteriorly to accommodate relatively large prey without disrupting the mesenteric supports for the rest of the intestinal tract. Second, the short distance between the esophagus and the beginning of the intestine would allow the seawater being drunk for osmoregulatory purposes to pass through the stomach along a very short pathway with minimal contact or interference with gastric digestion.

The stomach wall is composed of tissue layers similar to those of other vertebrates. The innermost lining is a columnar epithelium containing mucus-secreting cells and cells which are thought to secrete both pepsin and hydrochloric acid (HCl). There is some tendency for secretory cells to occur in the anterior (cardiac) part of the stomach. In a few fish the epithelium near the pylorus appears to be nonsecretory, has a rich blood supply and may be absorptive in function. Layers of striated muscle extend into the stomach wall from the esophagus, although a few fish have striated muscle in the pyloric (intestinal) region of the stomach.

4. The Intestine

While the intestine is basically a simple tube originating with a pyloric valve at the end of the stomach and ending at an iliocaecal valve at the vent, there are a great variety of anatomical specializations found in various species of fish. Salmonids have numerous pyloric caeca, which are spaghetti-sized blind tubes leading off from the intestine near the stomach (usually called midgut). Their histological structure and enzyme content appear identical to that of the adjacent intestine, suggesting that pyloric caeca serve to increase the surface area of the intestine. Some pyloric caeca have sphincter muscles at their bases. Larger and fewer outpocketings of the intestine include rectal caeca (absorptive), a typhlosole (a lengthwise dorsal infolding which increases surface area) and a variety of foldings and ridgings of the mucosal epithelium (to increase secretory and absorptive surface area). Villi (finger-like projections into the intestine of mammals) have never been seen in fish. There may be a few fish having a ciliated epithelium in parts of their intestine. Much of the internal elaborations as well as much of the profusion of epithelial development inside of fish intestines is greatly reduced during periods of starvation. The external diameter and length of the intestine also decrease during starvation. A rectal area of the intestine can usually be distinguished posteriorly by the decreased numbers of secretory (zymogen) cells and increased numbers of mucus-producing cells which can be seen histologically.

5. Innervation of the Digestive Tract

The digestive tract receives branches from three cranial nerves: facial (VII), glossopharyngeal (IX) and vagus (X). The first two serve the mouth and esophageal regions while the vagus reaches all of the visceral portions. These nerves are usually considered as part of the parasympathetic (cholinergic) nervous system, which is typically stimulatory for visceral organs. In the few cases where vagal action has been tested by electrically stimulating the vagus nerve, the responses were mixed. In fishes with stomachs, stimulation caused contraction of the stomach, but only part or none of the intestine contracted. In stomachless fish, only part of the intestine contracted.

A single splanchnic (spinal) nerve has been identified anteriorly going to the stomach and intestine. Two posterior splanchnic nerves supply the rectum. These appear comparable to the mam-mammalian sympathetic (adrenergic) system and are generally inhibitory.

In addition to these extrinsic nerves, there are also intrinsic nerves entirely within the gut (Fig. VI-2). These nerves form several networks (plexuses) which are thought to be the anatomical basis for true reflexive peristalsis. Physiological tests on an isolated trout intestine using standard neurotransmitter and inhibitor substances suggested that peristalsis resembled that in higher vertebrates.

In the few instances investigated, no nervous control of gastric, hepatic or pancreatic secretion was found, but it is too early for any firm conclusions. In general, knowledge about the innervation of visceral organs of fish is minimal.

6. Accessory Digestive Organs: Pancreas and Gall Bladder

Whereas the pancreas of sharks and rays is large and distinct, the pancreas of most teleost fishes is diffuse and not readily observable during gross dissection. A diffuse pancreas consists of small beads of pancreatic tissue scattered through the mesenteries of the fish, each supplied with an artery, vein, nerve and pancreatic duct. Usually the ducts eventually coalesce and join the bile duct to form the common bile duct before entering

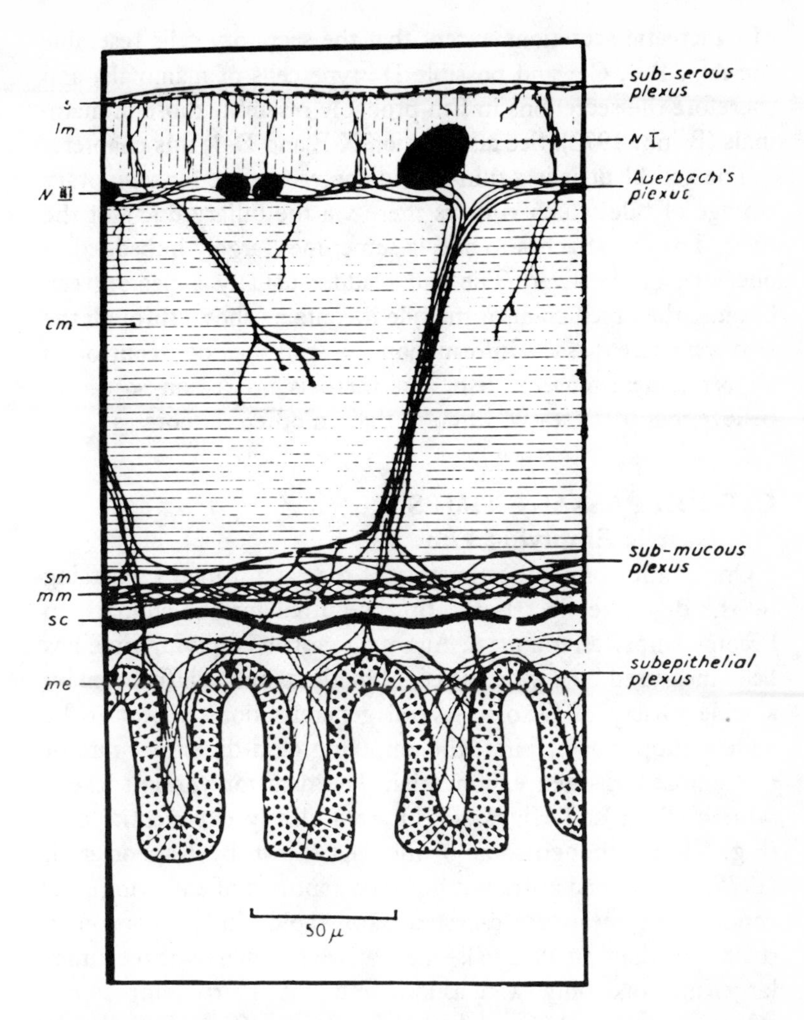

Fig. VI-2. Nervous elements in the stomach wall of trout: me - mucosal epithelium; sc - stratum compactum; mm - muscularis mucosa; sm submucosa; cm - circular muscle; lm - longitudinal muscle; s - serosa; NI, NIII -two types of nerve cells. (From Kapoor, et al., 1975, with permission of author and publisher.)

the upper part of the intestine, although two species of cottids have separate bile and pancreatic ducts.

Most of the functional information on the role of the pancreas in digestion is derived from histological and electron microscopic studies. There is no comprehensive understanding of the nature of pancreatic secretions except that the secretory cells resemble the A–, B–, C– and possible D–type cells of mammals, and therefore the secretions in fish probably resemble those in mammals (Brinn, 1973). See also section X-E and D-3, this chapter.

The gall bladder is a thin-walled contractile sac for temporary storage of bile which collects there via the biliary ducts of the liver. The sac attaches to and is sometimes largely imbedded in one lobe of the liver. The gall bladder usually appears green because the green color of the bile it contains shows through the semitransparent sac. Information on gall bladder control is almost nonexistent for fish, but there is no reason so far to believe that it is greatly different than in other vertebrates.

C. FOOD PASSAGE TIMES

1. Gastric Emptying Time

One of the common estimators of the rate of food processing by the digestive tract is the time required for the stomach to become empty after a meal. Although gastric emptying time has been measured by a wide variety of methods and is influenced by a wide variety of factors, several generalizations appear to be widely supportable. First, the emptying (and digestion) rate of the stomach is often exponential. The data for young sockeye salmon (Brett and Higgs, 1970) are a clearly exponential case (Fig. VI-3), although some of the data shown by Kapoor et al. (1975) approached a straight line (constant rate of emptying). Second, larger meals are digested more rapidly in proportion to their size than small meals—i.e., a meal which is three times larger may take only twice as long to digest. Third, temperature strongly controls the rate of gastric emptying (Fig. VI-4). Meals with a high fat content are digested most slowly, and final emptying of the stomach may be delayed by the presence of indigestible material such as crustacean exoskeletons (chitin). Longer intervals between meals usually mean that larger meals will be ingested if sufficient food is available, although sockeye salmon will not eat a maximal size meal below 10 °C or above 20 °C. In contrast to salmonids, neither the size of the meal nor the interval between meals in sunfish seemed to affect the rate of gastric digestion.

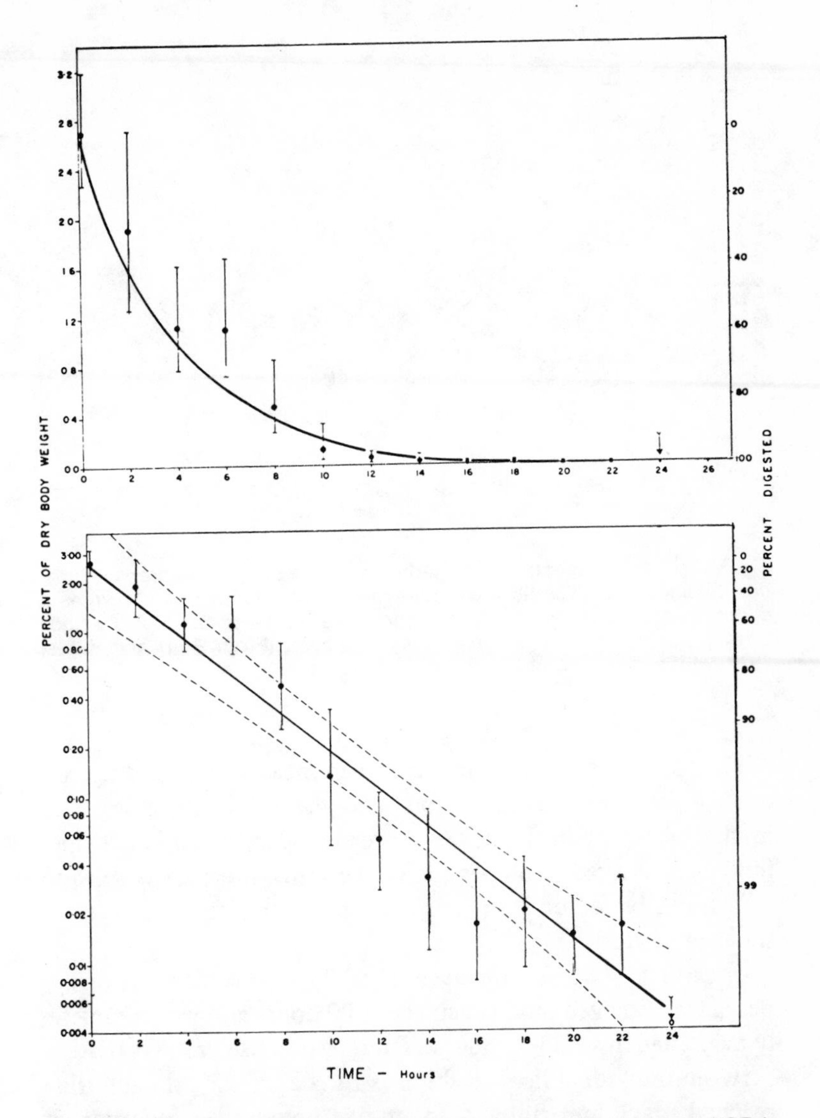

Fig. VI-3. Rate of decrease in the amount of organic matter (mean ± 2 SE) in the stomachs of fingerling sockeye salmon at 20°C. The curve in the upper graph was drawn from the transformed data in the lower graph, using the line of best fit. The 95% confidence limits on the line are shown as dotted lines in the lower graph. (From Brett and Higgs, 1970, with permission of author and publisher.)

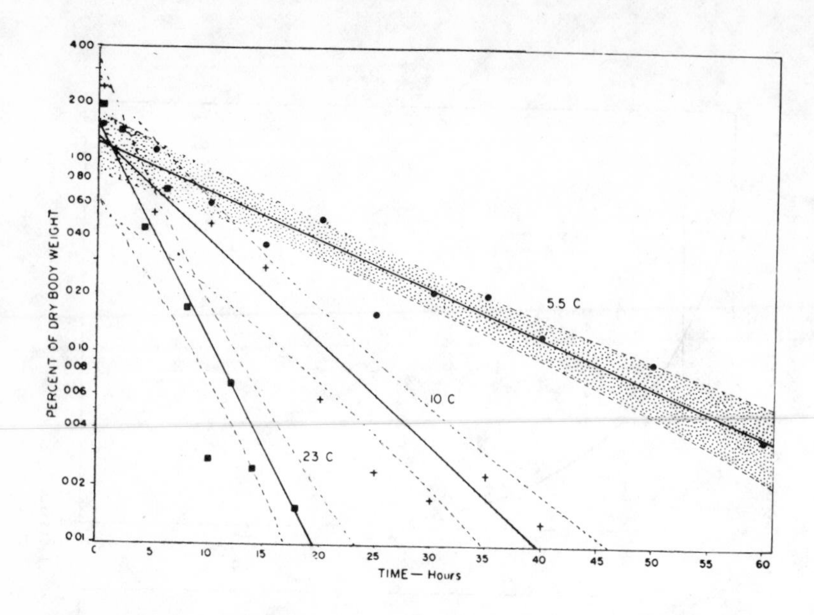

Fig. VI-4. Rate of digestion in fingerling sockeye salmon at three acclimation temperatures. The 95% confidence limit on each line is shown as a stipled band. The food ingested is shown as the percentage of dry body which remains following ingestion at time zero. (From Brett and Higgs, 1970, with permission of the author and publisher.)

2. Use of Fecal Markers

The total food passage time has been measured by adding colored or inert material to the food and waiting for its appearance in the feces. Typical markers have included chromium oxide, food dyes and radioisotopes such as cerium. Alternately, a group of fish can be fed uniform amounts and then sampled at various times afterward.

In pinfish *(Lagodon rhomboides)* at 24°C, the first major appearance of tagged food began about 20 hours and was completely evacuated after 28-37 hours. There was considerable variation between individual fish. In skipjack tuna at 23-25°, the intestine reached maximum fullness about five hours after ingestion of smelt. The stomach emptied at an initial rate of about 10%/hr, becoming empty in 12 hours and the intestine was empty after 14 hours. In general, tuna eat more food (up to 15% of body weight/day) and digest it more rapidly than is known for any other fish.

3. Gut Motility and Stasis

The peristaltic action of the intestine has been studied in several species by use of an isolated piece of intestine bathed in saline and attached to an apparatus to record its contractions. In tench *(Tinca vulgaris)*, Ohnesorge and Rauch (1968) showed that peristalsis was stopped by adrenalin and stimulated by acetylcholine. Peristalsis was also stimulated by stretching the intestine by inflating it with water, with the size of the peristaltic contraction being proportional to the degree of stretching (Fig. VI-5, D, A). Other studies in the same laboratory used electrical stimuli to produce peristalsis but failed to show any definite effects of adaptation temperature on peristaltic rate or amplitude. Similar pharmacological studies with similar results used isolated gut preparations from brown trout *(Salmo trutta)*.

The tench and brown trout results were both interpreted as showing that peristalsis in teleost fish is a true reflex resulting from the activity of an intrinsic nerve plexus in the gut wall. Even though striated muscle appears in some parts of the intestinal tract of some fish, the activity seen in the isolated gut preparations was entirely that of longitudinal and circular layers of smooth muscle. Thus, the peristaltic activity of the teleost intestine appears typically vertebrate.

Another form of food transport besides peristalsis also probably occurs. Scattered reports indicate the presence of cilia in the intestine of several teleosts, especially small or larval fish. In pond smelt *(Hypomesus olidus)* larvae, Iwai (1967) found scattered ciliated cells in the posterior half of the gut. Since the cilia were found to transport small particles posteriorly, Iwai suggested that the cilia might have had a subsidiary role in transporting particulate food, particularly if peristalsis were weak. Similar cilia also occurred in the gut of a small Japanese salmonid, the ayu *(Plecoglossus altivelis)*. In both cases there was a gradual disappearance of ciliation with age until no cilia were found in the adults. Although there has been some dispute as to whether the cell projections were cilia or microvilli in various species, Iwai (1967) used electron microscopy to show that both were present in the pond smelt (Fig. VI-6).

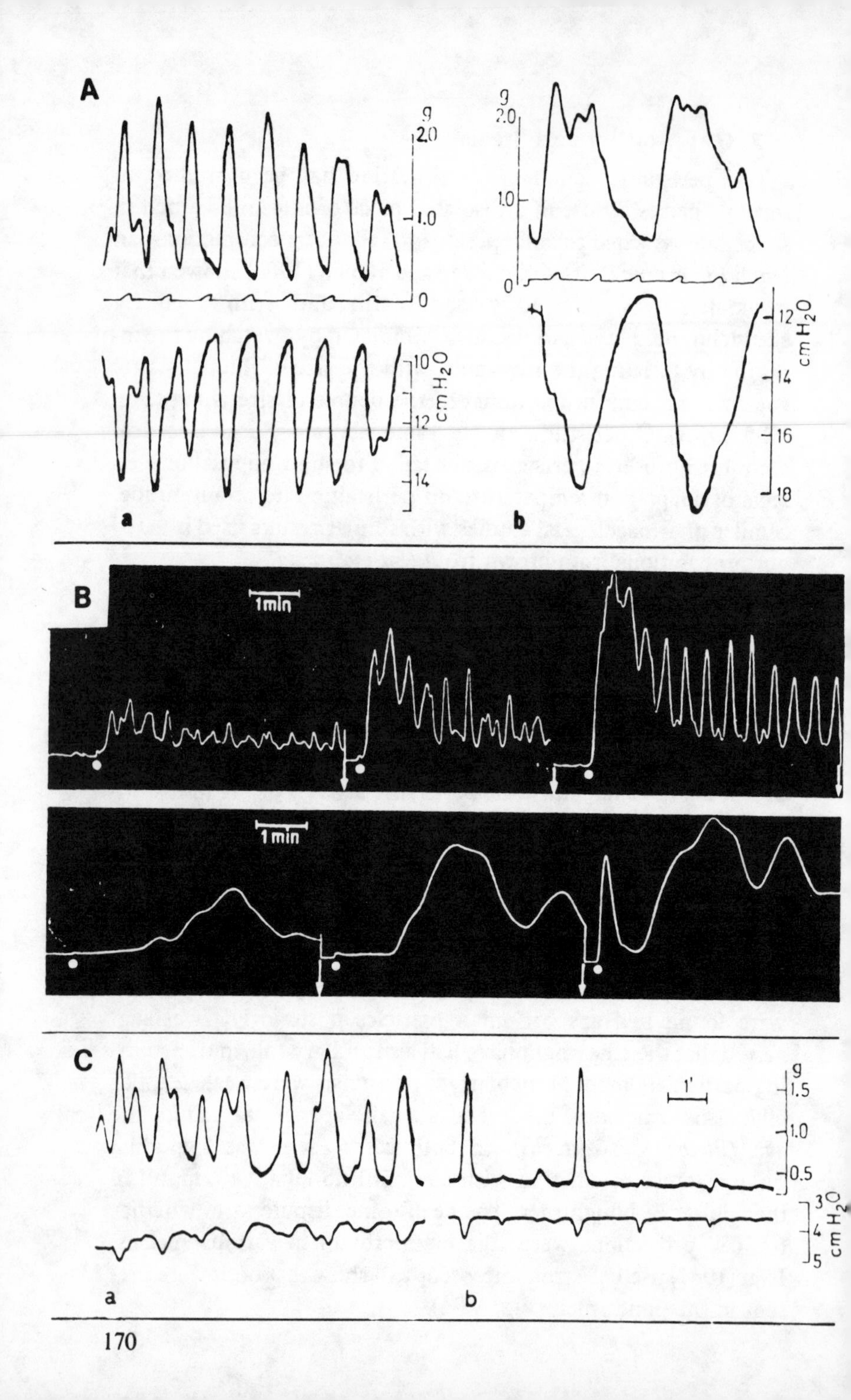
A
g
2.0
1.0
0
10
12
14
cm H2O
a
g
2.0
1.0
0
12
14
16
18
cm H2O
b
B
1 min
1 min
C
g
1.5
1.0
0.5
1'
3
4
5
cm H2O
a
b

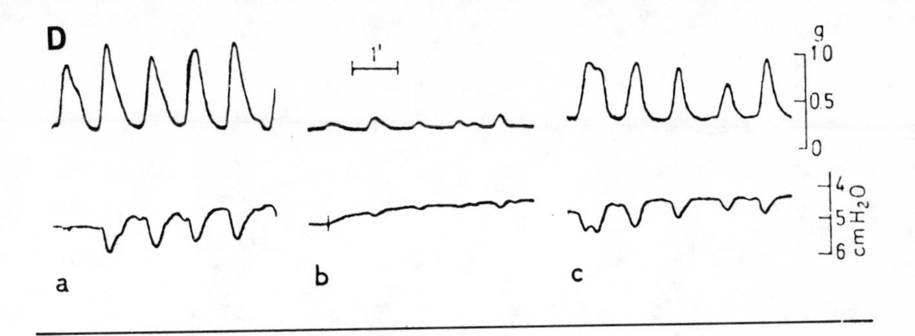

Opposite page and above: Fig. VI-5. Records of peristaltic contractions of isolated gut preparations from tench (*Tinca vulgaris* [= *Tinca tinca*] showing responses typical for vertebrates. In each of the figures below, the upper trace shows activity of the longitudinal muscles with tension in grams and the lower trace shows activity of the circular muscles with their tension in cm H_2O of internal pressure. Time marks all represent one minute. A. Part a, normal contractions of an isolated gut having an internal pressure of 10 cm H_2O and in Part b, the stimulatory effect of increasing the pressure to 12 cm H_2O. B. The stimulatory effect of acetylcholine. Dots indicate time of administration (l. to r.) 10^{-7}, 10^{-6}, 10^{-5} g/ml acetylcholine. Arrows show washing with fresh saline. C. Effect of atropine, an inhibitor of acetylcholine. Part a, control and in Part b, under the influence of 5×10^{-7} g/ml atropine. D. Effect of adrenalin. Part a, control, Part b, under the influence of 10^{-8} g/ml adrenalin; Part c, recovery after washing by replacement of the saline bath. (From Ohnesorge and Schmitz, 1968, and from Ohnesorge and Rauch, 1968, with permission of author and publisher.)

The cessation of peristalsis is called gut stasis. It is a common happening in mammals during stressful situations but has been only suggested in salmonids. Stroud and Nebaker (1974) suggested a relationship between disease outbreaks and gut stasis as being a mechanism by which pathogens in the gut could penetrate the gut wall with the help of the fish's own digestive enzymes. In my laboratory, changes in the plasma ions of marine salmonids suggested that gut stasis resulted from severe scale loss. Once the gut had stopped transporting Mg^{++} -rich rectal fluid, then the seawater being ingested for osmoregulatory purposes would be completely absorbed and produce levels of plasma Mg^{++} sufficient to block the myoneural junctions in skeletal muscles. The descaled fish eventually died of suffocation when the muscles operating the opercular pump became paralyzed. Control fish which had been similarly stressed except for the scale loss (they had severe O_2 debts) lost up to 15% of their

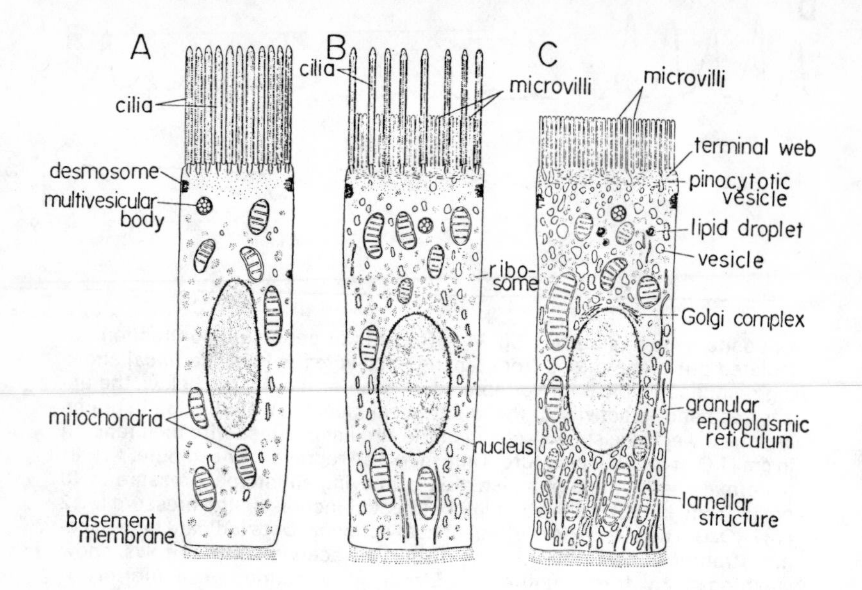

Fig. VI-6. Diagrammatic drawings of epithelial cells in gut of pond smelt larvae. (A) Typical ciliated cell. (B) Ciliated cell having both cilia and microvilli. (C) Columnar absorptive cell with microvilli forming a striated border. (From Iwai, 1967, with permission of the author.)

body weight and their body length shrank by up to 5%, but they survived.

D. DIGESTIVE SECRETIONS

1. Mouth and Esophagus

Only a few fish secrete digestive products into the mouth cavity, including the pharnyx. Most fish secrete mucus to protect the epithelium which lines the mouth and which has scattered taste receptors. Mucus may serve as a lubricant for chewing coral in certain scarid fishes. In general, the more abrasive the food the greater is the mucus production. Mucus may serve as food for the young of herbivorous, mouth-brooding cichlids.

The esophagus wall is often folded and ridged, sometimes in elaborate patterns. These elaborations often produce mucus in large amounts. Esophageal pouches may store food or grind food and produce mucus. Gastric-like secretory cells occur in the posterior esophagus of mullet *(Mugil)* and two cottids.

2. Gastric Secretions

The secretions in the stomach typically include mucus, hydrochloric acid (HCl) and the proteolytic enzyme pepsin. Histological examination of stomach epithelium shows only two types of secretory cells—a goblet cell which produces mucus, and a cell filled with secretory granules which are hypothesized to produce both pepsin and HCl. Pepsin has optimum activity at a pH level of about 2 and may have a secondary pH optimum around 4 in some fish. The amount of pepsin produced is strongly influenced by temperature, being reduced at either end of the temperature range. Production of acid is proportional to the size of the meal as well as to temperature. Distension of the stomach seems to be the stimulus for starting gastric secretion. Stomachless fish produce no HCl or pepsin.

The acidification of the stomach contents varies with the type and quantity of food. Since most food materials have a buffering action, more HCl is required for larger meals. With solid food such as fish prey, the required pH optimum may be achieved only in the outermost layer of the food mass. In the few species investigated, increased stretching of the stomach produced a proportional increase in the amount of acid. In many respects the amount of acid secreted into the stomach is more important than the amount of pepsin because the enzyme will not work effectively without the proper pH. In this regard, it would seem that the drinking of seawater (alkaline) for osmoregulatory purposes would hinder or even prevent gastric digestion. There are several possible solutions: not drinking seawater during gastric digestion; producing additional acid to acidify the seawater; routing the seawater through limited parts of the stomach by having the esophagus and pylorus close together; and digesting food, especially if solid, within a layer of mucus, pepsin and acid on the surface of the food. It has also been suggested that the evolutionary development of some stomachless fish species was another way to solve the problem, since the alkaline intestine should not be particularly influenced by seawater. None of these possibilities has been tested to date.

3. Intestinal Secretions

The intestinal secretions of fish contain a large number of enzymes which include all three major classes—proteases, lipases and carbohydrases—which hydrolyze the three respective classes of foods. Mucus is also secreted, and presumably HCO_3^- is secreted by the pancreas to neutralize the gastric HCl, as in mammals, but the latter has not been looked for in fish. There is some suggestion from histochemical investigations that the relative quantity of pancreatic and intestinal secretions may differ in different species.

The secretory cells of the intestine appear to follow the classic vertebrate pattern. The fish intestine has no villi but does have deep folds (rugae) in its wall. Secretory cells form in the depths of the folds, migrate to the crests of the folds and discharge their secretions. In goldfish given heavy doses of X-irradiation (Hyodo, 1964), the deterioration of the intestinal epithelium occurred over a two-week period because of the lack of cell division at the bottom of the folds (Fig. VI-7). The resulting thin-walled gut may not have been greatly different from the atrophied gut seen in non-feeding salmon during the spawning migration.

Trypsin is the predominant proteolytic enzyme of the intestine and is active between pH 7 to 11. The source of the trypsin is difficult to localize, but probably most comes from the pancreas with some from secretory cells in the intestinal wall, including the pyloric caeca. In general, pyloric caeca are histologically identical with the adjacent intestinal wall. Trypsin in fish probably requires activation by another enzyme, enterokinase, from the intestinal wall, as in higher vertebrates. A special trypsin which is active at acid as well as the normal alkaline pHs has been reported in salmon. There are probably other proteases present in the intestine, including an exopeptidase and a cathepsin (which is found in most tissues). Proteolytic activity seems

Fig. VI-7. Histopathological changes in intestinal epithelium of goldfish irradiated with 8 kr. A and B - normal; C - 3 days after irradiation; D - 7 days after irradiation; E - 9 days after irradiation; F - 11 days after irradiation; G -11 days after irradiation; H - 13 days after irradiation. The sequence shows the importance of rapid cell reproduction in maintaining the mucus and enzyme secretions of the intestine. (From Hyodo, 1964, with permission of author and publisher.)

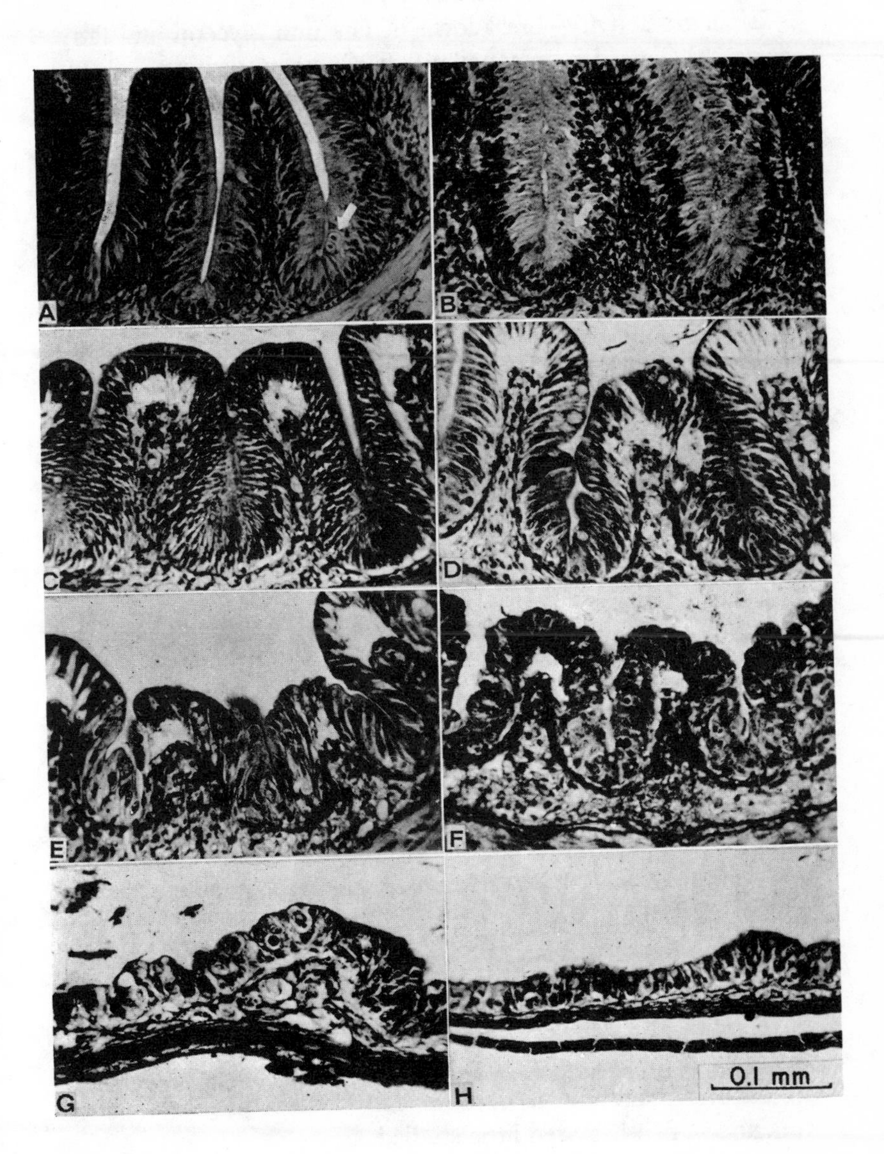
A
B
C
D
E
F
G
H
0.1 mm

strongest in predatory species which ingest large amounts of protein and weakest in herbivorous species.

Lipolytic activity—breakdown of fats into glycerol and fatty acids—has been demonstrated to occur in various extracts of pancreas, liver, intestine and pyloric caeca of fish in a fashion comparable to that of pancreatic lipase in mammals. Few details are known.

Carbohydrases abound, particularly in herbivores. Adult carp intestine showed activity by maltase, sucrase, lactase, melibiase, cellobiase, and a glucosidase. Amylase was present in rainbow trout and several other predators, but at levels lower than in carp or several other herbivores. Carbohydrases generally seem to have higher activity levels in herbivores than in predators, presumably because of the higher levels of dietary carbohydrate.

Bile enters the upper intestine from the gall bladder and facilitates the digestion and absorption of dietary lipids and lipid-related substances such as fat soluble vitamins (A, D, E and K). Bile is not an enzyme but a mixture of organic and inorganic salts produced by the liver as products of the catabolism of hemoglobin and cholesterol. Data on the inorganic composition of gall bladder bile from selected species of fish (Hunn, 1976) is shown in Table VI-2. If comparable to human bile, fish bile should also contain bile salts, bilirubin, cholesterol, fatty acids and lecithin.

Since bile is eliminated continuously from the liver as well as serving in digestion, the green color of mucus seen in the intestine of migrating (non-feeding) salmon, either smolts or adults, presumably is from bile. A condition observed in marine salmonids in which the gall bladder and the bile duct become filled with solidified mucus has been tentatively identified as being related to diet and appears to be reversible when the diet was changed. The cause and mechanism were unknown, but might have included excessive reabsorption of water from the bile by the wall of the gall bladder or excessive levels of Ca^{++} which would precipitate mucus from the gall bladder. In general, bile function in fish has been little studied.

4. Absorption

Absorption is the process by which various components of food are transferred from the intestine to the blood. The process has been studied negatively in fish--anything that did not appear in the feces was assumed to have been absorbed. The degree of concentration of an inert marker (e.g., chromium oxide) which was nondigestible and nonabsorbable provided similar estimates of absorption or digestibility. There is wide variation in the proportion of the ingested food which is absorbed depending on its digestibility. Food containing large amounts of plant material or large amounts of ballast such as mud may have absorption levels of less than 20%. Values typical for salmonids average around 80% for typical hatchery diets. The digested and absorbed food components which are immediately excreted by the kidneys or gills should also be included in the non-absorbed fraction of the ingested food (see next section below on specific dynamic action).

The mechanics of absorption are largely unknown. In mammals there are two routes of absorption. Carbohydrates and proteins pass through the gut epithelium into the blood stream. Lipids, if hydrolyzed to glycerol and fatty acids, behave similarly, but intact lipids which are reduced to small droplets (chylomicra) pass into the lymphatic ducts in the intestinal villi as an indirect route into the blood stream. In fish there is a lymphatic system, but its presence in the gut wall is minimal or perhaps even nonexistent and there are no villi in any fish looked at so far. However, lipids are fairly easy to identify histologically, and there have been several observations of the intestinal epithelium being lipid-rich after a meal. One paper suggested for rainbow trout that absorption of lipids from the gut used both the hepatic portal and lymphatic routes. In bream there is an increase in circulating lymphocytes during digestion which was thought to assist in absorption (Smirnova, 1966). Absorption of carbohydrates and proteins has not been studied.

E. SPECIFIC DYNAMIC ACTION

Upon ingesting a meal, the O_2 consumption of a fish increases markedly without any increase in activity. The events producing this increase in O_2 consumption are called specific dynamic action (SDA). One might suppose at first that the increased energy consumption relates to the synthesis of the enzymes, mucus and other digestive fluids needed for digesting a meal, but this is not

Species	Na+	K+	Ca++	Mg++	Cl-	pH
Oncorhynchus kisutch Coho Salmon	219.7 (±4.3)	14.8 (±3.4)	12.7 (±0.1)	2.7 (±0.4)	2.3 (±0.7)	7.26 (±0.06)
Salmo gairdneri Rainbow trout	277.5 (10)	7.6 (10)	18.1 (10)	2.7 (10)	3.3 (10)	7.6 (14)
Salmo salar Atlantic Salmon	243.6 (P)	11.8 (P)	32.9 (P)	4.9 (P)	6.2 (P)	--
Salvelinus namaycush Lake Trout	252.3 (P)	7.2 (P)	19.9 (P)	4.9 (P)	6.2 (P)	--
Esox lucius Pike	227.0 (±19.8)	3.4 (±0.4)	28.6 (±7.8)	7.3 (±2.4)	--	-- ±0.31
Cyprinus carpio Carp	267.8 (±33.1)	7.7 (±1.6)	29.2 (±9.8)	12.1 (±5.2)	1.4 (±0.4)	6.82 ±0.15
Ictalurus natalis Yellow Bullhead (Catfish)	197.1 (±21.2)	1.4 (±0.5)	11.9 (±1.4)	4.1 (±0.5)	--	8.17 ±0.12
Aplodinotus grunniens Freshwater Drum	150.1 (±2.1)	4.5 (±1.6)	9.3 (±1.6)	6.1 (±1.9)	11.6 (±5.6)	--
Pomoxis nigromaculatus Black Crappie	143.1 (±15.6)	3.7 (±0.9)	4.8 (±2.0)	2.6 (±0.7)	35.4 ±19.2	--
Man (from Guyton: *Medical Physiology*, 1962)	145	5	5	--	100	--

Table VI-2. Concentrations of inorganic ions (mEq/l) and pH of gall bladder bile from selected species of fish. All species were in freshwater. Data gives mean ± S.D. or P = pooled sample from 4-7 fish. Potassium levels appear to be characteristically low in catfish species (here *Ictalurus natalis*, also occurs in channel catfish) and chloride ion appears the most variable. In part this variability may be explained through known active transport of Cl^- by the wall of the gall bladder. (Adapted from Hunn, 1976.)

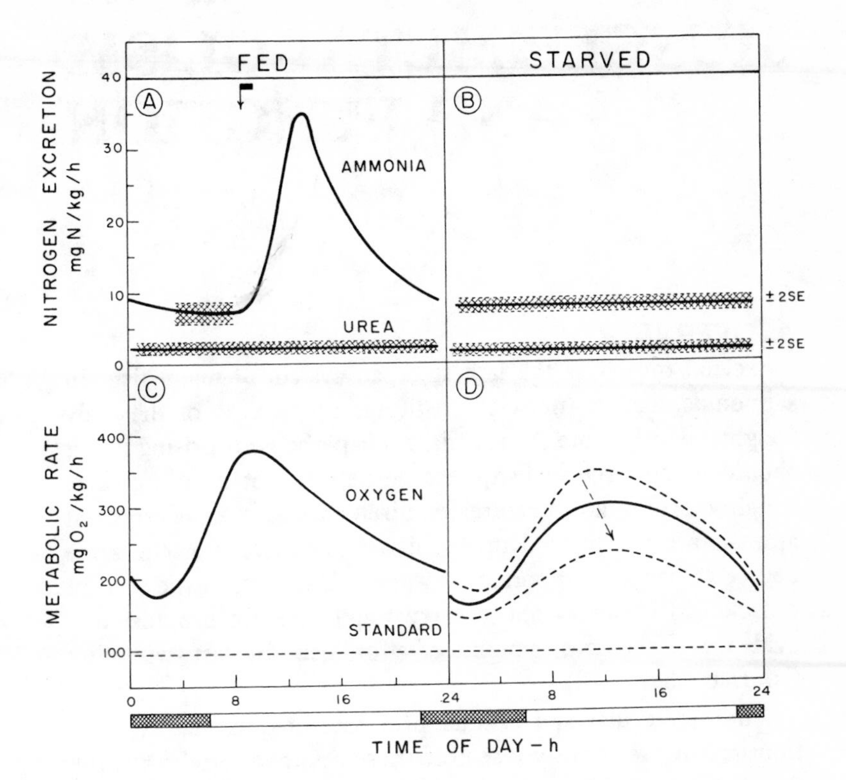

Fig. VI-8. Pattern of changes in rates of nitrogen excretion (A,B) and oxygen consumption (C,D) of sockeye salmon, shown when being fed daily (A,C) and after starving for 22 days (B,D). The average trend for oxygen consumption of starved fish appears as a solid line with broken lines indicating the general pattern of decrease as starvation progressed. Temperature 15°C. (From Brett and Zala, 1975, with permission of author and publisher.)

the case. The enzymes are already present intracellularly prior to eating and are only released during digestion.

The size of the SDA relates to the amount of protein in a meal and to the proportion of that meal which is used for energy rather than for growth. Protein that is to be used for energy must first have the amino group ($-NH_2$) removed and excreted as ammonia (NH_3), both processes requiring an input of energy. Thus the pulse of increased O_2 consumption is followed by a pulse of increased ammonia excretion (Fig. VI-8).

Chapter VII: MUSCLE METABOLISM AND FUNCTION

A. OVERVIEW

Skeletal muscle is the largest organ system of many fish. In salmonids, skeletal muscle constitutes up to 60% of the body weight, slightly more than in man. It is perhaps surprising that it should be so large in fish, because fish do not have the same weight-carrying and postural requirements as man. Terrestrial animals are not buoyed up by air to anywhere near the same degree as fish are supported by water. That most of the skeletal muscle of fish serves not to carry weight but for propulsion is evidence of the great amount of effort used by fast-swimming fish (see Chapt. I-B).

While molecular and cellular processes of muscular contraction in fish presumably resemble those of other vertebrates, the gross anatomy of fish muscle differs markedly from that of the land vertebrates which have become such familiar preparations in physiology laboratories. The "typical" vertebrate muscle attaches directly to a bone on one end and tapers down to a distinct tendon on the more mobile end. Fish muscle, on the other hand, forms transverse vertical layers on either side of the axial skeleton and only rarely attaches directly to a bone or to any discrete tendon. Further, the transverse layers (myomeres) make anterior-posterior zigzags into somewhat flattened W-shapes which become even more exaggerated next to the skeleton than on the surface (Fig. VII-1). The functional basis for the W-shape of these myomeres is not entirely clear, but it presumably serves to spread out the contraction of any one myomere over a much longer portion of the body length than just the length of the myomere. The myosepta (the white layers of connective tissue between the myomeres) appear to serve the same function as tendons, so their sloping attachment to the skeleton probably is the means of translating the muscle contraction (more or less parallel to the axial skeleton) into a sideways (bending) force for swimming.

The directions in which the muscle fibers run through the myomeres are not always parallel to the axial skeleton. Thus the bending force applied to the axial skeleton during swimming may result from a combination of the angular orientation of the myomere and myosepta as a unit and the independent angular orientation of the fibers within the myomere. Alexander (1969) showed an evolutionary sequence in the development of angular orientation. Sharks have angular myomeres, but most of the fibers run parallel to the backbone. In primitive teleosts, including eels and salmonids, the fibers near the head are angled outward, then parallel in midbody, and then inward posteriorly toward the tail to give an arched "muscle trajectory." In more advanced species such as perch, rockfish, tunas and mackerels, the combination of complex myomere shaping and fiber angles up to 40° from the body axis produce a spiral muscle trajectory. Some of these fish revert to a more primitive (salmonid or even shark-like) arrangement of myomeres in and near the caudal peduncle, presumably because parallel fibers have more tensile strength for transmitting power to the caudal fin than angled ones. In extreme cases such as tunas, mackerels and sierra, there are bony tendons, without which the muscles in the caudal peduncle presumably would tear during contraction of the larger muscles anteriorly. The advantage of angled fibers and a spiral trajectory appears to be greater speed of contraction.

Another feature of gross anatomy of fish muscle is the presence of layers of dark (red) muscle in regions which carry on sustained activity—usually swimming. Fish which swim most of the time, such as salmonids or tunas, have a layer of dark muscle just under the skin. One can assume that natural selection over millions of years would put dark muscle in the position of best mechanical advantage or leverage on the caudal fin, so the peripheral position of dark muscle should be indicative of how the muscle functions. It might seem that this layer of dark mus-

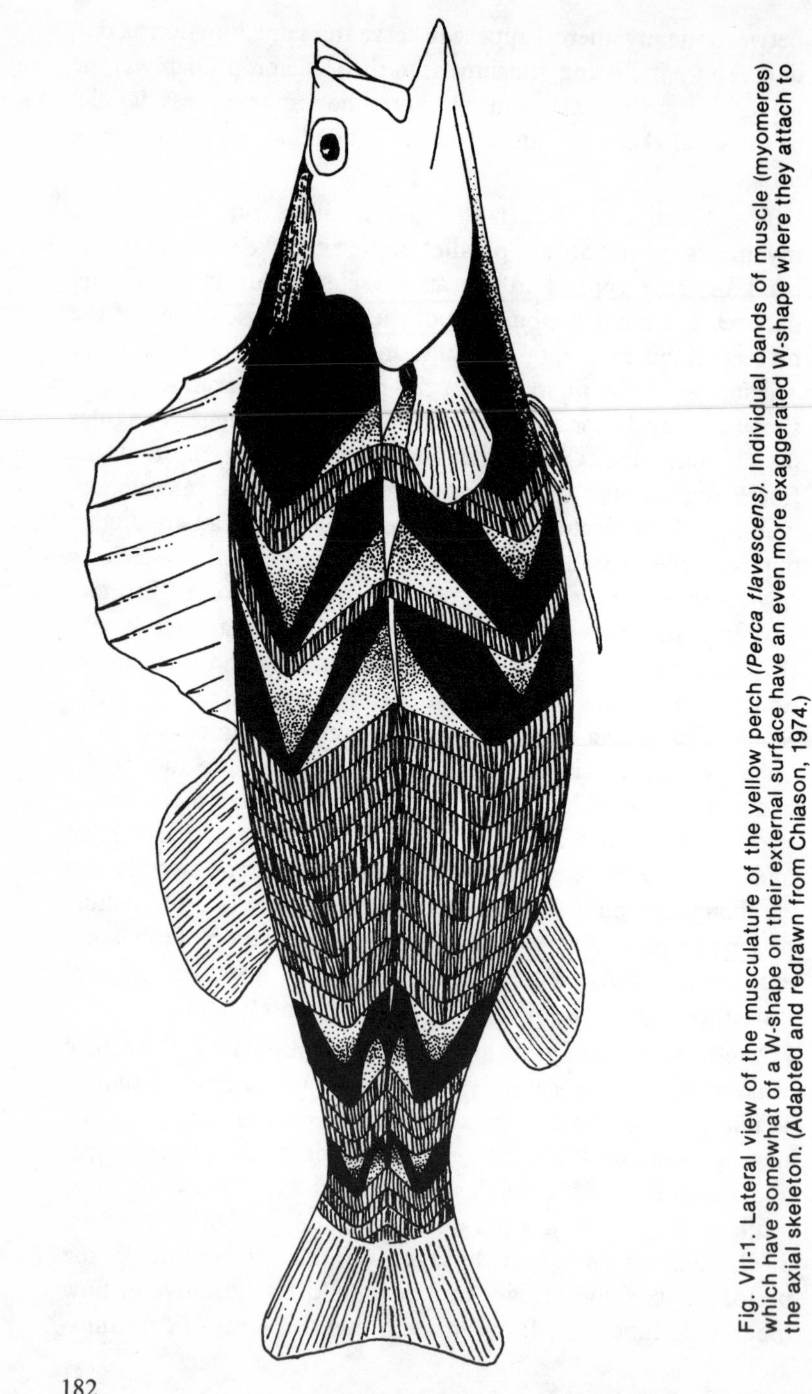

Fig. VII-1. Lateral view of the musculature of the yellow perch *(Perca flavescens).* Individual bands of muscle (myomeres) which have somewhat of a W-shape on their external surface have an even more exaggerated W-shape where they attach to the axial skeleton. (Adapted and redrawn from Chiason, 1974.)

cle simply shortens the distance from the back corner of the head to the caudal fin, but this is not the case. Vertical cuts completely through the muscle to the skeleton reduced the swimming ability of a rainbow trout only slightly (L. Smith, unpublished). Thus muscle contractions in peripheral layers transmit their force to the axial skeleton quite locally and are not dependent on a long series of muscle fibers connected end-to-end all along each side of the body, although this may be somewhat true for red muscle where the fibers lie all parallel to each other in a straight strip of muscle.

In fish such as pile perch and shiner perch (Embiotocidae), whose main propulsion is by pectoral fins, then the dark muscle relates to the pectoral rather than the caudal fin. Seahorses and pipefish (Syngnathidae) propel themselves primarily with rapid undulations of the dorsal fin and reportedly have dark muscle associated with them. Thus dark muscle occurs wherever there is vigorous, sustained activity.

The blood supply to light and dark muscle tissue also differs. Smith and Bell (1975) found extensive capillary beds associated with the dark muscle layer of salmon. Also, the blood supply to the dark muscle either became more dense toward the tail or the dark muscle layer became thicker toward the tail or both. The blood supply to the white muscle was rather minimal by comparison to that of the dark muscle. This is consistent with the general view that metabolism of dark muscle is aerobic, while white muscle soon becomes anerobic during strenuous activity (see below).

B. COMPOSITION AND METABOLISM OF LIGHT AND DARK MUSCLE

The inorganic composition of Atlantic salmon muscle at various stages of the life cycle is shown in Table VII-1. In general, these numbers demonstrate the low Na^+ and Cl^- and high K^+ characteristic of the intracellular compartment (shown in Fig. I-2). The increasing water content during and after spawning is characteristic of non-feeding salmonids as they consume lipids and some proteins for energy and replace them with water.

Stage	Conditions	% Water	Na^+	K^+	Ca^+	Cl^-
Smolt	Freshwater	--	16.3*	210*	3.14*	2.60*
Smolt	2 weeks seawater	79.2	33.3	185	3.55	--
Adult	Seawater	67.6	28.1	206	2.84	6.12
Adult	Freshwater, estuary	64.1	29.7	194	2.95	--
Adult	Freshwater, spawning	77.6	66.4	261	2.84	--
Adult	Freshwater, post-spawn.	80.5	45.5	199	2.22	--
Adult	⅔ seawater, 4 wks.	77.8	54.7	200	2.16	--
Adult	Seawater, 3 hrs.	77.9	80.7	205	2.99	--
Adult	Seawater, 2 days	82.2	36.7	238	1.94	--

Table VII-1. Inorganic composition of muscle tissue of Atlantic salmon *(Salmo salar)* at various stages of the life cycle and in various envrionments. Units for ions are in mM/kg water except that which indicates mM/kg weight of muscle. (Adapted from Parry, 1961.)

The organic composition of light and dark muscle differs in relation to differences in their metabolism at different activity levels. Light muscle is characterized by high levels of glycogen as the primary energy source, while dark muscle has high levels of lipid. In jack mackerel *(Trachurus symmetricus)*, Pritchard, et al. (1971) found that the failure to swim at any speed was associated with complete depletion of glycogen in light (white) mscle. Dark (red) muscle also showed longer term decreases in glycogen and slow decreases in lipid content, but only after extended periods of swimming at velocities less than the sustained speed threshold. High lactate levels were found in both types of muscle and did not seem related to fatigue at any speed, although this is contrary to the situation in salmonids noted below. The interpretation from this data was that light muscle was the primary locomotor organ near and above the threshold for sustained swimming speed and that dark muscle was the primary locomotor organ at lower speeds in combination with unknown levels of effort by light muscle.

There have been numerous and extensive studies in salmonids on muscle (usually light muscle) fatigue and the associated accumulation of lactate—the end product of the anerobic production of energy from glucose or glycogen. Edgar Black, an earlier worker who published prolifically in the 1950s and early 1960s, found an association between blood lactate levels in excess of about 150 mg/100 ml blood with delayed mortality in adult salmonids following severe exercise such as might occur during transport, capture-and-release fishing or tagging. More recently, Mearns (1971) observed blood lactate levels in juvenile salmonids in excess of 500 mg/100 ml blood but saw no mortality. Muscle lactate levels usually exceeded the blood lactate several fold. In Black's studies where the blood lactate approached 200 mg/100 ml blood, the muscle lactate went as high as 700-900 mg/100 g muscle. One of Black's students, Don Stevens, decided to see what would happen if the blood lactate went as high as the muscle lactate and injected into the blood of a rainbow trout sufficient sodium lactate to produce that condition. The fish became rigid as in rigor mortis (unpublished data) and died. Whether this was really rigor mortis (insufficient chemical energy remaining in muscle cells for them to relax) or some kind of pH effect was not determined, but it demonstrated the importance of the muscle tissues' not dumping all of their lactate into the blood stream at once. Since Stevens' experiment in the 1960's, I have seen similar symptoms in occasional salmonids after severe stress in which premortal rigor began posteriorly and moved progressively forward until the fish died when the opercular pump stopped. I presumed that this resulted from excessive lactate, but did not make lactate measurements.

Large accumulations of lactate obviously need to be dealt with cautiously and promptly. The most significant alternative is to oxidize the lactate—i.e., repay the oxygen debt. By oxidizing about one-fifth of the lactate to provide energy and to form H_2O and CO_2, about four-fifths of the lactate can be reformed into glycogen. Oxidation of blood lactate takes place mostly in the liver, while lactate which remains inside the muscle cells can be oxidized there. Oxidation requires several hours. Another alternative could be to excrete the excess lactate to avoid the pH problem and pay the metabolic price of throwing away over 90% of the energy in the original glucose molecule. Mearns showed in my laboratory that lactate appeared in urine during high levels of blood lactate but that the quantity was only a few percent of the tôtal lactate build-up and therefore was not a significant energy "leak." Loss of lactate across the gills was also possible, but not investigated, and there are some early reports of lactate being metabolized directly in the gills.

C. EFFECTS OF EXERCISE ON MUSCLE

The effects of muscular training on athletes has long been recognized and studied. Similar logic, when applied to the migratory salmonids which make spectacular athletic performances in terms of distances travelled and overcoming obstacles to their upstream migration, would suggest that the exercising of salmon raised in hatcheries should improve their rate of survival and return to the fishery. A number of experiments to test this idea have been performed by several investigators, but most results have suggested that there was no improvement in survival rate of exercised fish. Burrows and his co-workers (Thomas, et al., 1964) showed that exercised fish had greater swimming stamina than non-exercised fish, but the survival rate comparison of the exercised and non-exercised fish was confused by the reduced growth rate (due to using more energy for swimming and less for growth) of the exercised fish. The fish which are largest at time of release generally have the best survival rate, so the larger size of the non-exercised fish apparently balanced any improvement in survival produced by the exercise. The only information that I have seen which supports the beneficial effects of exercise was an informal report by Swedish workers who found that exercise approximately doubled the return rate of some hatchery-reared Atlantic salmon over a 7-year period.

The metabolic effects of exercise appear to be much the same in fish as in man. Hammond and Hickman (1966) found that physical conditioning of rainbow trout enable them to swim longer at a given speed, to tolerate higher levels of plasma and muscle lactate before fatigue occurred (refusal to swim), and to remove the accumulated lactate faster than in unexercised fish.

Exercise also increases the size of the muscle fibers in fish similarly as in man. Greer-Walker (1971) exercised coalfish (*Gadus virens,* similar to cod) for two months at various swimming speeds. At 0.93 L/sec the dark muscle fibers increased in diameter by 55% with no change in the light muscle fibers. At 2.01 L/sec, both dark and light muscle fibers increased in diameter by 61% and 32% respectively. Exercised fish gained 13% less weight during the two months exercise period than non-exercised controls. During 130 days of starvation, cod *(Gadus morhua)* lost 30% of their body weight and muscle fiber diameters decreased 15% for dark muscle and 40% for light muscle. These data support the generalization that only dark muscle is working during low-speed (cruising) swimming but that white muscle plays increasingly important roles at higher speeds.

D. INNERVATION OF SWIMMING MUSCLES

The smooth, undulatory motions of swimming would seem difficult to achieve with the highly segmented arrangement and short fiber length characteristic of fish muscle. In part, smoothness of contraction comes from displacement of the myomere from a flat sheet of muscle into the sloping W-shaped arrangement already discussed and shown in Fig. VII-1. Another basis for smoothness, however, lies in the multiple, overlapping innervation of spinal nerves for each myomere and muscle fiber. Hudson (1969) found that each muscle fiber in a sculpin *(Cottus scorpius)* received from two to five nerve axons from each spinal nerve and received an equal complement of fibers from each of four adjacent spinal nerves. While it could not be determined whether all of the muscle fibers had single or multiple nerves and single or multiple endings for each fiber, the multiplicity of connections would readily spread out the effect of any single nerve impulse over several myomeres. Thus a short wave of nerve impulses traveling down the spinal cord would produce a broad smooth wave of muscular contraction in either case. It also appeared probable that a wave of muscular contraction passing down one side of a fish's body was accompanied by a simultaneous wave of relaxation on the opposite side of the body.

E. SPECIALIZED MUSCLES

1. Pectoral Fin Muscles

In sharks the myomeres near the base of the pectoral fins extend onto the fin sufficiently to move the fin in an anterior-posterior plane and to change the angle of attack (amount of lift) slightly. In contrast, in fish such as live-bearing surfperches (Embiotocidae), in which the pectoral fins serve as the routine means of locomotion, both mobility of the fin and the activity level of the muscles at the base of the fin are greatly increased. Such fins can be moved in a figure-eight motion like a bird wing, paddled forward or backward, paddled individually or together or in opposition, and are also used for braking, turning and equilibrium. Such flexibility of movement requires much specialization—perhaps increased innervation of the region around the base of the fin, increased number of nerve endings in the cerebellum to coordinate the increased activity and certainly an increase in the independence from each other of many sections of myomeres which formerly contracted as an entire unit. However, I know of no such study detailing the muscular anatomy and functional specialization of pectoral musculature, although an isolated pectoral abductor muscle has been used to test the toxic effects of organophosphate pesticides on cholinergic neuromuscular transmission (Schneider and Weber, 1975). Also, the muscles associated with the embiotocid pectoral fins are dark muscle and there is no dark muscle in the caudal fin musculature.

2. Sound-producing Muscles in Swimbladders

A number of fish make humming, grunting or other low-pitched sounds. Such species include the toadfish *(Opsanus tau)*, squirrelfish *(Holocentrus rufus)* and the northern midshipman *(Porichthys notatus)*. The mechanism in each case is a pair of striated muscles located symmetrically on and completely attached to the wall of the swimbladder. Sound is produced by rapid muscular contractions of the muscles, which causes resonant vibrations of the swimbladder in much the same fashion as you produce sound in a taut balloon by dragging a sticky finger across the surface.

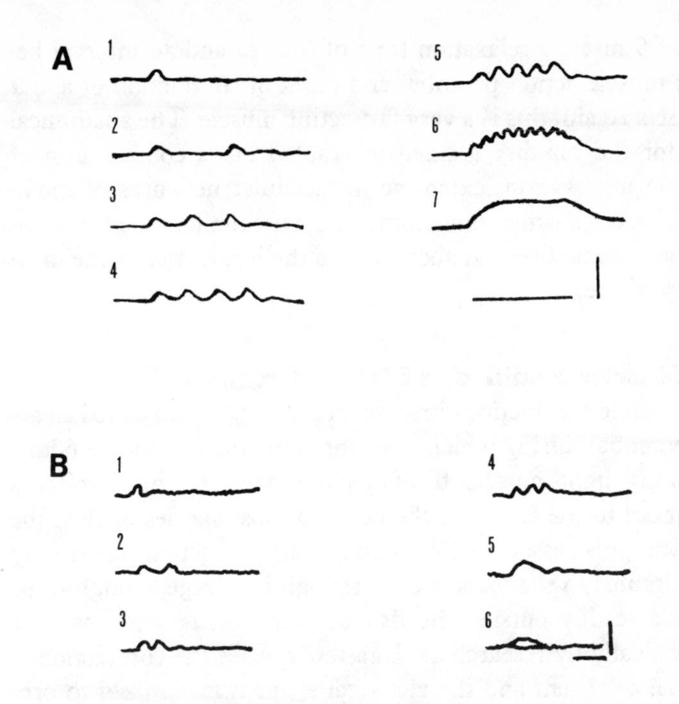

Fig. VII-2. Tension recordings from muscles of A squirrelfish. A. Sound-producing muscle: 1 - single stimulus to nerve; 2 - two stimuli at 26/sec; 3 -three stimuli at 52/sec; 4 - four stimuli at 63/sec; 5 - stimulation at 112/sec; 6 - stimulation at 189/sec; 7 - stimulation at greater than 200/sec. Duration of stimulus train was 50 msec in each case. B. Tension recordings in white dorsal muscle: 1 - single stimulus; 2 - two stimuli at 24/sec; 3 - two stimuli at 42/sec; 4 - three stimuli at 50/sec; 5 - stimuli at 100/sec; 6 - superimposed traces, one at 100/sec, the other a single stimulation. In both A and B, vertical calibration = 50g, horizontal calibration 50 msec. (From Gainer, et al., 1965, with permission of the publisher.)

The physiological significance of sonic muscle is not that it produces sound, but that it can contract and relax fast enough to produce sound. Gainer, et al. (1965) showed that squirrelfish sonic muscle could respond individually to stimuli with a frequency of 112/sec and still showed some individuality to the muscle contractions at 200 stimuli/sec. In contrast, light skeletal muscle showed some lack of complete relaxation at 42 stimuli/sec and considerable lack of relaxation (fusion or summation of contractions) between pulses at 50/sec. (Fig. VII-2A & B). In a similar paper on sonic muscle of toadfish, Gainer and Klancher (1965) described the contraction as having an average contraction time of 5 msec, a relaxation time of 8 msec and an interval between muscle action potential and onset of contraction of about 0.5 msec. Again, this is a very fast-acting muscle. The anatomical basis for the rapidity seemed to result from a combination of multiple innervation, extensive intracellular networks of membranes (sarcoplasmic reticulum) and an orientation of the individual muscle fibers at about 45° to the lengthwise of the muscle as a whole.

3. Muscles Modified as Electric Organs

All muscle contractions involve an electrical event called an action potential during which a resting potential of about −60mV in resting light muscle flip-flops to about +5 mV (voltages referenced to the inside of the cell). In most species of fish, the electrical pulse associated with muscular contraction serves only as a stimulus to the muscle even though it is large enough to be detected readily outside the fish by some predators as well as electronically by researchers. In a few species the contraction is minimal or absent and the electrical event is maximized to produce significant voltages and current flow. These species are taxonomically unrelated and in six different families including two skates, a stargazer, several gymnotid eels (knifefish), a catfish and the electric eel of the Amazon River. Also, in different fish the electric organs have evolved from different muscles. In the knifefish the nerve to the muscle evolved into electric organs and the muscle degenerated entirely. Further, the skates and the stargazer are marine; the remainder are freshwater. Thus the evolution of electric organs probably occurred independently in each of these instances. Also, it has occurred only in fish, presumably because only the aquatic environment has sufficient electrical conductivity for electrical signals to be purposeful.

The basic mechanism for producing electricity in an electrocyte (also called electroplax, electroplacque and electroplate) involves a broadly flattened cell which develops resting potentials similar to those of other muscle cells (Fig. VII-3A). Upon stimulation of one face of the cell, the polarity of one face reverses momentarily and current flows (Fig. VII-3B). Although no muscular contraction occurs, much energy is used at higher

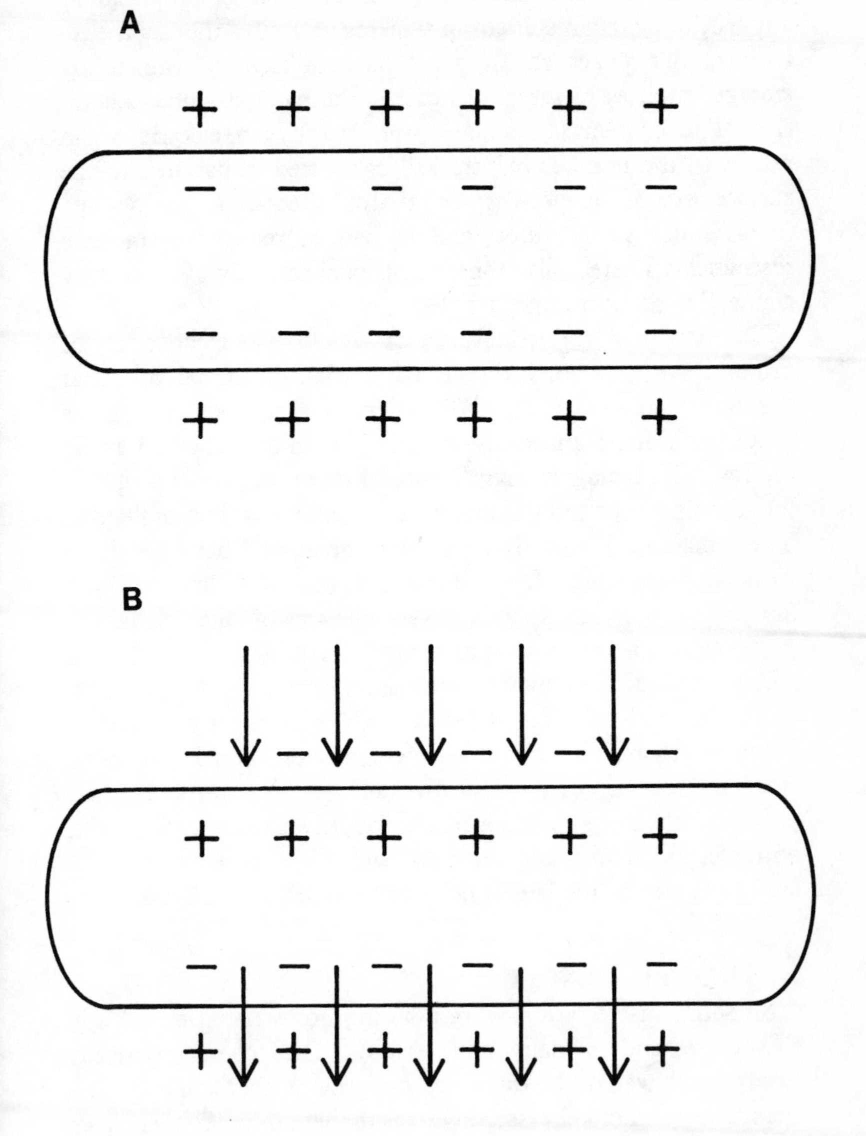

Fig. VII-3. Diagram representing how electroctyes produce electrical current. A. Resting (polarized) electrocyte, no current flowing. B. Active electrocyte with upper surface depolarized and current flowing.

voltage levels and fatigue can occur as in any other muscle when glycogen supplies become exhausted. The voltage produced by any particular fish results from stacking of up to a thousand electrocytes like poker chips, which is equivalent to connecting storage batteries in series to make the individual voltages additive. The current (amperage) produced by an electric organ relates to the number of "stacks" connected in parallel, to the surface area of the electrocytes involved (hence the wafer shape to maximize surface area) and in part to reduced membrane resistances. There is also some complex channeling of current by connective tissue around the electrocytes.

The actual voltages produced vary widely. The electric eel can produce over 500 volts, the electric catfish up to 300 volts and the electric ray *(Torpedo)* about 60 volts. The electrical power produced is also significant—about 10 kw in the eel and 1 kw in the ray. The biological significance of these outputs is probably for stunning prey and definitely for defending against predators. The remainder of the fishes already mentioned have relatively weak outputs—from a few volts down to tens of millivolts. These low voltage outputs show variable phasing of high frequency pulses which serve as a sophisticated electro-orientation system and as an electrocommunication system between individual fish.

The amount of information available on electric organs is much larger and more sophisticated than can be covered here. Readers desiring further information might consult Bennett (1971) as a starting point. Readers should also note that all fish with electric organs, and some without, have an electro-sensory system as part of the lateral line system (Chapter VIII-E-4).

F. SMOOTH MUSCLE

Smooth muscles are slow-responding non-striated muscles in vertebrates and in many invertebrates. They characteristically occur in the wall of the digestive tract, the urinary and gall bladders, ducts and blood vessels. Nerves to these muscles come from either the sympathetic or parasympathetic nervous systems or often from both. A distinctive feature of the smooth muscle is that it can change its length drastically with only minimal changes in the amount of tension it can produce. This is of ob-

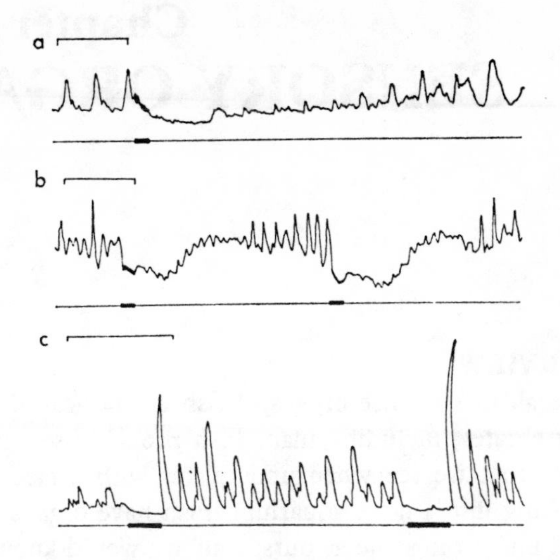

Fig. VII-4. Typical smooth muscle contractions as seen in the stomachs of three fishes, also showing inhibition produced by intracranial stimulation of the vagus nerve. a - *Scyliorhinus canicula* at 15.5°C. At the bar, stimulation of both vagi with 10 V, 1 msec pulses at 4/sec for 60 sec. b - *Conger conger* at 16.0°C. Both vagi stimulated at 8 V, 0.5 msec pulses at 1/sec for 90 sec at the first bar and at 2/sec for 120 at the second bar. All time markers are 5 min. (From Campbell, 1975, with permission of author and publisher.)

vious importance in maintaining gut peristalsis even when the gut becomes distended with a large meal, for example. A further distinctive feature of smooth muscle is that contractions can be stimulated by nerves or by stretching or may be spontaneous.

The functions of smooth muscle in fish have been studied mostly with regard to control of gut and duct motility and mostly in one laboratory. Fig. VII-4 is from Campbell (1975) and shows typical peristaltic contractions of the stomach in several species of fish and their subsequent inhibition via stimulation of the vagus (10th cranial) nerve. The slow contractions and the long period of slow recovery from inhibition are all typical of smooth muscle of any vertebrate. The review by Campbell (1970) on the autonomic nervous system of fish contains considerable material on smooth muscle.

Chapter VIII: SENSORY ORGANS

A. OVERVIEW

In general, many sense organs of fish are typical of those of other vertebrates, including man. Fish probably see much the same view that we see when underwater with a face mask or when looking into a large aquarium. Fish have organs for taste and smell much the same as ours; thus we would know what a hot, soapy bath tastes like if we had taste buds on the outside of our body as many fish do. Fish have no inner ear (cochlea) as elaborate as ours, but apparently hear similarly in the lower part of our hearing range. The lateral line system of fish serves as a dynamic pressure detector, especially at very low frequencies (well below human hearing—down to one cycle per second (Hertz) or less) in a way not found in humans. Humans and fish have similar organs to detect fluid motions in their respective semicircular canals, however. Thus there is a considerable degree of commonality between sensory systems of man and fish.

There are also some significant differences between sensory organs in man and fish. The typical fish eye works as poorly in air as human eyes do underwater with water directly against the cornea—both are unable to focus clearly. The specialized lateral line organs of some fish detect electrical fields in the sea or, in the case of electric fish, fields which they produce with their own electric organs (see Chapter VII). Electrosensory systems do not occur at all in man, of course. Fish maintain normal three-dimensional orientation in poor visibility while humans have considerable difficulty in keeping oriented while SCUBA diving in low visibility or while flying airplanes in clouds. Humans operate rather poorly in three-dimensional fluids when their visual cues are severely limited.

Further, we have no idea of whether the brain of a fish interprets the input from its sensory organs similarly as we do, for example, since much of the significance of sensations is determined in the brain, not in the sensory organs. For example, pain is a sense which we can reasonably assume that fishes have—an awareness of malfunctions or injury—but we have no way to understand whether they suffer because suffering is entirely an interpretation of pain by the brain. As long as we remain unable to communicate with fish (or any other animal), there will be no physiological basis for studying all of the ramifications of pain or any other sensation. Thus there is considerable basis for our awareness of how fish perceive their world, but it needs to be tempered with objectivity and knowledge of the limitations of the parallelism between fish and man.

There are physiological as well as philosophical problems in interpreting the electrical events which occur at various points in the neurosensory systems of any animal. Recordings of electrical changes inside of individual receptor cells show many changes which are not transmitted across the junction (synapse) to the nerve cell—i.e., the receptor changes do not attain the nerve threshold. Once into nerve pathways, receptor information may be integrated in varying degrees, depending on the sensory system, before the information reaches the brain. For example, integration occurs in the bipolar cells of the fish retina and in the olfactory bulb. There are instances in which the recording method produces summations of electrical events from whole sensory organs—the electroretinogram from the eye, for example. These extracellular electrical recordings appear quite different from intracellular recordings from individual receptor or nerve cells.

Finally, recordings can be made from portions of the brain devoted to a particular sensory system—optic lobe, olfactory lobe, etc. These electroencephalograms (EEG) show a high level of integration which makes responses to discreet stimuli difficult to interpret. The olfactory EEG is an example of this kind of response. Detection of an odorous substance is demonstrable, but whether the odor is an attractive or a repulsive one cannot be determined physiologically, only behaviorly. At the EEG level, the electrical events recorded show only a small part of a complex electrical field within the brain lobes. Particularly at the brain level, only empirical association between the sensory stimuli and the electrical recording can be made. A specific sensory input could stir up quite an "electrical storm" in the brain, but the fish might decide to ignore the information and do nothing about it, for example. On the one hand, a physiologist might attach great significance to such a large neural response while a behaviorist would say nothing happened because there was no overt response. To a degree, both would be correct. While such an extreme difference in interpretation of neurosensory information would be unlikely, it illustrates the hazards of interpreting the significance of neurophysiological data.

B. THE EYE

1. Anatomy

The anatomy of a typical fish eye appears in Fig. VIII-1. A fish eye is typical of vertebrate eyes in its general configuration but differs in specific details. The lens in fish is spherical rather than ovoid as in humans. In land vertebrates there is a major difference in refractive index between air and the cornea so that the cornea does more than half of the focusing of light rays. In fish the difference in refractive index between the cornea and water is so small that the cornea does very little focusing, making a powerful lens necessary. The most powerful simple lens is a spherical one, as seen in fish eyes. The spherical lens also requires a different mechanism for focusing light rays on the retina. In man the lens is thickened slightly for close vision, but in fish the spherical lens can get no thicker, so instead, it must be moved as in a camera. The lens appears to be typically in a forward position for close vision and is moved nearer the retina and slightly downward by the retractor muscle for distant vision. The travel distance is probably small since a fish eye compares to a camera with an extremely wide angle lens and great depth of focus. The slightly flattened shape of the eyeball and the lens projecting in front of the iris also contribute to the wide angle capabilities of the eye.

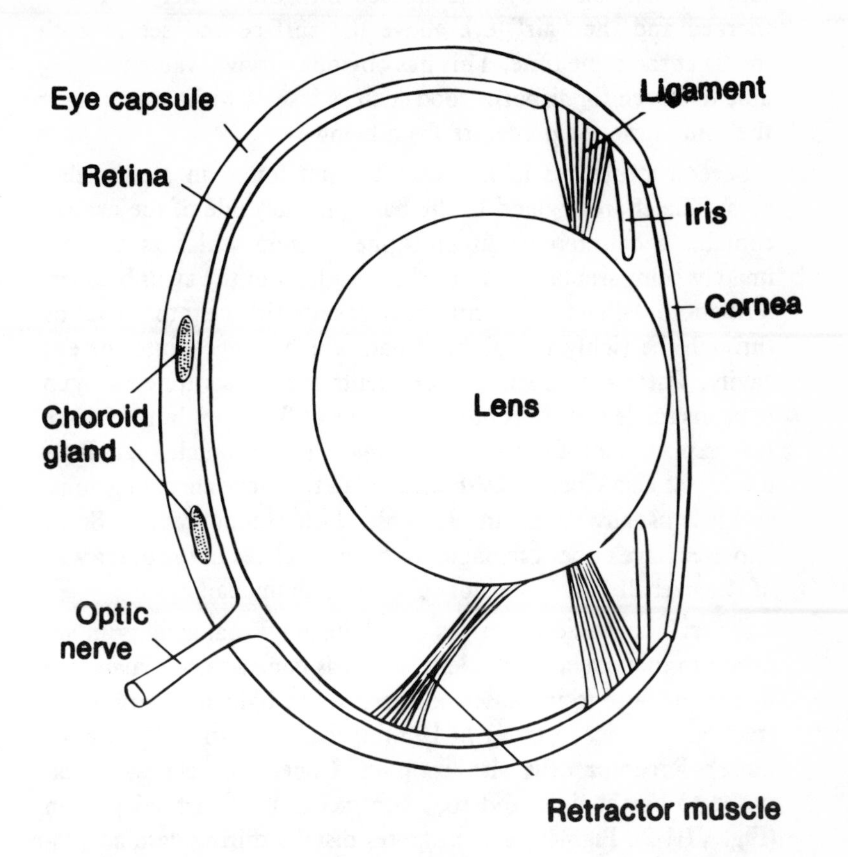

Fig. VIII-1. Diagrammatic vertical section of a typical teleost eye showing only the major structures. (From Wedemeyer, et al., 1976, with permission of TFH Publications.)

For those few fish which need to see in air, there are focusing problems to solve. In air a curved fish cornea becomes effective as a lens and causes light to be focused too strongly (in front of the retina). One solution to this problem would be to have a nearly flat cornea. A more complex solution occurs in the four-eyed fish *(Anableps)*, in which each eye is divided into two separate portions for vision in water and vision in air. The eyes project upward from the head and, since the portion for air vision is on top, the fish can rest at the surface with the "water" eye submerged and the "air" eye above the surface and see in both media at the same time. This has obvious survival value in being able to search for airborne food such as insects while watching at the same time for predators from below.

Several structures in fish eyes are not found in mammalian eyes. The choroid gland in the back (medial) side of the eyeball contains a countercurrent multiplier system which is approximately comparable to that of the gas gland in the swimbladder. The choroid gland apparently provides nourishment to the retina through the richly vascularized part which projects into the eye cavity. There have been measurements of supersaturated oxygen tensions made just in front of the retina of fish eyes, but there are few explanations for the physiological role of such high P-O_2 levels (see also Chapter IV-F-2). Not all fish have choroid glands, and it is usually absent in fish which lack pseudobranchs. Some fish eyes have either cartilaginous or bony plates in the outer wall of the eyeball which are not found in mammalian eyes.

Several mechanical changes occur during visual activity in addition to movement of the lens. The iris contracts or expands in direct response to increases or decreases in light intensity. Contraction is typically faster (2-15 min) than relaxation (30 min or more). Receptor cells also contract. Cones contract when exposed to bright light and rods contract during dark adaptation (Fig. VIII-2). Pigment also migrates distally during dark adaptation and may uncover a reflective layer (tapetum) in others. (The tapetum does not occur in some fish and is nonocclusible in others.) These photomechanical changes in the retina serve to maximize the sharpness of vision in bright light and maximize the capture of even stray light rays (with attendant blurring of vision) by the rods during vision in dim light. The reflective layer usually has its reflective cells arranged perpendicular to the normally-focused light rays so as to minimize light scattering and thus fuzziness of vision. It should be pointed out that light rays come from the bottom of Fig. VIII-2 and must pass through small blood vessels and layers of nerve cells before reaching the receptor cells.

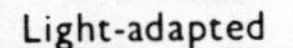

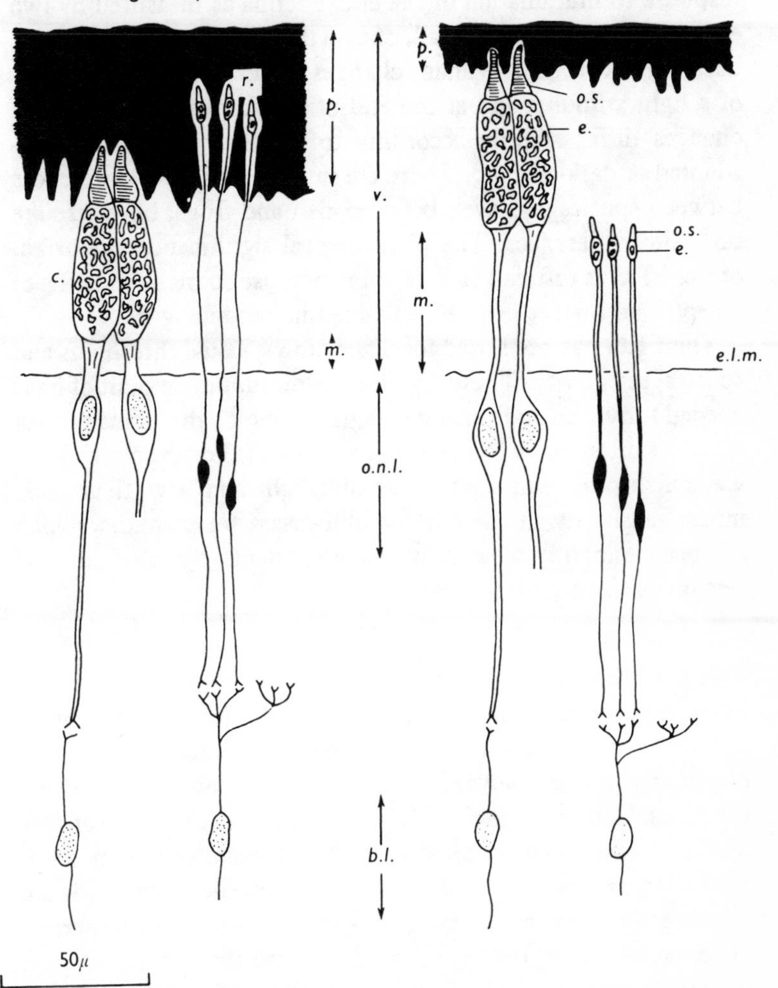

Fig. VIII-2. The retina of an adult herring drawn semi-diagrammatically in the dark- and light-adapted state. *p* - width of pigment; *v* - width of visual cell layer; *m* - length of cone myoids; *b.l.* - bipolar layer; *o.n.l.* - outer nuclear layer; *e.l.m.* - external limiting membrane; *e* - ellipsoid; *o.s.* - outer segment; *c* - cone; *r* - rod. (From Blaxter and Jones, 1967, with permission of authors.)

2. Electroretinogram

The electroretinogram (ERG) is the summed electrical response to illumination of the entire retina as measured by two electrodes outside the eyeball, one on the cornea, the other on the back of the eye. Small voltage changes (<1mV) occur at the onset of a light stimulus and at the end of the stimulus. The voltage changes differ slightly according to whether the eye is light-adapted or dark-adapted. There are small quantitative differences between species and even between fish and frogs, but there are no major differences. The physiological significance and origin of the ERG is difficult to determine because so many millions of receptor and nerve cells are involved in producing it.

The ERG has been used for determining visual thresholds and spectral sensitivity. These data for the minimum amount of light needed for vision and the wavelengths to which the retina is most sensitive often relate to the ecology of a fish. For example, deep-water fish often can see in very dim light and have their maximum sensitivity in the blue or blue-green wavelengths, which are predominantly what is available to them. (See also the next section on S-potentials.)

3. S-Potential

The S-potential is the voltage change recorded by intracellular electrodes from bipolar (also called horizontal) cells which overlay the retina (shown in Fig. VIII-2, the lowest cells in the diagram). Compared to the ERG, the S-potential is at a relatively low level of integration because any one bipolar cell connects to only a few receptor cells. On the other hand, the S-potentials are not directly from the cones, as was thought by their discoverer, G. Svaetichin, in 1953. His students eventually named these potentials after him when the origin of the signals was finally clarified.

S-potentials occur in great variety, both from different cells in the same retina and in retinas from different species (Fig. VIII-3). The method by which such records are obtained involves stimulation of an isolated, darkened retina with pulses of light having equal length and equal intensity, but of different colors (wavelengths are noted across the bottom of the recordings

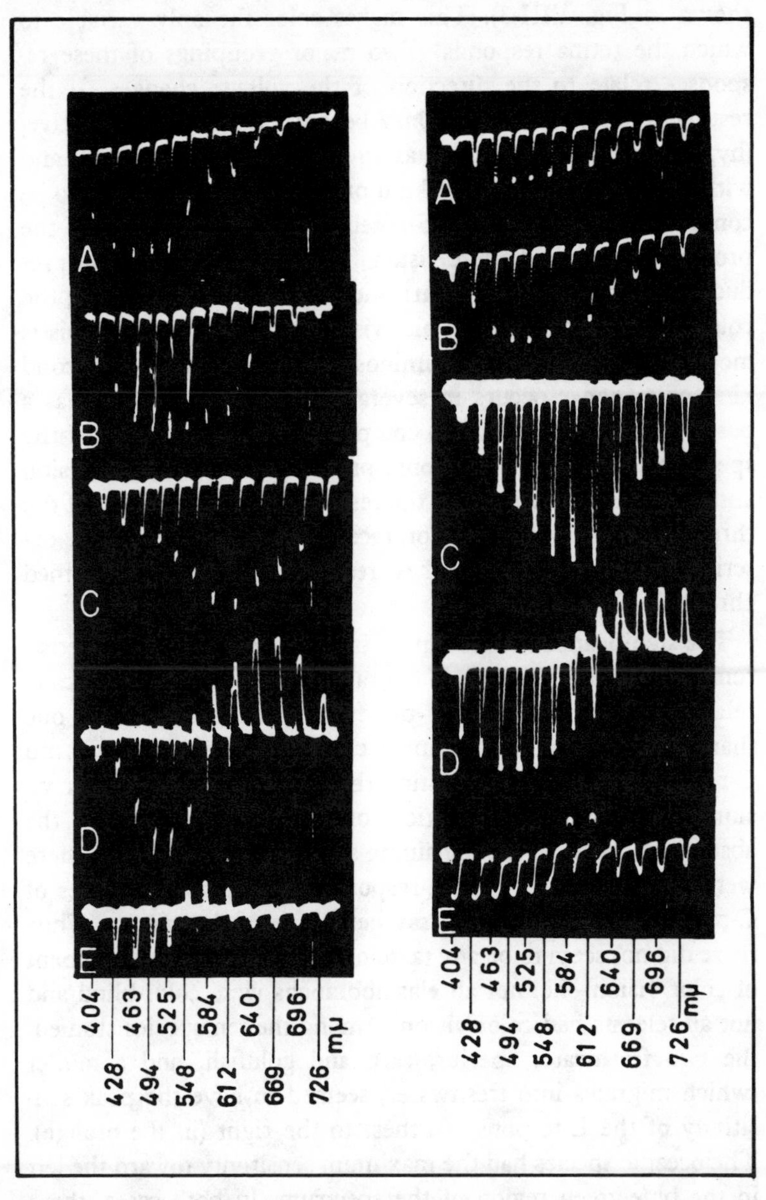

Fig. VIII-3. S-potentials recorded from bipolar cells of teleost retinas in response to equal-intensity pulses of light of different colors. Left column -carp. Right column - goldfish. A, L_1-type; B, L_2-type; C, L_3-type; D, C_1-type; E, C_2-type. Vertical calibration, 5-mV. Color of light shown in wavelength of mμ (nm) from blue to red (l. to r.). (From Tamura and Niwa, 1967, with permission of author and publisher.)

shown in Fig. VIII-3). This makes color the only variable to which the retina responds. Two major groupings of these responses relate to the direction of the voltage changes. If the resting potential of about -40mV becomes increasingly negative, (hyperpolarized) by as much as an additional -20 to -30 mV and with all colors of light, that kind of bipolar cell is thought to be connected to rods and the S-potential to represent part of the process of black and white vision. The color of light which produces the greatest hyperpolarization is interpreted as being the color at which black and white vision is most sensitive. This is most commonly termed a luminosity (L) response. The second kind of response occurs in several variations but always has a positive-going (depolarizing) component in some portion of the spectrum. This type of response probably relates to color vision and the different varieties of the response come from each of the three different kinds of color receptors (cones) which are described in the next section. These responses are commonly termed chromatic (C) responses.

The significance of S-potentials in the functioning of the retina remains unclear, although Tamura and Niwa (1967) empirically related them to the ecology of several species of fish. In one shark, a porgy and the Japanese chi dai *(Evynnis japonica),* no C-response occurred, suggesting that these fish lacked color vision. Histological examination of the retina confirmed the absence of cones. In the remaining seven species examined, there were up to three types of L-responses and up to four types of C-responses, with the sting ray being one of the species. Thus there did not seem to be any taxonomic sequence of development of color vision—i.e. not all elasmobranchs were color-blind and not all teleosts had color vision. Among the ten species studied, the two freshwater species, carp and goldfish, and a mullet (which migrates into freshwater) seemed to have the peak sensitivity of the L-response furthest to the right (in the orange). The oceanic species had the maximum sensitivity toward the left in the blue-green region of the spectrum. In both cases, these sensitivities were interpreted as adaptations to match the predominate color of light found in those respective environments. The greater number of C-responses found in some fishes were interpreted as providing greater discrimination of colors, but no ecological significance of this capability was suggested.

In large part, the sensitivity of the retina to particular colors of light depends on the pigments found in rods and cones. The initial response of photoreceptors to a light stimulus is a chemical one in which a pigment bleaches and partially decomposes. The color of the pigment determines the color of light which is maximally absorbed—e.g., a red pigment would absorb green maximally—but exactly how the photochemical decomposition of the pigment produces a nerve response is unknown for any vertebrate.

There have been many comparative studies made of vertebrate photopigments and some generalizations have been established. There are two broad groups of photopigments, rhodopsins and porphyropsins. The opsin portions of the pigments are proteins and combine with either one of two retinenes (derivatives of vitamin A) to produce the two groups. The photochemistry of visual pigments in fish is comparable to that in other vertebrates, but the interesting feature about fish has been the great variety of pigments—more than all of the rest of the vertebrates put together. The literature on visual pigments of fish is thus too extensive to review adequately here and is even only briefly reviewed in Hoar and Randall (Vol V, Chapts. 1, 2, 3; 1971).

4. Responses of Individual Photoreceptors

Responses of individual receptors have not been widely studied because of technical difficulties in penetrating the cells with microelectrodes. Some of the receptors are too small to be penetrated by even the smallest micropipet-type intracellular electrode. In the few cases where responses from cones (which are larger than rods) have been obtained from fish retinas, there were three peak sensitivities identified which correspond to the maximal light absorption peaks for the pigments found in that retina (Fig. VIII-4), although there are problems in making direct correlations between type of pigment and type of response. While it has been possible to mark the cells from which recordings were made and then identify them histologically, there has

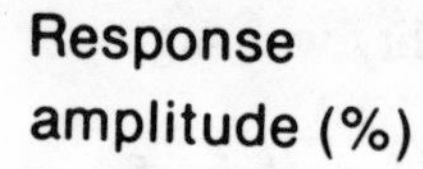

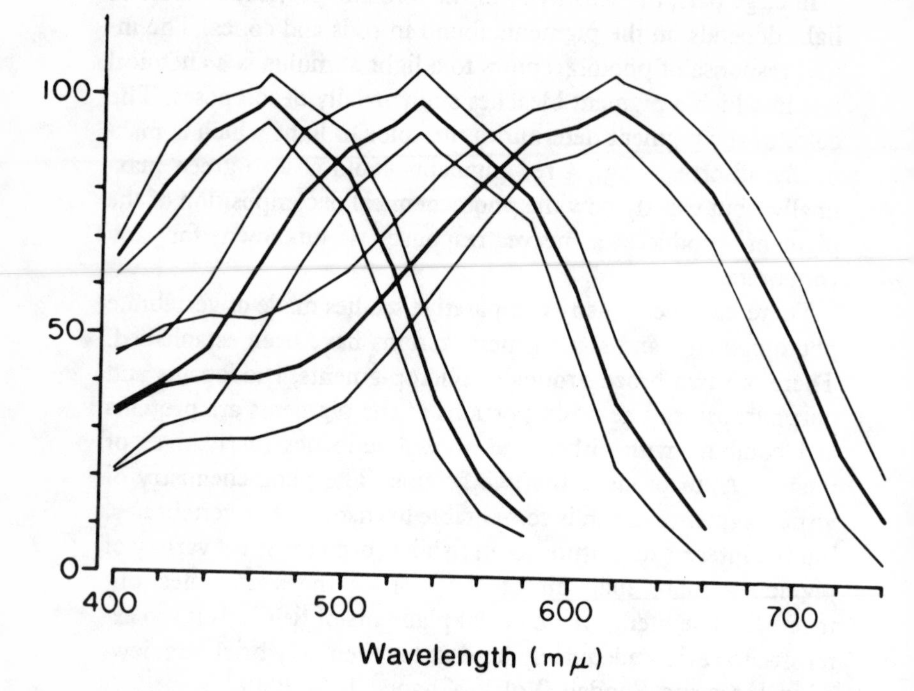

Fig. VIII-4. The averaged response spectra and standard deviation curves of three cone types in carp. This kind of evidence suggests that color vision arises from the combination of three color responses in somewhat the same fashion as color pictures are printed in three colors. (From Tomita, 1971, with permission of author and publisher.)

been no way to extract pigment from only the one receptor cell or even from only one kind of receptor cell.

There have been problems of interpretation of the responses recorded from cones and from bipolar cells. Most of the problems relate to the increased negativity (hyperpolarization) of the responses while the eventual transmission of the information on a nerve fiber results from decreased negativity (depolarization). Exactly how the hyperpolarizing response can induce a nerve impulse is unknown.

C. THE PINEAL ORGAN

The pineal organ is a puzzling structure lying on the dorsal midline of the brain, partly posterior to and partly between the forebrain lobes. Its finger-like shape rests immediately under the roof of the cranium. Histological and histochemical examination of the organ in sockeye salmon suggested the presence of photoreceptor cells and some secretory (possibly endocrine) cells (Hafeez and Ford, 1967). This is in general agreement with findings in other fishes, in amphibians and in reptiles.

The functions of the pineal organ remain a bit vague and difficult to assess. As a photoreceptor, it probably serves to detect changes in day length (photoperiod). This seems quite workable in juvenile salmonids, in which both the skin and the skull over the pineal organ are translucent. The skull thickens in older fish but may not be totally opaque. How much light is required to stimulate the pineal photoreceptors apparently hasn't been tested. Other functions may relate to adjustments in body coloration with changes in illumination level, although the lateral eyes also function in this respect. Surgical removal of the pineal does not produce any life-threatening changes.

D. CHEMOSENSORY ORGANS

1. Olfactory Organs

In general, the olfactory organs are comparable to the nasal organs of smell in man—same general location, innervation and role. On the other hand, the nostrils of fish only rarely (e.g., lungfish) open into the mouth cavity and may have either one or two openings externally. In the case of the salmonids' single nostril on each side, a transverse flap directs water inward on the foreward side and outwards on the other side of the opening, somewhat comparable to the flow-through arrangement produced by having two openings on each side. Further, fish olfactory organs deal with odors dissolved in water while mammalian organs function in air, although the latter organs may simply have a thinner layer of fluid over the sensory surface.

The nerve tract from the olfactory organ to brain usually has either one of two configurations. In one pattern (e.g., skate) the olfactory nerve (cranial nerve I) leads from the sensor to an enlargement called the olfactory bulb which then connects to the brain via the olfactory tract. The anterior part of the forebrain is usually called the olfactory lobe, although usually it is not clearly demarcated from the forebrain. In the other general pattern, (e.g., salmonids, carp, hake) the olfactory bulb and olfactory tract merge into the forebrain, sometimes smoothly, sometimes with a faint constriction between the bulb and forebrain. The olfactory electroencephalogram (EEG) described below for salmon were recorded from such an olfactory bulb of the forebrain.

2. The Olfactory Electroencephalogram (EEG)

Electrical recordings of olfactory activity can be obtained from the surface of the olfactory bulb. The usual procedure for doing this involves immobilizing the fish with Flaxadil (or other nerve-muscle-blocking agents similar to curare), providing for respiratory water flow, removing the skull over the olfactory bulbs, positioning the electrodes and then running the water to be tested into the nares (top of the fish's head is out of water).

It was found early during olfactory investigation of salmon spawning migrations (Hara and co-workers, several papers, starting in 1965) that the olfactory EEG clearly showed a stronger response to home stream water than to other water sources (Fig. VIII-5). Later investigations suggested a more complicated situation however. The first report came from experiments on adult fish which had already returned to a hatchery which was supplied from a small creek representing only about 1/200th of the total flow of the river at its mouth. When diluting the home water with distilled water, the home water response disappeared after about a 65-fold dilution—i.e., the salmon presumably could not detect the creek odor when it entered the mouth of the river. Water from the mouth of the river, however, produced an EEG response as strong as creek water from the hatchery. This was interpreted to mean that the salmon were following a trail or sequence of odors during their homing migration rather than just one odor.

Another approach to understanding the role of olfaction in salmonid homing involved putting an artificial odor (morpholine) into the hatchery water a few weeks before downstream migration, then introducing the same odor into a different stream a few

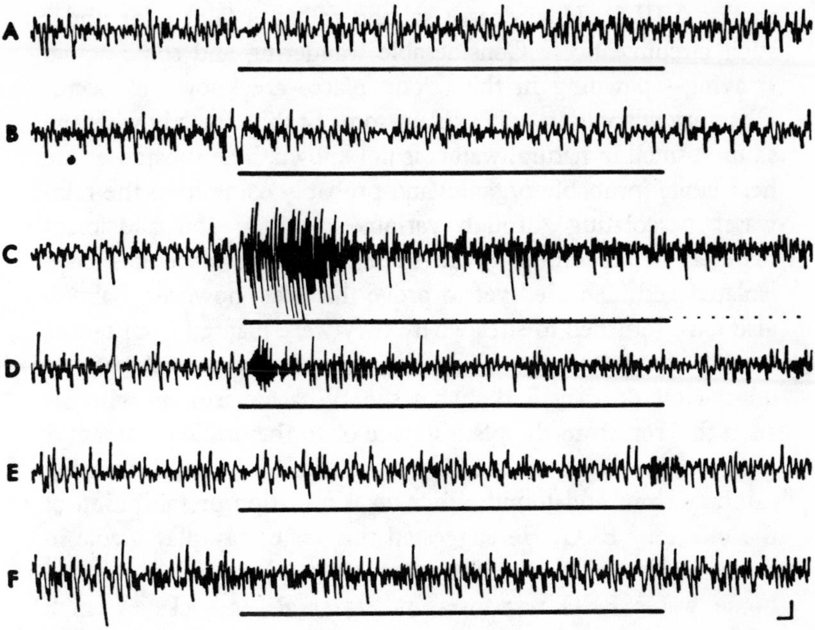

Fig. VIII-5. Effects of infusion of different waters into the olfactory receptor upon EEG patterns in the olfactory bulb of an adult male silver salmon *(Oncorhynchus kisutch).* A - distilled water; B - tap water; C - water from University of Washington's spawning pond (home water); D - water from Lake Washington (Seward Park); E - Green Lake (Seattle); F - Lake Sammamish water. All waters are from different parts of the same watershed, both upstream and downstream from the home water. Vertical calibration, 50µv; horizontal calibration, 1 sec; temperture, 10.2°C. (From Hara, et al., 1965, with permission of author and publisher. Copyrighted 1965 by the American Association for the Advancement of Science.)

miles away when these same fish returned to spawn. Cooper, et al. (1976) found that nearly 90% of that particular stock of coho salmon returning from Lake Michigan returned to the new river with the artificial odor. These fish also showed a characteristic home water EEG response to the artificial odor. These and many other experiments suggest the major importance of olfactory cues for guiding the spawning migration of Pacific salmon in coastal and river waters.

The olfactory story is not completely understood yet, by any means. The olfactory response is an adaptive one—i.e., the response decreases or even stops with continued stimulation (as in Fig. VIII-5). How does a salmon follow a faint odor under such circumstances? Considerable wandering and some actual straying—spawning in the wrong place—are known to occur. Olfactory adaptation may explain some of this. Exactly what the salmon smell in natural waters is not known. The substances are heat labile (probably organic) and probably come from the rainwater percolating through various layers of soil and plant material before entering the stream. No substance has been isolated and identified yet to prove this idea, however. Salmon also have returned to sites where they were planted from tanker trucks—sites where there is no obvious unique source of water or odor. How do they find such a site? What determines whether such fish return to the planting site or to the original hatchery? Hara (1974) tested morpholine as an olfactory stimulant using rainbow trout and found either no stimulation or inhibition of the olfactory EEG. He suggested that taste may play a role in homing. Bodznick (1975) was unable to demonstrate specific home water EEG responses in Frazer River sockeye which related to their making choices of which major tributaries to ascend. He considered it possible that the home water response does not develop until salmon reach their spawning grounds or hatchery. What do these fish follow? Fascinating questions with no answers, yet.

Negative olfactory responses are known, but not fully understood, either. Rinsing your hands (without soap) in a stream will cause migrating adult salmonids to retreat downstream, often for some distance. The substance has been isolated as L-serine, an amino acid. Rinsings from bear skin, sea lion skin and a deer foot produced the same response. Yet when testing the electrical response of the olfactory epithelium to many amino acids, the response to L-serine resembled that of the other amino acids. However, salmon were not repelled by the other amino acids. Salmon apparently recognize several mammals as desirable to avoid, but the olfactory basis for it remains unclear. Fish have been shown behaviorally to recognize body odors from other fish, some of which produce alarm responses, but I did not find any EEG information about such responses.

3. Gustatory Sensors

Gustatory organs (taste buds) in man occur on the tongue. In contrast to the complexity of our sense of smell, they respond to only four simple flavors—salt, sweet, sour and bitter. Thus, most of the flavors of food, etc., combine olfaction and gustation. Fish have gustatory sensors which appear microscopically similar to those of man. They are usually not located on the tongue and may occur inside or outside of the mouth. Fish with barbels often have concentrations of taste buds there, making preparation of isolated barbels practical for studying their responses to different chemical stimuli.

Barbels respond to simple chemical stimuli, similarly as taste buds in man, but not to exactly the same chemicals. Fujiya and Bardach (1966) studied isolated barbels from one freshwater and two marine species and found faint responses to glucose and sucrose, while quinine (bitter) and acetic acid (sour) produced a strong response. The amino acid cystein also produced a strong response, but not the related molecules cystine or homocysteine. Adaptation to most stimuli occurred in 10-20 sec., except for the response to salt in the two marine species, in which there was a constant output which was proportional to the salt concentration. In other experiments, Konishi and Hidaka (1967) found that polyvalent cations such as Fe^{++}, Ca^{++} and Mg^{++} generally depressed the responses of taste buds, but that polyvalent anions perhaps enhanced the responses to monovalent salts. These responses to combinations of salts and valencies were most apparent at low concentrations. The responses to salts seen at higher concentrations lacked the enhancement characteristics and may have involved different receptors.

Salmonids have no barbels but have taste receptors on the front of the snout and in the palatine organ on the roof of the mouth. Sutterlin and Sutterlin (1970), using Atlantic salmon parr, recorded electrical activity from the cranial nerves (VII, IX, respectively) which innervate these sites while washing each of the test sites with a great variety of solutions. External receptors on the snout responded equally to NaCl and KCl but less to $MgCl_2$ and even less to $CaCl_2$. They also responded to mineral and organic acids but not to neutral amino acids or sugars. The palatine organ, on the other hand, responded to both the neutral amino acids and sugars and to strongly ionized substances—salts, acids, etc.

Heavy metals had some unusual effects. Treatment of the external receptors on the snout with dilute solutions of Hg^{++} or Pb^{++} followed by rinsing with distilled water reduced the response of the receptors to other stimulatory substances. Once the response was reduced or blocked by Hg^{++}, it remained blocked for extended periods of rinsing and testing. However, a 10-sec. rinsing with dilute $CuSO_4$ restored the normal response. Further, the receptors did not respond electrically to either the mercury or the copper solutions, and yet fish would reject food pellets that had been treated with the same dilute Hg^{++} solution. The response to the food pellets probably involved the palatine receptor rather than the snout or the nasal receptors.

In addition to localized taste receptors in and around the mouth, other organs have been identified on the surface of the body which appear to be chemosensors. These can be found on fins, in the skin and as part of the lateral line system and occur in both elasmobranchs and teleosts.

E. THE LATERAL LINE SYSTEM

The lateral line system of fish provides a kind of information that is not available to man, a fact which makes interpretation of the responses of the system more difficult than if we had a comparable system. On the other hand, the lateral line's basic sensory unit, the ordinary neuromast organ, appears to operate on the same mechanical principles as the mechanoreceptors in the semicircular canals and the inner ear in man. Thus the basic physiological processes of the neuromast organs of fish have been studied extensively by biomedical researchers, but the manner in which fish use their complete lateral line system is less well known.

1. Location and Structure of Neuromasts

In sharks and rays and in some primitive fish the neuromast organs project from the body surface in many places. Most fish, however, cover their neuromast organs to some degree, either

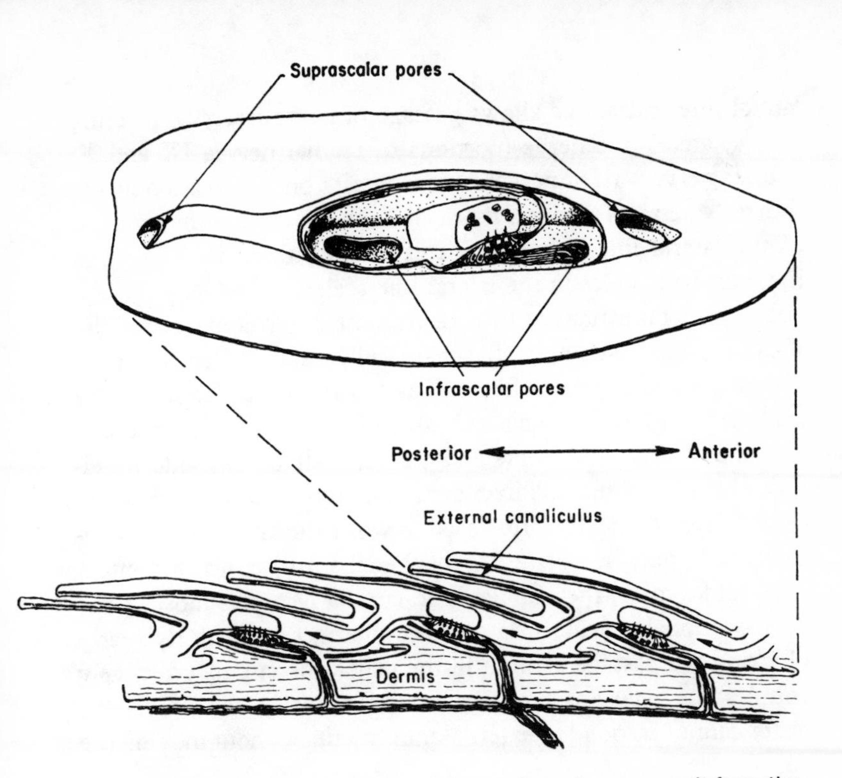

Fig. VIII-6. General anatomy of lateral line scale and sensory unit from the lateral line of salmonids. The lateral line thus consists more of a series of overlapping chambers, each with a neuromast organ, than a straight tube as is sometimes represented. (From Wedemeyer, et al., 1976, with permission of TFH Publications.)

partly enclosed in a groove or completely enclosed inside a tube with periodic openings to the surface. The enclosing of the neuromast organs serves to limit the directions from which water movement can occur and to increase their sensitivity by reducing the amount of water movement needed to cause a response. Enclosed neuromast organs (i.e., the typical lateral line system) extend along the middle of the side of the body for part or all of the length of the body behind the gill covers. Usually the lateral line is single, but it branches in a few species. In many fish the lateral line also extends forward onto the head, in which case it usually branches. In salmonids the lateral line neuromasts occur inside of special lateral line scales having chambers with four openings (Fig. VIII-6). Two of the openings connect to join lateral line scales, one allows passage of nerve fibers connecting to the lateral line nerve (portions of cranial nerves IX and X; cranial nerve VIII innervates neuromasts on the head) and the fourth penetrates through the scale to the outside. What you see of the lateral line in salmonids externally, then, is the row of these external holes in the lateral line scales.

The general structure of a neuromast organ appears in Fig. VIII-7. The basic unit, the hair cell, supports an array of stereocilia and one kinocilium. The kinocilium differs from the stereocilia by having nine longitudinal fibers located peripherally plus a pair of central fibers and a basal body on one side which give the kinocilium a directional polarity—i.e., stereocilia are radially symmetrical in cross-section while the kinocilium is symmetrical in only one plane. All of the cilia project upward into a gelatin-like mass, the cupula. The cupula occupies most of the cross-sectional area of the lateral line chamber or canal. By nearly blocking the passage of fluid in the canal, the cupula effectively couples the neuromast organs to the water in the canal—the water cannot respond to external movements without moving the cupula.

2. Function of Ordinary Neuromast Organs

Neuromast organs function by combining two electrical processes. First, neuromast organs produce a spontaneous output which remains constant in the absence of stimuli. Second, a receptor potential decreases or increases in response to external stimuli and modifies the spontaneous output. The resulting signal to the brain has oscillations in the number of nerve impulses per unit time which mimic the oscillating motions from wave action, tailbeats, sounds and other displacement motions in a fish's external environment. Expressed in electronic terms, the spontaneous output acts as a carrier frequency for frequency modulation (FM) by the receptor potential. Thus neuromast organs differ from most other receptors which respond only to the presence (increase) of a stimulus—the neuromast cells also respond to a decreasing stimulus.

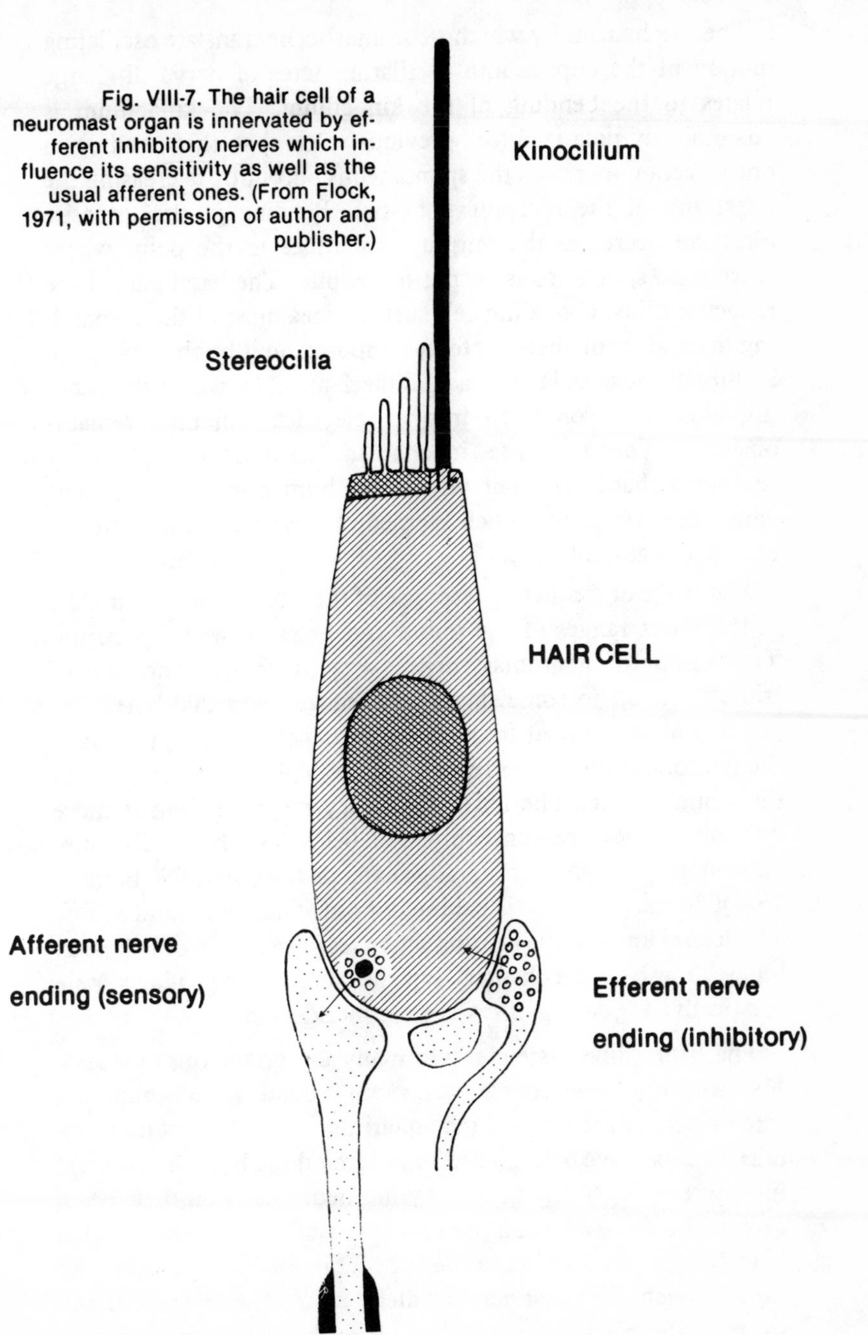

Fig. VIII-7. The hair cell of a neuromast organ is innervated by efferent inhibitory nerves which influence its sensitivity as well as the usual afferent ones. (From Flock, 1971, with permission of author and publisher.)

The mechanism by which neuromast cells translate oscillating motion of the cupula into oscillating rates of nerve discharge relates to the bending of the kinocilium. The kinocilium is anatomically polarized (see previous section) so that bending in one direction increases the spontaneous output by decreasing the negativity of the receptor potential. Bending in the opposite direction decreases the output, sometimes to the point where there is no spontaneous output for awhile. The basal part of the kinocilium just above the cell surface does most of the responding to bending or shearing forces imposed on it by the sliding to-and-fro of the cupula. The actual mechanism by which the bending motion is converted into an electrical potential remains obscure. The response mechanism in kinocilia probably resembles that in the hair cells of the human inner ear, an idea which has stimulated much biomedical research on neuromasts of fish and amphibians.

The range of frequency response of the lateral line system falls in the lower ranges of the sound and pressure wave spectrum. The output of neuromast organs exactly follows the rate of stimulation up to somewhere between 100 and 200 hertz, depending on species. At frequencies greater than this upper limit, the neuromast output corresponds to some harmonic fraction of the stimulus rate. The lower frequency response limit is more difficult to measure—probably well below one hertz. By comparison, the human hearing range extends from about 50 hertz to 15,000 hertz, putting a fish at the bottom of our range and below. The lateral line is not a hearing organ (see discussion of fish hearing, section F-3, this chapter), but the comparison provides some perspective for an organ system not found in man.

The lateral line system serves many important roles for fish. Fish with their eyes covered can swim around without bumping into objects or the sides of the aquarium. The fish's motion produces a bow wave or pressure wave around the head. The lateral line system, especially in fish having neuromasts on their head, detects the bow wave and the reflections of the bow wave which bounce off objects ahead of the fish. This ability has been called distant touch. The manner in which fish in schools space themselves uniformly in three dimensions probably relates to sensing some position in relation to the bow wave or other wake waves of the fish ahead. This probably also reduces swimming effort for all but the leading fish, perhaps by as much as 40%. Migrating mullet, for example, travel considerably faster in groups of ten or more fish than in groups of five or fewer. Adult salmon migrating upstream take advantage of every back eddy or other favorable turbulence to reduce the effort required to reach the spawning ground. Downstream migrants seem to identify the fastest, strongest currents which will speed their trip to the sea. Such activities must depend heavily on information from the lateral line system.

3. Electroreception: Specialized Neuromast Organs

In addition to the ordinary neuromast organs just described, some neuromast organs lack the ordinary cupula and hair cells. Furthermore, the sensory cells of these neuromast organs are deeply buried in the skin tissues and connect either to the lateral line canal or to the skin surface (most common) by a jelly-filled tube. These specialized neuromast organs can respond to the low-level electric fields produced by muscle contraction, by electric organs and by seawater moving through the earth's magnetic field. The combination of electric signals produced by an electric organ (Chapt. VII, Section E-3) and electrosensors produces a radar-like sensory system which can also serve for communication between fish (reviewed by Hopkins, 1974).

The sensitivity of the electroreceptors in electric fish varies widely with species, size of the individual and the conductivity of the water. Amazon electric eels *(Electrophorus electricus)* required a minimum imposed signal strength of from 2 to 30 mV/cm before the electroreceptors responded with a train of 1-15 nerve impulses. They discriminated changes in voltage of from 1.5 to 5 mV/cm, however, at voltages above the minimum (Hagiwara, et al., 1965). In weakly-electric (gymnotid) fish, the sensitivities ranged from 0.25 uV/cm in 2000 Ohm water to 0.5 uV/cm in 10,000 Ohm water. The amperages produced by the fish varied from as little as 41.6 uA/cm in a 6-cm fish to over 1200 uA/cm in a 22-cm fish. These sensitivities and field strengths produced a capability to detect objects within the fish's electric field at a distance of about one body length and to communicate electrically with other electric fish up to a distance of 5-6 body lengths (Knudsen, 1975). It appeared that fish which produce weaker electric fields have appropriately more sensitive electroreceptors, since the electrosensory field of the electric eel ranges from 20-50 volts whereas that of the gymnotids rarely exceeds a few millivolts.

Some non-electric fish may also sense electric fields. Both Atlantic salmon and anguillid eels detected voltages in the same microvolt range that occurs in the ocean as a result of ocean currents moving seawater through the earth's magnetic field and generating an electrical current. By also detecting the polarity of the electrical current and keeping it to one side, fish could swim downstream in the ocean currents (Rommel and McCleave, 1973). Pacific salmon have been shown to swim predominately downstream as an active orientation and not just to drift with the current (Royce, et al., 1968). Swimming downstream in the North Pacific currents accounts for much of the migration route of Pacific salmon and suggests that the electric sensitivity seen in laboratory experiments may function as an electro-orientation system in the real world.

F. FUNCTIONS OF OTOLITHS AND SEMICIRCULAR CANALS

1. Location, Structure and Innervation

The receptors of the labyrinth—a collective name for the semicircular canals plus the associated chambers and otoliths—closely resemble those of the lateral line in both embryological origin and functional mechanics. Their location, orientation and innervation differ sufficiently from that of the lateral line so that their sensory information serves several purposes quite different from that of the lateral line.

One anatomical difference between the lateral line system and the labyrinth is that the fluid-filled tubules and pouches of the labyrinth are nearly or completely isolated from the outside environment. Another difference is the location of the labyrinths deep inside the head on each side of the medulla with the fluid-filled ducts and chambers arranged with a precise orientation to the body axis. The labyrinths also contain calcareous otoliths (statocysts) which have no counterpart in the lateral line system. Finally, the labyrinth receives the eighth cranial nerve exclusively, while the lateral line uses portions of the ninth and tenth cranial nerves. Thus the labyrinth and lateral line organs differ markedly in their gross anatomy, even though their sensors are essentially alike.

2. Role of the Labyrinth in Orientation

Orientation (equilibrium)—maintaining a prescribed relationship to gravity—involves two groups of receptors in fish. One group of labyrinth receptors concerns accelerations, particularly rotation or changes in direction. Other receptors report the direction of gravity. The combination of dynamic and static information from two different parts of the labyrinth enables fish to continue normal activities at night or during poor visibility when human orientation would be severely hampered by lack of visual light. For example, SCUBA diving in total darkness can produce severe disorientation because the visual component of equilibrium is absent and the human semicircular canals cannot do the total job alone. On the other hand, fish also have problems. Herring schools usually disperse at night and regroup each morning because of their sensory inability to maintain orientation to the other fish, although their straight-and-level swimming ability probably remains unaffected at night. Their schooling behavior apparently requires visual input as well as lateral line and labyrinth information.

The dynamic part of the labyrinth sensory system resides in the semicircular canals (Fig. VIII-8). The fluid contained in the canals tends to lag behind any motion of the fish's head. Motion of the fluid inside one or more of the canals stimulates hair cell sensors which are strategically located in bulges at the base of the canals. The hair cells resemble the neuromast organs in the lateral line sysem. While the lateral line sensors on the side of the body orient only in the anterior-posterior direction and those on the head in many directions, the semicircular canal sensors orient specifically to the axis of each of the three canals. Thus one or more patches of hair cells can detect motion in any one or a combination of the canals.

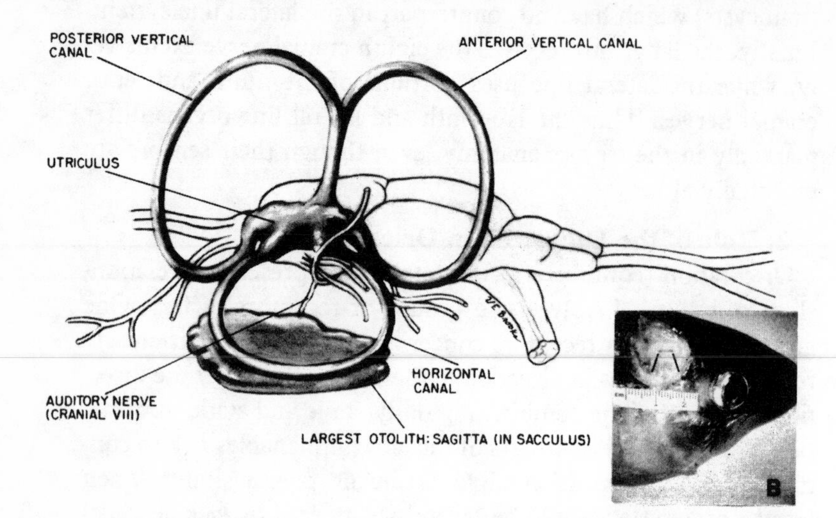

Fig. VIII-8. The appearance, innervation and general relationship of one set of semicircular canals to the brain in coho salmon *(Oncorhynchus kisutch)*. The left side (not shown) is a mirror image of the right side. (From Smith and Bell, 1976, with permission of author and publisher.)

The static detection system consists of bony structures (otoliths) resting in fluid-filled pouches which are lined with sensory hairs. The otoliths, being heavier than water, rest on whatever portion of the pouch is downward with respect to gravity. The sensory hairs report the position of the otoliths to the brain, where the information probably is compared to that from the semicircular canals to determine whether the otolith is responding to gravity or centrifugal force (acceleration).

Each labyrinth usually contains two small pouches (utriculus, lagena) and a single larger pouch (sacculus). Each pouch contains one otolith which fills most of each cavity. Additional pouches or expansions of the semicircular canals contain patches of sensory hairs but no otoliths. The smaller otoliths may be composed of particles of calcium salts embedded in a gelatinous matrix, but the larger otoliths are solid and often are elaborate in shape. Concentric layers in the larger otolith allow estimation of the fish's age—one major layer equals one year, although minute daily layers have also been identified. The multiplicity of pouches and otoliths suggest a multiplicity of functions or at least a division of labor within one function.

Recordings of electrical activity from nerves serving each pouch have shown that some otoliths serve as gravity sensors, some serve in hearing and all respond to linear acceleration to some degree. Elasmobranchs, lampreys and hagfish have been investigated most often because their cartilaginous skulls make access to the nerves easier than in bony fish. Electrical records from nerves to each pouch showed that the utriculus most often served as the gravity receptor (see next paragraph on righting response), while the sacculus served for hearing. In some herring-like fish the reverse seemed to be true. Since all of the labyrinth systems could potentially respond to gravity, acceleration and vibration, having different otoliths performing different combinations of duties in different fish should cause little surprise.

The righting reflex serves to keep fish upright—a particularly important factor in fish having their swimbladder below their center of gravity (e.g., salmonids). Such fish, once they begin to rotate, require an immediate, vigorous response with all fins to prevent turning belly-up. The fins move to counteract any rolling motion. If responding to being rolled to the right, for example, the dorsal fin tilts to the right, the paired fins move downward on the right and upward on the left, and the caudal fin may rotate to the right on top and left on the bottom. The eyes, on the other hand, attempt to remain level: the right eye rotates upward and the left eye downward. Cutting the nerve to the utriculus often abolishes this response. The response also occurs in fish without swimbladders, although perhaps not so emphatically.

3. Role of the Labyrinth in Hearing

Whether fish have a sense of hearing has been discussed scientifically for about 150 years, but its presence was not confirmed until the advent of modern electrophysiological techniques.

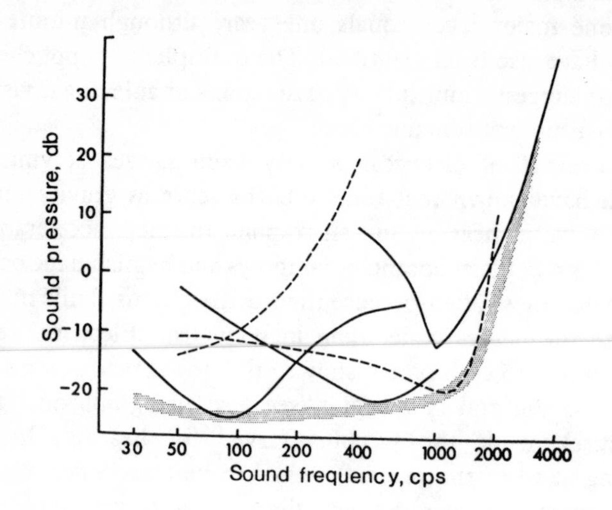

Fig. VIII-9. Sound pressure thresholds versus pitch for five single receptor units in the auditory system of herring. Two of the curves are drawn broken for clarity. The broad, shaded curve tentatively represents an overall audiogram for herring. (From Enger, 1967, with permission of author and publisher.)

Recordings from the auditory nerve showed impulses which remained synchronized with sound vibrations up to frequencies considerably greater than those to which the lateral line responded. Until obtaining this kind of evidence, however, physiologists believed that fish might lack a sense of hearing because they lacked the hearing organ (cochlea) found in the inner ear of humans and other higher vertebrates. It required considerable insight into the physics of underwater acoustics (often resulting from sonar research) before people even began to understand how fish hear.

The basic problem which fish have in using a hair cell is that fish are transparent to sound. This means that the bodies of fish resemble water sufficiently closely that sound waves pass through water or fish equally well, just like the fish wasn't there. This further means that the whole fish vibrates in synchrony with the sound waves in the water. The problem arises because the hair cells respond to bending, which usually results from displacement. If there were perfect synchrony of vibration there would be no displacement. However, otoliths vibrate slightly differently than the rest of the fish's body because their density differs considerably from that of water. The differential in density and resonance causes sufficient displacement to stimulate the hair cells. Since any density discontinuity enhances such diplacement, the presence of air bubbles such as the swim bladder also enhances displacement and therefore hearing sensitivity (see section 4 below). Thus fish can hear, although they lack the organ of hearing found in humans.

The range of frequencies and loudness within which fish respond varies widely with species and individuals, with circumstances (size of the test tank, for example) and with the ambient noise level. A typical audiogram as determined for herring appears in Fig. VIII-9. The upper limit of about 2000 hertz is typical of many fish, although the range among species is wide—from about 400 to over 4000 hertz. The lower limit of 30 hertz in herring also appears typical (range is perhaps 25-100 hertz). The exact limits remain difficult to determine because the sensors continue to respond somewhat outside of these frequencies with increased strength of stimulation. The responsiveness of the hairs slowly fades away over a span of frequencies rather than ending abruptly at a fixed frequency.

4. Mechanical Amplification of Hearing

Since fluid displacement is the basic stimulus for hearing, any means which increases displacement will enhance a fish's ability to hear low-level sounds. Maximal enhancement occurs with maximum difference in density, so most enhancement schemes involve placing an air bubble (density of air is approximately 800X less than water, while bone is only 2-3X heavier than water) as near to the labyrinth as possible. In addition to providing a density differential, an air bubble provides much greater elasticity than bone and thus readily changes size with sound pressure waves—i.e., bubbles are good resonators (displacement enhancers). Smaller bubbles resonate best with high frequency sound and large bubbles resonate at low frequencies. Bubbles held captive inside of fish resonate much less than bubbles free in the water column. Sonar researchers find that free bubbles produce an echo about ten times larger than swimbladders of the same size. Even working at only 10% of their potential effectiveness, swimbladders form a major class of mechanical amplifiers for hearing in fish.

The amplification process requires that the sound waves be changed from far-field to near-field status. Near-field sound waves involve both pressure and displacement, while far-field sound consists only of pressure waves. Displacement waves extend for about one-sixth of a wavelength away from the source of the sound. Since sound travels about 4500 feet (1370 meters) per second in water, a sound of 100 hertz has a wavelength of about 45 feet (13.7 meters) or a sound frequency of 450 hertz has a wavelength of about 10 feet (3 meters). The near-field would thus extend out to about 7.5 and 1.7 feet (2.3 and 0.5 meters) from the source, respectively. Most fish, then, have their labyrinths well within the near-field of their swim bladders.

Resonators such as swimbladders radiate their near-field displacement waves in a spherical fashion—i.e., the amount of displacement decreases rapidly with increasing distance from the source. Maximum sensitivity of hearing comes from maximum closeness of the resonator to the labyrinth. Different groups of fish have maximized this closeness in a variety of ways. Herring and anchovies have tubular extensions of the swimbladder which extend anteriorly close to the labyrinth. An air-breathing eel, *Symbranchus,* holds air bubbles in the roof of the mouth so that they probably enhance the functioning of their hearing organ just above the bubbles as well as serving for respiration. A whole group (order) of fish, the Ostariophysi (which includes the carp), has a series of bones (the Weberian apparatus) connected to the anterior end of the swimbladder which mechanically transmits the vibrations of the swimbladder to the back of the cranium. That a wide variety of fish have some sort of hearing enhancement suggests a survival advantage for having good hearing.

Chapter IX: CENTRAL NERVOUS SYSTEM

A. OVERVIEW

The brain performs three primary functions. First, the input portion of the brain receives and interprets the information from all of the sensory organs, internal as well as external. Second, the output portion of the brain sends coordinated commands to all parts of the body, either as nerve impulses or hormones. Most commands involve activity (stimulation), although some commands produce inaction (are inhibitory). The third, integration, is between these two aspects of brain function. Integration ranges at one extreme from simple, nearly automatic reflexes such as those which regulate heart rate and breathing, to complex learning at the other extreme. This third aspect of brain function varies most among vertebrates, being minimal in protochordates and hagfish and maximal in mammals and man. Bony fish are perhaps somewhere near the middle of the range of integration capability.

In an evolutionary context the vertebrate brain rates as a conservative organ. Brains change very little, even when most other organs have changed greatly. As an example, consider the brain of sole and flounder in which larval fish swim erect and adults rest on the substrate and swim on their sides. The metamorphosis from larva to adult involves drastic changes in the position of the eyes and to a lesser extent of the nostrils. This, in turn, involves rerouting of the optic and some other cranial nerves. The brain, however, remains in its "original" position—i.e., lies on its side in the adult—along with the semicircular canals. In the latter case, the interpretation of semicircular canal input certainly must change—downward in the adult must

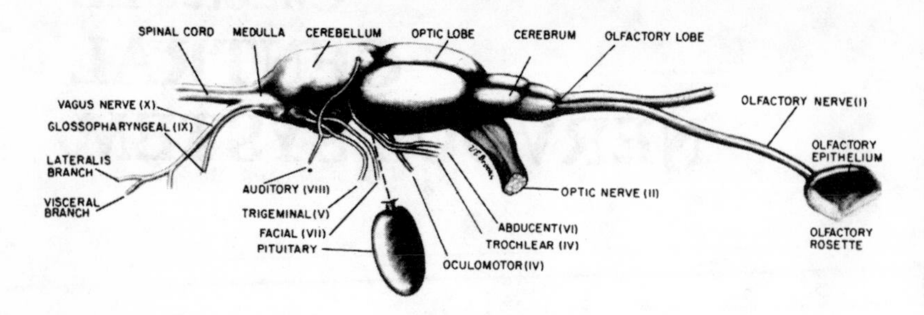

Fig. IX-1. Brain and cranial nerves of an adult coho salmon, seen from the right side. (From Smith and Bell, 1975, with permission of authors and publisher.)

be signified by the otolith resting on the side of its sensory pouch rather than resting ventrally. The conservation of the brain perhaps prevents or minimizes a number of related evolutionary changes in the skull, the musculature of the jaws and opercula and hence the configuration of the entire head.

On the other hand, the brain is not totally unresponsive to changes in form and function elsewhere in a fish's body. In fish with large eyes, for example, the additional receptors in the eye require additional sites in the brain to receive the additional information. Thus the optic lobes of fish with large eyes may be 20-30% larger than the optic lobes of fish with small eyes. Fish with electroreceptors in the lateral line system may have a few extra bulges or ridges on each side of the medulla where the lateral line nerve (part of cranial nerve X) enters. The specialized or emphasized functions of the brain can thus be recognized anatomically in some fish as relatively subtle differences in its size and shape.

The typical configuration of a teleost brain resembles that of other similar vertebrates including sharks on the more primitive side and frogs on the more advanced side. The typical lobes of a fish brain appear in Fig. IX-1, using salmon as an example. The brain is basically a tubular structure with swellings and thickenings along its length which produce the lobes. Seen from the dorsal side, the swellings include the cerebral hemispheres (merged with the olfactory lobes anteriorly in salmon), the optic lobes and the single cerebellum. Seen from the side, the underlying brain stem is visible as the diencephalon beneath the optic lobes and the medulla which begins beneath the cerebellum and merges posteriorly with the upper portions of the spinal cord. The obviously tubular nature of the brain in very primitive fish (hagfish, lamprey) and its large cavity have largely disappeared in teleost fishes. They have only a small, irregular passage communicating through the brain stem to the relatively small cavities inside each lobe of the brain. The trend for more and more brain tissue in the lobes and less and less cavity continues through all of the higher vertebrates, but the basic arrangement of lobes changes relatively little.

B. PROJECTION OF SENSORY INPUT

The most complete information on sensory projection comes from research on humans because of the ease during brain surgery of stimulating the brain directly while patients describe their sensations. The concepts thus developed involve a spatial arrangement in the brain lobes of nerve endings from all the various sensors. Since all nerve impulses are basically alike except for frequency, the only way for the brain to know whether the impulses relate to vision or hearing or taste is for the brain to "know" by its nerve connections in the brain what kind of sensor produced the nerve impulses. Further, for the brain to interpret a visual image rather than simply the presence or absence of light, the brain requires that its nerve endings have a spatial arrangement comparable to that of the sensors. This spatial representation (projection) of the external sensors in the brain explains such things as the phantom limb phenomena in amputees. Nerves cut at the stump ends of amputated limbs often become irritated and produce the sensation of coldness or itching. Because the brain portion of the limb's nerve tracts still include a projection of amputated limb, the brain misinterprets the location of the itching, leading to the sensation that the absent limb itches rather than the irritated stump. As a child, I once saw a man absent-mindedly scratching his wooden leg and thought it absolutely hilarious. Only many years later did I learn the physiology involved and empathize with the man's inability to ease the itching of a leg that existed only in his brain as a then irrelevant sensory projection.

The concept of sensory projection clearly applies to fish as well as to humans, even though a fish's capability to describe its sensations to us is extremely limited. The idea that brain lobe size relates to the quantity of its sensory input (discussed above) assumes that sensory projection occurs. Measurements of electrical activity in specific brain lobes also correspond to stimulation of appropriate sensory organs.

Vision, olfaction and hearing seem to be the sensations most readily identifiable as electrical activity in the brain lobes. Enger (1957) recorded electrical activity (electroencephalogram or EEG) from various brain lobes of cod. In fish resting in the dark, the cerebral hemispheres produced slow (alpha) brain waves at less than 7 hertz, while the optic and acoustic lobes produced electrical signals of from 8 to 13 hertz. The cerebral activity was unaffected by photic or acoustic stimulation, but electrical activity in the corresponding sensory lobes increased into the range of 14-32 hertz. Electrical activity in the optic lobes followed a flickering light stimulus up to 40 hertz. In the acoustic lobes on the side of the medulla, electrical activity faithfully reproduced the acoustic stimuli up to about 100 hertz (Fig. IX-2).

The amplitude of electrical activity in the olfactory lobes also corresponds to stimulation of the nasal epithelium. This is well known as the home-water response in Pacific salmon (Chapt. VIII, part D-2, and Fig. VIII-5). Thus three of the major senses have been localized in specific lobes of the fish brain.

C. CEREBRAL AND CEREBELLAR FUNCTIONS

The cerebral hemispheres (forebrain) of higher vertebrates are the site of learning, and it appears reasonable to extend this generalization to fish. Kaplan and Aronson (1967, 1969) studied the learning abilities of *Tilapia* from which they had removed various portions of the brain. The learning situation required that the fish would swim through a hole in a partition dividing their tank in half within a few seconds after a light was turned on or they received an electric shock. Fish with intact brains learned to escape the electric shock by swimming through the hole

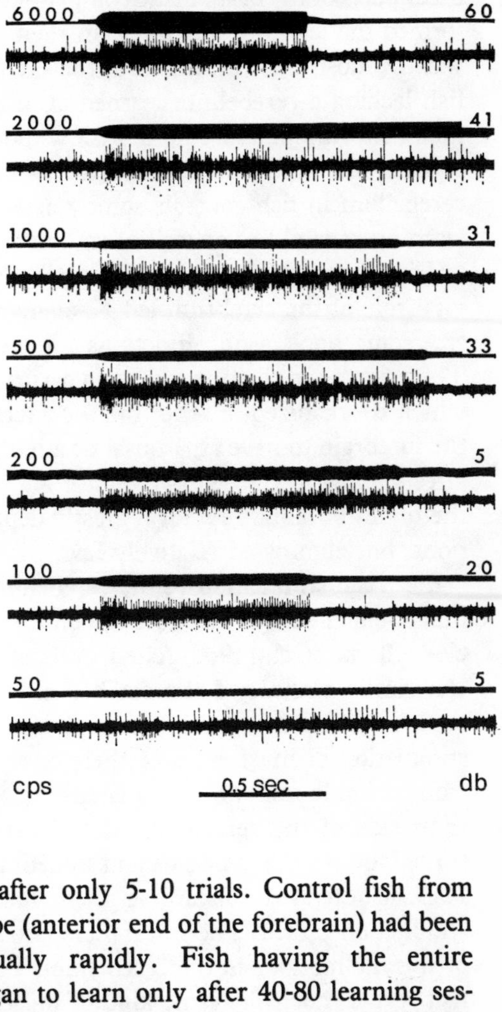

Fig. IX-2. Multi-unit responses in the acoustic region of the medulla to pure tone stimulation with various frequencies and pressures in herring. Upper line of each pair represents the hydrophone signal; lower record is the response in the medulla. Sound frequency (in cycles/sec) and pressure (in dB ref. l μBar) are indicated on the left and right sides, respectively, of each record. (From Enger, 1967, with permission of author and publisher.)

90-100% of the time after only 5-10 trials. Control fish from which the olfactory lobe (anterior end of the forebrain) had been removed learned equally rapidly. Fish having the entire forebrain removed began to learn only after 40-80 learning sessions and never reached the performance level of control fish. Some fish never learned to avoid the shock. Fish which learned to avoid the electric shock and then had their forebrains removed lost their learned response and relearned it very slowly.

The cerebellum of higher vertebrates serves for muscular coordination and proprioception (identifying the location and tension of muscles). The cerebellum of fish appears to serve a broader range of functions. In the experiments (above) by Kaplan and Aronson, the results suggested that in fish without forebrains other parts of the brain made compensatory adjustments and performed the learning processes. In their opinion, the cerebellum was the most likely candidate as the site for this learning because fish lacking a cerebellum learned even more slowly and poorly than fish without a forebrain. Fish without a cerebellum also had equilibrium problems for a day or two after surgery. Thus the cerebellum in fish controls some relatively essential part of the learning capabilities as well as muscular coordination.

The relatively large role of the cerebellum in learning and non-vital role of the cerebrum led Kaplan and Aronson to hypothesize some non-specific functions for the cerebral hemispheres. They suggested that the forebrain acts as a facilitator of behavior which is organized in other brain centers. They also considered the forebrain to have an arousal or attention-focusing role which affects many of the sensory, motor and associative processes of the brain. Forebrain removal mostly depressed a variety of functions, but eliminated relatively few.

Electrical stimulation of the cerebellum can provide information about the cerebellar role in control or coordination of muscles. Clark, et al. (1960) (cited by Kaplan and Aronson, 1969) electrically stimulated the cerebellum of goldfish, sunfish and catfish and observed changes in swimming. Fish responded to stimulation of most areas of their cerebellum by turning away from the side that was stimulated—e.g., fish stimulated on the right side of the cerebellum would turn left. Sometimes they turned toward the same side and sometimes they swam in spirals. Fish also tended to reverse their turning pattern after the stimulation ceased. The researchers also found some localization of specific motions in the cerebellum. Clearly the cerebellum of fish serves much the same kind of muscular coordination functions as in mammals, even though it may perform other functions, such as learning, as well.

D. OTHER FUNCTIONAL CENTERS IN THE BRAIN

Higher vertebrates have several clusters of brain cells called centers, each of which controls and coordinates a specific function. These centers are well studied in mammals and man, but little studied in fish. Hirano, et al. (1972) measured the drinking rates of eels in seawater after isolation of various lobes. They could sever the entire brain between the cerebrum and optic lobes or between the optic lobes and the cerebellum without any effect on drinking rate. The eels stopped drinking and died soon after having their vagus nerves (which attach to the posterior medulla) cut. Thus control of osmoregulation resides somewhere in the medullar and cerebellar portions of the brain. Hammel, et al. (1969) found that they could control the temperature preference of an arctic sculpin *(Myoxocephalus)* by changing the temperature of the anterior part of the cerebrum. Other centers will also eventually be located—e.g., a respiratory center probably exits in the medulla.

E. AUTONOMIC NERVOUS SYSTEM

The autonomic nervous system is one of the less known parts of the nervous system of fish. There still remain some major questions about its anatomy or even its existence for some organs, which makes it difficult to discuss functions of the system. In man, the autonomic nervous system contains two subdivisions—sympathetic (associated with thoracic and lumbar spinal nerves) and parasympathetic (from cranial and sacral spinal nerves). The sympathetic nervous system produces catecholamines (is adrenergic) at its nerve endings, while the parasympathetic nerves produce acetylcholine (are cholinergic). In bony fish, the autonomic nervous system lacks such clear anatomical and functional divisions although there are both adrenergic and cholinergic nerves. Campbell, one of the few people working consistently in the area, recognizes cranial, upper spinal and middle spinal components of the autonomic system in fish. This brief discussion is based mostly on Campbell's (1970) review of autonomic functions.

The anteriormost portion of the autonomic system in fish consists of motor fibers derived from the oculomotor (III) cranial nerve which secrete acetylcholine (Ach) and cause constriction of the iris. The largest and best known part of the teleost autonomic system emanates from several branches of the vagus (X) nerve and serves many branchial and visceral organs. The list of possible organs innervated includes the gills (vasoconstriction of blood vessels), possibly the stomach, the heart (adrenergic nerves accelerate the heart, cholinergic nerves slow it) and the swimbladder (regulation of blood flow and membranes at one or several sites each). The higher spinal portion of the autonomic system regulates the stomach and intestine (cholinergic endings enhance peristalsis, adrenergic nerves slow or stop it) and the urogenital ducts. The middle spinal portion controls the skin chromatophores to a unique degree. In most animals chromatophores respond over broad areas, but particularly in flatfish the autonomic system controls them in sufficiently small patches to enable flatfish to match the texture of their background (e.g., sand or small gravel vs. large gravel) as well as the color. Thus the autonomic system of fishes is known to control many of the vegetative (slow, routine, automatic) functions of the body.

A number of other functions *presumably* operate under the autonomic system, but such control has *not* been demonstrated in fish. The functions include contraction of both biliary and urinary bladders, stimulation of intestinal and pancreatic secretion and contraction of the spleen. The autonomic system in sharks and rays resembles that of bony fish, except that in elasmobranchs the autonomic system also receives connections from additional cranial nerves (V, VII, IX).

F. PHYSIOLOGY OF ELECTROFISHING

Electrofishing uses an electrical current to guide fish into a metallic scoop net (electrode) and may eventually result in the fish's complete immobility (electroanesthesia). Electrofishing equipment commonly uses DC current and usually allows for pulsing the current at various intervals and pulse durations and reversing the polarity as well as changing the strength. Electrofishing has become the method of choice for sampling fish living in areas where underwater obstacles make other fishing gear ineffective because electrical fields can draw fish out of hiding. Effective use of electrofishing gear is a considerable art because of the variability of the conductivity of different waters, species differences and the experience needed to predict the response of the fish to the electric field at various distances from the electrode.

The physiological basis for the behavior of fish in an electric field has been developed mostly from interpretation of the behavior of the fish (Vibert, 1963; 1969). He described the behavior of fish at various voltage settings and polarities. With the fish facing the positive electrode, 14 volts inhibited swimming, 18 volts stimulated coordinated swimming, 40 volts produced narcosis (electroanesthesia), 90 volts stimulated spinal reflexes directly (his interpretation) to give pseudo-forced swimming, and 150 volts produced tetany (rigidity) via direct stimulation of the muscles. Responses differed somewhat with reversed polarity. When the fish faced the negative electrode, 14 volts caused tetany-like swimming (interpreted as facilitation of motor axons in the brain). With increased voltage settings, brain control failed and actual tetany occurred, first (his opinion) through stimulation of motor areas in the brain, then spinal nerves and finally the muscles themselves, as before. The difficulty of determining the voltage actually experienced by the fish makes many of Vibert's interpretations difficult to substantiate with laboratory neurophysiological experiments, although he described some nerve transection experiments as being done in moderate electric fields of approximately 100 mV/cm.

In spite of the apparently drastic and pervasive effects of electrofishing on fish behavior, it appears relatively unstressful as judged by several stress indicators (see discussion of stress response, Chapt. XII-B). Schreck, et al. (1976) found that plasma glucose in rainbow trout increased significantly above control levels three hours after electroshocking. Preliminary experiments on coho salmon in my laboratory showed an immediate and continuing increase in blood glucose and decrease in plasma chloride following electroshock, although neither change was severe. Anesthesia by electroshock was once proposed as being less stressful than unbuffered tricaine methanesulfonate (MS-222) but was never adopted widely.

Chapter X:
ENDOCRINE SYSTEM

A. OVERVIEW

The endocrine system of fish must be viewed in relation to the endocrine systems of other vertebrates because fish are typical vertebrates in many respects. Some hormones, such as thyroid hormone and catecholamines, have remained remarkably similar in all vertebrates. On the other hand, endocrine systems of fish also show great diversity, perhaps more so than all of the land vertebrates together. Thus, when fish differ from higher vertebrates, you should not assume that fish necessarily represent a "lower" or more primitive condition than tetrapods, only a different one. Fish have faced a wide variety of environmental problems, have had long evolutionary periods for developing adaptations to them and have responded to these problems in a variety of ways. Viewing fish only as primitive ancestors of the tetrapod vertebrates grossly oversimplifies their physiology and underutilizes the richness of comparative information available from fish endocrine systems.

The endocrine system serves roles complementary to and often together with that of the central nervous system. The brain coordinates many bodily functions having a response time range of milliseconds to seconds, while the endocrine organs respond mostly over the time range of minutes to days. Although nerve activity is primarily an electrical event, it also involves production of chemicals (neurotransmitter substances) at nerve junctions (synapses). Endocrine organ activity may involve nerve transmission, but emphasizes production of chemicals—hormones. The effects of a single pulse of hormones entering the blood stream lasts for as much as an hour or more, in contrast to a much shorter duration for the effects of a single nerve impulse. While nerve impulses can go to highly specific sites, hormones usually travel via the circulatory system and potentially can influence the entire body. However, most hormones also have highly specific effects, in spite of their general distribution, due to the limited distribution of their specific receptor sites. Because of its emphasis on secretion and its generally widespread influence, the endocrine system has often been described as serving for chemical coordination via the circulatory system.

However, the chemical and neural control systems appear less and less independent as research on them continues. At one time the pituitary gland was considered a highly independent "master" controller of all the other endocrine glands. Then a portal system of blood vessels was described as carrying secretions from the floor of the brain (hypothalamus, immediately above the pituitary) down into the pituitary. Now most pituitary hormones are known to be controlled by secretion of releasing factors produced in the hypothalamus. The hypothalamus, in turn, appears to be mostly controlled by the neural activity of a portion of the brain near the optic nerves called the preoptic nucleus. The preoptic nucleus contains nerve endings from most other parts of the brain, including the sensory lobes. In addition, the three widespread neurotransmitter chemicals common in most nervous systems have turned out to be not too different from some hormones, and some endocrine organs once thought to be controlled only by pituitary hormones are now known to have their own nerve supply. Thus there is much overlap between neural and hormonal control systems.

The endocrine system integrates many external conditions—salinity, temperature, photoperiod—to control osmoregulation, metabolism, reproduction and migration via combinations of several sensors and several endocrine organs. For example, information on background color and texture travels from the eyes to the optic lobes and then somehow become translated into the combination of neural and hormonal activity needed to coordinate chromatophores so that they produce camouflage instead of just random colors. Also fish "recognize" some circumstances as stressful and respond by producing catecholamines and corticosteroids (see Chapter XII on stress response). A maximal stress response involves both the autonomic nervous system and at least two levels of endocrine gland functions. Thus, the integrative role of the endocrine system necessitates that it merge its functions not only with those of the nervous system, but with many other systems as well.

The interrelationships within the endocrine system also are many and complex, but often follow two principles. First, many of the responses involve two endocrine glands—the pituitary and some other endocrine gland under the control of the pituitary. Second, the hormone from the second gland often inhibits production of the pituitary hormone, a process called inhibitory feedback. Combinations of inhibitory feedback systems form an elaborate set of checks and balances which normally prevent excessive or "runaway" responses. Thus the endocrine system controls itself as well as controlling other organ systems.

The study of endocrine systems of fish attracts much interest and many researchers, so the literature of fish endocrinology is very large, much larger than can be reviewed in detail in a book of this size. The large number of papers on fish endocrinology makes the use of review articles necessary for everyone but full-time endocrinologists. Reviews which I found useful included those of Bern (1967), which is out of date only for lack of information on the ultimobranchial gland (discovered later), of the series of chapters in *Fish Physiology* (Hoar and Randall, 1970, Volume 2 and part of Volume 3) and of a series of papers published in the *American Zoologist*, Volume 12, No. 3 (1973) and Volume 17, No. 4 (1977). Figure X-1 shows many of the multiple actions and interactions within the endocrine system of fish (excepting the ultimobranchial gland), and Table X-1 catalogs the hormones presently known in fish. These are offered both as an overview and a summary for this chapter.

B. THE PITUITARY

The pituitary gland hangs beneath the hypothalamus portion of the diencephalon (brainstem below the anterior edge of the optic lobes) by a narrow stalk of nervous tissue and blood vessels. It rests on the midline in a pocket in the parasphenoid bone. In salmonids, the location is just at the posterior margin of the hard paland immediately in front of the junction of the first gill bars in

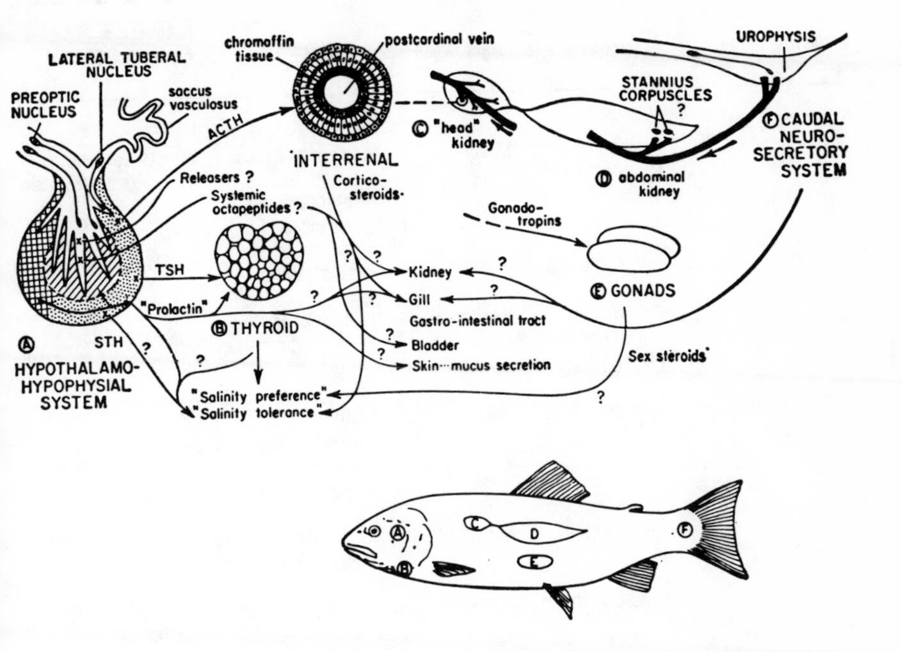

Fig. X-1. This diagram of possible endocrine factors in osmoregulation in teleosts shows most of the endocrine organs and their location in salmon. Arrows show supposed interactions between hormones and the many question marks emphasize the areas of insufficient information. One additional hormone, calcitonin, and its source, the ultimobranchial gland located on the transverse membrane between the liver and the heart, do not appear in this diagram because they were discovered later. (From Bern, 1967, in *Science* 158: 455-462, copyright 1967 by the American Association for the Advancement of Science, by permission.)

the roof of the mouth. In a 10-inch (25-cm) trout, the gland is about one-eighth of an inch (3 mm) in diameter and contains several histologically-identifiable regions. Recently it has also been possible to identify histochemically some of the hormones produced by various regions of the pituitary (Figure X-2). The following discussion covers each of the pituitary hormones briefly.

Prolactin (also called paralactin) functions predominantly in freshwater osmoregulation and therefore affects many organs. The most general role involves increasing the retention of certain

Endocrine organ	Secretion	Target Organ	Effects	Feedback or Antagonist Effects
Pituitary gland Pars distalis (rostral)	Prolactin (paralactin)	Many	Growth, mucus production, ion retention, melanin production, decreased water permeability	Effects sometimes inhibited by thyroxin
	Adrenocortico-tropic hormone (ACTH)	Interrenal gland	Stimulates cortisol production	Inhibited by Cortisol
Pars distalis (caudal)	Somatotropin (STH)	Many	Stimulates growth, appetite, prevents liver hypertrophy	
	Thyrotropin (TSH)	Thyroid gland	Stimulates thyroxin production	Inhibited by thyroxin
	Gonadotropin(s)	Gonads	Stimulates egg and sperm production	Inhibited by thyroxin, sex hormones from gonads
Pars intermedia	Melanocyte stimulating hormone (MSH)	Melanocytes	Dispersion of melanocytes	Melatonin from pineal organ causes concentration of melanocytes
Neurohypophysis (Pars nervosa)	Oxytocin	Gills, kidney	Osmoregulation, contraction of smooth muscle.	
	Arginine vasotocin, Isotocin	Blood vessels	Osmoregulation, smooth muscle.	
Interrenal Gland	Cortisol, cortisone, corticosterone	Bladder, gut, kidney, gills	Stress response, osmoregulation	Effects are often antagonistic to prolactin effects
Chromaffin Cells	Epinephrine, norepinephrine	Circulatory system	Increase heart rate, vasodilation or vasoconstriction, increased blood glucose	

Table X-1. Endocrine organs, secretions, and effects typical of teleost fishes.

Endocrine organ	Secretion	Target Organ	Effects	Feedback or Antagonist Effects
Corpuscles of Stannius	Hypocalcin	May decrease blood calcium	?	
Ultimobranchial Gland	Calcitonin	Lowers blood calcium in mammals, may regulate Ca^{++} excretion by fish gills	Skeleton ?	
Caudal Neuro-secretory system (urophysis)	Urotensin I-IV, Acetylcholine	Increased salt retention in kidney and gills, increased GFR, may cause contraction of reproductive smooth muscle	Kidney, gills	
Endocrine Pancreas	Insulin	Increased permeability to glucose, prevent excessive breakdown of glycogen	All cells	
	Glucagon	Mobilizes glycogen and fat	All cells	
Renin-Angiotensin system (Juxtaglomerular cells)	Angiotensins, Renin	Controls blood pressure, may regulate aldosterone production, may assist adaptation from FW to SW	Kidney	
Thyroid	Thyroxin	Increased O_2 consumption, migratory restlessness, reproductive behavior	Metabolic rate, brain	

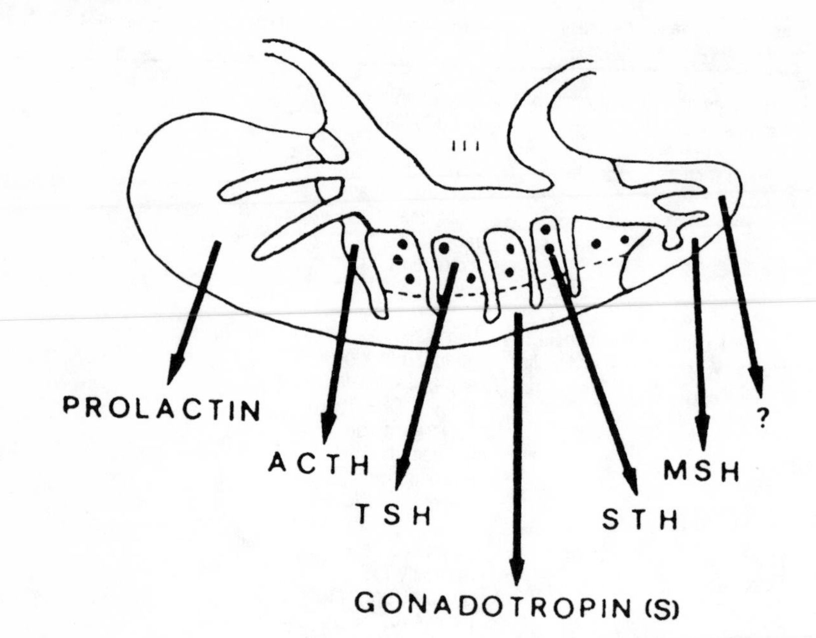

Fig. X-2. Diagrammatic representation of a mid-sagittal section through a representative teleost pituitary gland *(Xiphophorus)* to indicate the various regions and the hormones produced. OC - optic chiasma; III - third ventricle; ACTH - adrenocorticotropic hormone; TSH - thyroid stimulating hormone; STH - somatotropin; MSH - melanocyte stimulating hormone. See also Table X - I. (From Schreibman, et al., 1973, with permission of author and publisher.)

ions by the epithelia of the gills, kidneys and urinary bladder and decreasing the permeability of external surfaces to water. For example, removal of the pituitary from eels increased the rate of ion loss from their gills and, in several other species, caused death from ion losses. However, extracts from fish pituitaries caused no response in either mammalian mammary glands or in pigeon crops, the two standard bioassays for prolactin from land vertebrates. On the other hand, mammalian prolactin will replace part of the function of prolactin in fish whose pituitary glands have been removed. This differential of function caused earlier investigators to use the name paralactin for the fish hormone since it clearly differed from the mammalian prolactin. The present trend is to consider both of them as variants of prolactin.

A great many specific effects are known for prolactin. In fish, prolactin increases the glomerular filtration rate in the kidney. Prolactin also increases ATPase activity (see Section XII-C-1) in the kidney and urinary bladder, but decreases ATPase activity in the gills. Uptake of water from the intestine and the bladder decreases as well as inflow of water across the gills, presumably from decreased epithelial permeability. Na^+ uptake increases in the kidney and bladder while Na^+ efflux decreases at the gills. Mucous production increases at the gills, intestine and skin. Most of these effects are the opposite of those produced by cortisol, and several are inhibited by thyroxin. All of these effects are consistent with optimization of freshwater osmoregulation.

The name prolactin comes from the hormone's role of promoting lactation in mammals, where it has been studied extensively. In making comparisons among the lowest vertebrates, there has been some tendency to assume that prolactin should necessarily be involved in reproduction as in mammals. In fish, prolactin stimulates the production of epidermal mucus—hardly a reproductive function. However, in discus fish *(Symphysodon discus)*, young fish swarm about the female after hatching and feed on an epidermal mucous secretion which is stimulated by prolactin. While this somewhat resembles lactation, most endocrinologists interpret this as being an osmoregulatory function which accidentally became involved in reproduction rather than indicating a basically reproductive role for prolactin. In general, prolactin seems to have undergone more molecular modification and role changing than almost any other vertebrate hormone.

Somatotropin (STH) (also called growth hormone) acts, like prolactin, in a widespread fashion without any intermediate hormone gland being involved. In general, it stimulates appetite and growth and prevents liver hypertrophy. Bovine growth hormone injected intramuscularly once a week increased growth in coho salmon by 50-100% over controls in only 56 days, depending on ration level. In this same series of experiments, the authors considered the economic feasibility of enhancing growth in cultured salmon this way, but they concluded that it would be too expensive. The hormone was expensive and the labor to inject it too costly even for the considerable increase in growth obtained (Markert, et al., 1977). Relatively few experiments have been performed on STH compared to prolactin, and further progress awaits development of a quantitative assay for STH.

The neurohypophysis (that part of the pituitary gland embryologically derived from nerve tissue) produces two hormones in fish—oxytocin and arginine vasotocin. In mammals the equivalent two hormones are chemically similar—both are eight-unit peptides differing by only two amino acids on one end of the molecule. Both stimulate contraction of uterine (and other smooth) muscle and milk production to some degree. Mammalian vasotocin is also known as antidiuretic hormone (ADH) and stimulates reabsorption of water in the collecting tubules of the kidney. In fish the osmoregulatory role of both oxytocin and arginine vasotocin (AVT) seems to be emphasized, with AVT causing increased urine production (diuresis) in freshwater fish. Oxytocin influences osmoregulation by producing vasoconstriction in gill blood vessels. The role of AVT in contracting arteriolar smooth muscle to maintain or increase blood pressure (which occurs in some fish and not others) could also influence osmoregulation by increasing the amount of glomerular filtration. AVT appears not to be the agent causing ovarian and oviduct smooth muscle contractions in livebearing fish, because a third, related hormone, isotocin, is ten times more effective than AVT. Although all of these oxytocin-like hormones differ only slightly in chemical structure, these small differences have led to major differences in function. This in turn originally produced a confusing array of names as various investigators believed that they had different hormones when they looked in different animals. That situation is now largely sorted out, although many details remain to be settled (Pang, 1977; Chan, 1977; LaPointe, 1977).

Melanocyte stimulating hormone (MSH) comes from the posterior part of the pituitary and acts directly to disperse (enlarge) melanocytes in fish skin, thereby darkening a fish's coloration. Control of coloration in fish has turned out to be considerably more complex than just controlling MSH, however. There are several kinds of pigment cells (chromatophores), each of which is controlled separately. Further, a single type of chromatophore may respond differently to the same stimulus according to its location. Finally, MSH is not the only hormone which influences melanophores—melatonin from the pineal organ (Chapter VIII-C) causes melanophores to contract, giving a blanching effect on fish coloration.

The nervous system further complicates coloration control of fish. A rule of thumb says that fish which adapt to a change of background color in less than 10 minutes do so entirely or mostly by neural mechanisms. Fish which require longer than 10 minutes have primarily hormonal controls. Fish which adapt at intermediate rates have mixtures of neural and hormonal controls. Fish which change color over several days probably produce new cells and pigments rather than adjusting the dispersion of existing chromatophores. Thus fish cover the entire range from totally neural to totally hormonal control of coloration.

In salmonids, an injury may produce rapid darkening of sharply-defined areas of skin. This suggests a combination of neural and hormonal controls on salmonid melanocytes. My hypothesis is that normal control results from a balance of neural blanching and MSH darkening. Thus when nerve damage stops the neural blanching effect, then the full darkening effect of MSH follows in a few minutes. Sick salmonids are often darker in color than other comparable fish in the same tank.

The function of several pituitary hormones is to stimulate secretory activity in another hormone gland. These pituitary hormones include adrenocorticotropic hormone (ACTH), thyrotropin (TSH) and two or more gonadotropins. ACTH stimulates production of cortisol and other corticosteroids by the interrenal gland, whose widespread functions are discussed later in this chapter. The production of ACTH was shown in goldfish to be controlled by a hormone, corticotropin releasing factor, from the hypothalamus (in the floor of the brain immediately above the pituitary). TSH stimulates the thyroid gland to produce thyroxin or release it from storage. The hypothalamus probably secretes both stimulatory and inhibitory factors which control the production of TSH. The control of reproduction in fish has received little study compared to mammals or compared to other hormone systems in fish. Gonadotropins (perhaps only one gonadotropin in some fish) are known to stimulate gamete production by teleost gonads but have not yet been demonstrated in hagfish, lampreys or sharks and rays at the time of writing this. Gonadotropins are probably also controlled by hypothalamic secretions. Details of reproductive hormones are included in Chapter XI.

C. THE UROPHYSIS

The urophysis, also called the caudal neurosecretory system, is a neurosecretory site in the posterior portion of the spinal cord. The urophysis may be visible as a bulge on the ventral side of the spinal cord or may be identifiable only histologically (Figure X-3). A urophysis has been found in every species of fish examined, so it seems likely to be universal in fish.

Although the urophysis was discovered 150 years ago, its functions still remain controversial. Four secretions have been identified which were originally designated as urotensins I through IV. Dr. Howard Bern's assessment of their status as of April, 1977 follows (personal communication). Urotensin I has no known effect on fish, but lowers blood pressure in most types of land vertebrates. Urotensin II causes contraction of smooth muscle such as isolated strips of trout rectum and bladder, but has been shown not to be 5-hydroxytryptamine or acetylcholine (both are neurotransmitter substances), which also cause smooth muscle to contract. Urotensin III stimulates the uptake of Na^+ by gills and increases retention of Na^+ by kidney tubules. Urotensin IV probably is arginine vasotocin, but was identified as such only in some rainbow trout in Japan. Other investigators have looked but not found it in other species. In addition to urotensins, the urophysis of carp produces large amounts of acetylcholine. In general, urophysis secretions are usually thought of as relating to osmoregulatory functions with their greatest effect on the kidney.

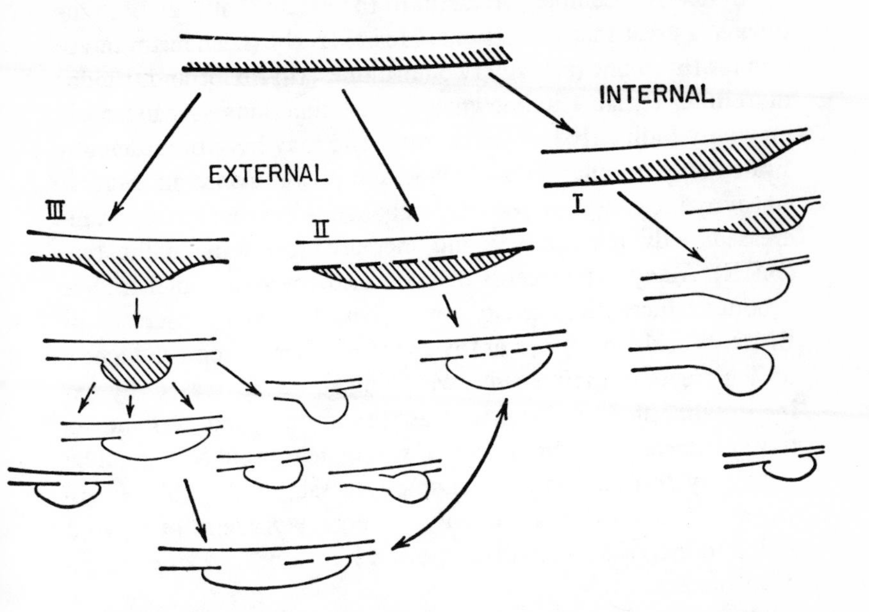

Fig. X-3. Varieties of urophysial structure in teleosts and possible derivation of types by ventral and caudal expansion of neurohemal tissue (hatched area) into or through meningeal membranes (heavy black line) to produce lobate urophyses. The "ancestral type" is an exaggeration of the elasmobranch situation and is encountered by some isospondylous teleosts (e.g. - the clupeoid, *Elops*). Type I is exemplified by *Conger;* type II by *Salmo;* type III by *Albula.* (From Bern, 1967, in *Science* 158: 455-462, copyrighted 1967 by the American Association for the Advancement of Science by permission.)

D. THYROID GLAND

The thyroid gland of salmonids consists of thyroid follicles scattered along the ventral aorta and in the isthmus muscles of the lower jaw. The follicles are often too small to be seen readily during gross dissection. Their blood supply comes from a branch of the coronary artery and is easily damaged during any experimental surgery near the heart. In some fish thyroid follicles also occur along the common cardinal vein, in the head kidney and even as far posteriorly as the vicinity of the spleen. The more dispersed condition is interpreted as primitive, and one author suggested that the localization had a selective advantage for controlling the blood supply to the gland.

Thyroxine resembles prolactin to the extent that they both influence a great many functions. However, the mechanism of action for thyroxine (actually two hormones, thyroxine and triiodothyronine) remains unknown, so some functions ascribed to it may only indirectly involve thyroxine or may have been demonstrated only in other animals. For example, in mammals increased thyroxine causes increased metabolism—increased activity, increased body temperature and increased basal metabolic rate (BMR). Many experiments on increased thyroxine in fish have produced increased activity, but most showed no increase in BMR. Weekly injections of thyroxine in Atlantic cod produced a 35% increase in their steady swimming speed. Increased thyroxine in salmonids produced increased swimming and jumping activity. Increased activity may also stimulate thyroid activity. The migratory restlessness seen in smolting salmonids probably involves increased thyroid activity. The general trend of thyroid action to increase metabolism is clear, but some of the details remain puzzling.

Thyroxine also influences growth, metamorphosis and reproduction. Captive sturgeon were successfully treated with thyroxine to alleviate ovarian degeneration. In general, thyroxine is required for gonadal maturation. An example of its influence on growth occurs in non-migratory alewives in Lake Erie which grow only half as large as their migratory relatives. The non-migratory fish appear to suffer from an iodine (and therefore thyroxine) deficiency as judged by their enlarged thyroid follicles (a goiter-like condition also called thyroid hyperplasia). Pacific salmon in Lake Michigan also seem to experience iodine deficiency problems, although not so severely. Still other functions of thyroxine have been shown to include effects on skin, the central nervous system, seasonal adaptation, temperature tolerance and osmoregulation.

E. ENDOCRINE ROLE OF THE PANCREAS

The mammalian pancreas produces two hormones, insulin and glucagon, which control the level of blood glucose. Glucose passes into cells only in the presence of insulin, which thus regulates blood glucose levels downward. The lack of insulin causes high blood glucose levels (diabetes mellitus) and a shift from carbohydrate metabolism to lipid metabolism. Glucagon increases blood glucose through glycogenolysis (see Chapter V-J-2) and also causes release of stored lipids from the liver. Thus regulation of blood glucose level in mammals normally results from balancing the opposing effects of the two hormones.

The reason for describing mammalian hormones is that the fish situation is unclear and may differ from mammals to some unknown degree. For example, blood glucose in humans averages 90 mg/100ml of blood and ranges from 140 mg/100 ml after a large meal to 60 mg/100ml after severe exercise. In rainbow trout, normal blood glucose wanders between 50 and 150 mg/100ml of blood and may not have any closely regulated norm. Blood glucose rises after a meal but may decline slowly for many days afterward if there is no additional food. Spontaneous diabetes (Sekoke disease) occurs in Japanese carp as a result of rancid fish oil in their diet and responds minimally to injections of mammalian insulin. Some investigators consider salmonids as "semi-diabetic" because of their poor regulation of blood glucose, and their emphasis on lipids for muscle metobolism. With poor regulation of blood glucose, some glucose might be expected to leak past the kidney tubules and appear in the urine, but apparently this does not commonly happen. Glucose in the urine is characteristic of both diabetic carp and diabetic mammals, however.

In discussing carp diabetes, Yokote (1970) described those degrees of response by fish to mammalian insulin. Some fish responded to insulin much like mammals—a moderate dose of insulin drastically lowered blood glucose levels and led to convulsions. Fish which responded thusly included brown trout, scup, menhaden, both common and bull's-eye mackerel, sea bass and skate. Some fish responded minimally or not at all, even when given massive doses of mammalian insulin. The daddy sculpin was the only fish with this lack of response which resembled some of the forms of insulin-resistant diabetes in mammals. The third group of fish responded to moderate doses of mammalian insulin with decreased blood glucose, sometimes achieving zero blood glucose, but without having any convulsions. These fish included the puffer, sea robin, toadfish, tench and hagfish.

The structure of the pancreas of fish may differ from that of mammals. In addition to their diffuse distribution, pancreatic islets in fish may have four types of secretory cells instead of the mammalian three (possibly four) types. However, the fourth type of cell in fish may be an artifact resulting from use of mammalian histochemical tests on fish tissues. Nerve cells are absent in pancreas tissue of mammals and birds, but have been seen in electron micrographs of pancreatic tissue of most lower vertebrates including conger and anguillid eels, catfish *(Ictalurus)* and swordtails *(Xiphophorus).* The nerve endings appeared adrenergic, but actual nervous control of pancreatic secretion remains largely undefined. See also Section V-B.

Glucagon in mammals mobilizes nutrients—glucose, amino acids, fatty acids—in time of metabolic need. It also stimulates prompt release of insulin and causes increased cardiac output. Histochemical studies in fish suggested the presence of glucagon, but that appeared to be as far as things had progressed at the time of writing. Gastrin granules (part of the system controlling release of pancreatic digestive secretions into the gut) may also have been present in the histochemical sections.

F. CALCIUM REGULATION: THE ULTIMOBRANCHIAL GLAND AND THE CORPUSCLES OF STANNIUS

Calcium regulation in teleosts is incompletely understood, so the mammalian situation will have to serve as a model, although it probably differs from the teleost situation in several respects. Mammals regulate their blood calcium level very exactingly as part of calcium's role in maintaining normal excitability of cell membranes in nerve and muscle tissue. Tetany (uncontrollable muscle contraction) occurs in humans when blood calcium decreases more than 30% below the norm of 10 mg/100ml blood. Depressed reflexes and general sluggishness begin at 12 mg/100ml blood and become increasingly pronounced at higher levels of blood calcium. The skeleton serves as a functionally versatile reservoir for calcium. Increased parathyroid hormone causes lowered blood calcium through deposition of calcium in bone, while decreased hormone levels lead to increased blood calcium via solubilization of bone to whatever extent that the calcium was not provided by the diet.

Calcium metabolism in teleost fish probably follows the same general scheme as in mammals, but with some significant differences. While fish probably regulate their blood calcium relatively effectively, the various tables of ion data in Chapter II show as much as a two-fold variability among species for apparently normal levels of blood calcium. Further, some species show considerable differences in normal blood calcium between freshwater and marine stages of their life cycle; others show no differential. In contrast to mammals, marine fish must cope with a calcium-rich environment, and apparently some fish cope better than others. Freshwater fish may or may not face a calcium-poor environment, depending on the water hardness and the calcium content of their diet. Fish obviously have a skeleton (and calcified scales in some species) to serve as a calcium reservoir, although there has been discussion about whether the acellular bone found in most bony fish is exchangeable with blood calcium or not. The present consensus of reviewers seems to agree that cellular bone (mammals, eels) and acellular bone (most teleosts) serve as a reservoir for blood calcium, although the role of a bone in fish may be much smaller than in land vertebrates.

Present data, although scanty, suggests that the ultimobranchial gland hormone, calcitonin, assists in excretion of calcium in fish living in calcium-rich seawater or perhaps some hard freshwaters. The gland and its hormone have been most studied in salmon as a source of calcitonin to lower blood calcium and to accelerate bone deposition in people suffering from weak, spongy bones (osteoporosis, which occurs particularly severely in elderly women). Copp and coworkers sorted through several tons of viscera at a salmon cannery, removed the transverse membrane (between heart and liver), clipped out the faint white streak on the membrane (which is all that is visible of the ultimobranchial gland) and extracted the calcitonin. Salmon calcitonin proved to be more effective and longer lasting in lowering blood calcium in people than any other preparation. The special effectiveness of salmon calcitonin probably resulted from the salmon hormone being sufficiently similar to the human one to be effective, but sufficiently different to slow down the normal enzymatic degradation processes. Copp finally injected salmon calcitonin into salmon and found *no* change in blood calcium levels. Presumably the calcitonin was doing something rather than nothing; perhaps it enhanced the excretion of calcium at the gills only enough to balance the passive inflow from seawater. Involvement of regulation of phosphate ion has also been suggested.

The corpuscles of Stannius, two white or yellow dots on the ventral surface of the posterior kidney, are even more puzzling than the ultimobranchial gland. No clearly defined hormone or active fraction has been isolated from them. Removal of the corpuscles of Stannius (Stanniectomy) in killifish caused hypercalcemia, and the intact corpuscles histologically appear more active in seawater-adapted killifish than in freshwater-adapted killifish. Such evidence suggests that the corpuscles of Stannius lower blood calcium. The mechanism of lowering blood calcium in European eels seemed to be increased bone deposition, while in another experiment on eels the urinary calcium level increased.

The puzzle continues and grows larger. The corpuscles of Stannius also influence other electrolytes, and other hormone glands affect calcium levels. Giving extracts from the posterior portion of the pituitary caused increased blood calcium in killifish. Stanniectomy in killifish disturbed their ion balance in ways that were largely remedied by injections of prolactin, which may have stimulated reproduction of mineralocorticoid hormones from the interrenal cells. Thus the corpuscles of Stannius may be part of the larger osmoregulatory system which includes the pituitary, the interrenal gland (discussed below), the urophysis and the ultimobranchial gland. Because of their apparent complexity of interactions with other hormone glands, the corpuscles of Stannius are expected to puzzle endocrinologists for some time.

G. INTERRENAL BODIES AND CHROMAFFIN CELLS

1. General

The adrenal system of fish consists of two glands, interrenal bodies and chromaffin cells, embedded in the tissues of the head kidneys; both produce two or more hormones having widespread effects. There are a number of parallels between the adrenal-like systems of fishes and the adrenal glands of mammals, but the parallelism should not be pushed too far. In both groups the hormones—glucocorticoids, mineralocorticoids, epinephrine and norepinephrine—serve similar, although not identical, functions. Both have pituitary control of their functions via ACTH from the pituitary (this Chapter, Section B). Both groups appear to have similar generalized stress responses (below and Chapter 12). However, mammals have discrete adrenal glands with functionally separate cortical (outer) and medullary (inner) regions. Teleost fish have similarly separated functions, but in two separate glands within the head kidney. The glands can be distinguished only histologically, making adrenalectomy experiments extremely difficult because the anterior head kidney is removed. Further, some adrenocortical hormones in fish have minor chemical and functional differences from those of mammals, probably reflecting the major differences in environmental challenges faced by mammals and fish. Finally, there are considerable data suggesting that the adrenal systems of all fish are not necessarily alike. Thus the literature on the adrenal hormones of fish is voluminous and still expanding, sometimes contradictory, often confusing and always relevant to a wide variety of fish functions. Figure X-4 suggests some of these widespread roles in osmoregulation and their interrelationships with other organs and other hormones.

2. Interrenal Bodies

The interrenal bodies (tissue) in fish consist of clusters of follicular (fluid containing) cells embedded in the head kidney. These cells produce several corticosteroid hormones which are all related to each other by virtue of being derived from cholesterol and differ primarily in their side chains (Figure X-5). The similarity of molecular structure of corticosteroid hormones probably also accounts for the overlapping of function between glucocorticoids and mineralocorticoids and between corticosteroids and sex hormones. Such cases as some male chinook smolts

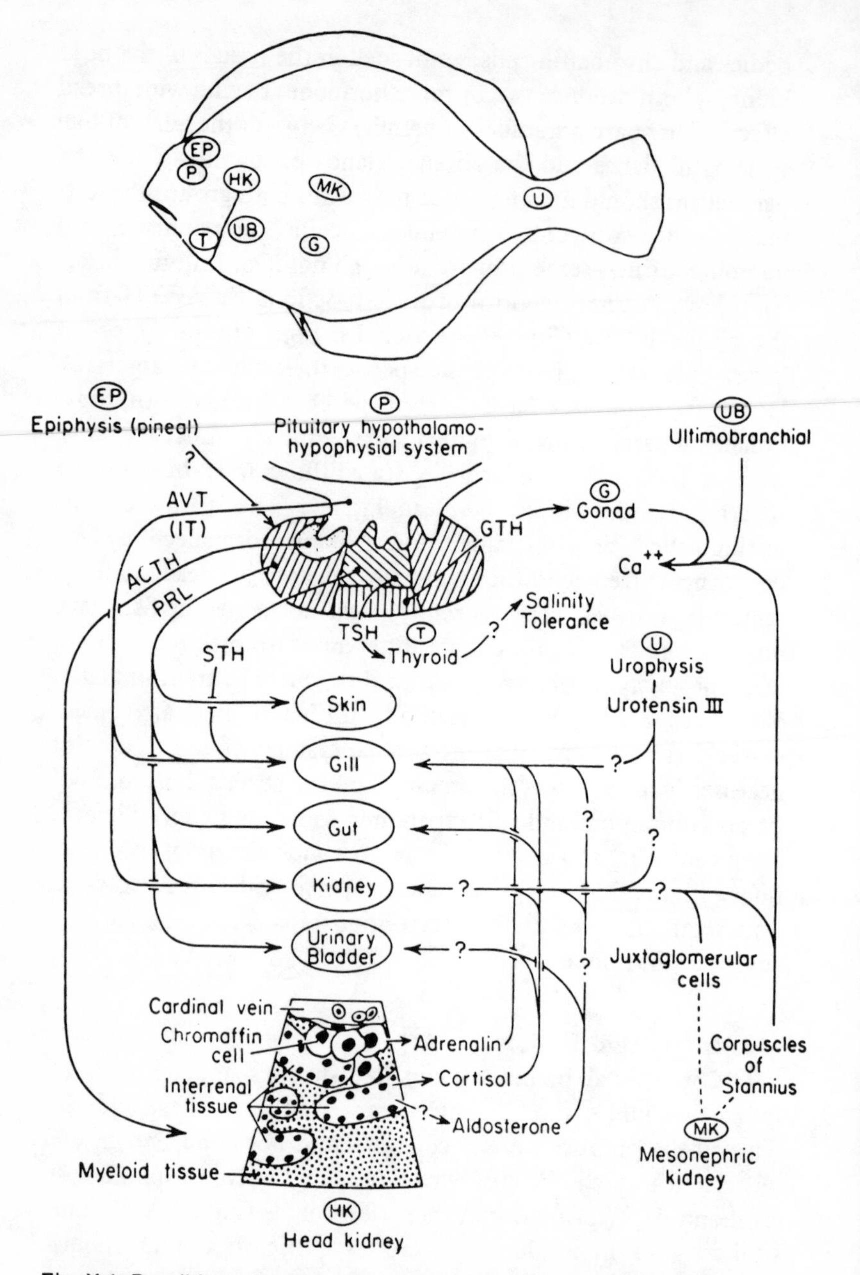

Fig. X-4. Possible endocrine involvement in control of osmoregulatory effector organs. Apparent diversity of control may reflect imperfection or complexity of hydromineral regulation. Compare this with Fig. X-1 and Table X-1. (From Johnson, 1973, with permission of author and publisher.)

Pregnenolone

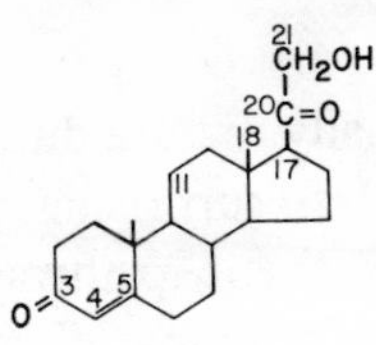

ADRENOCORTICO-
STEROIDS

SEX-RELATED
STEROIDS

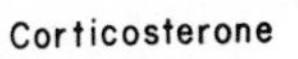

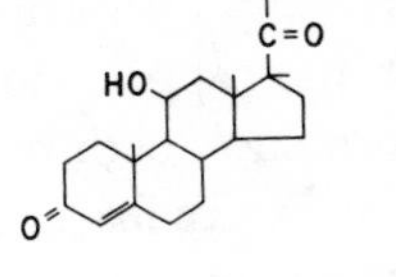

Estradiol

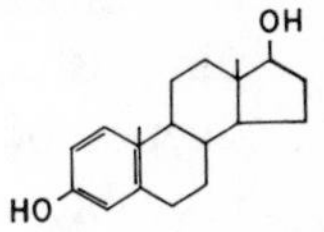

Cortisol

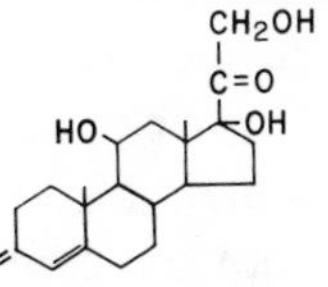

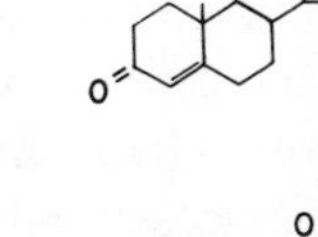

Progesterone

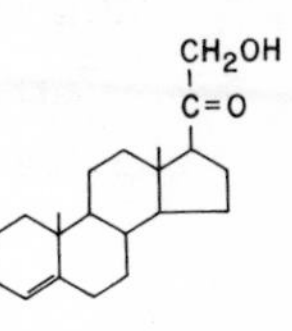

Aldosterone

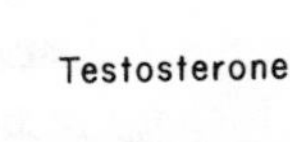

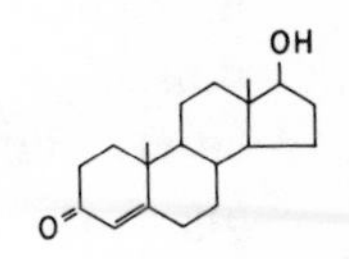

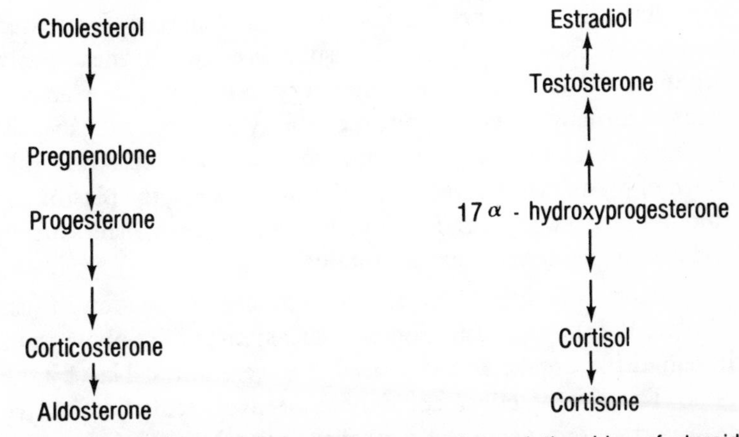

Fig. X-5. Chemical similarity, derivation and interrelationships of steroid hormones. The closeness of chemical structure and location of production suggest possible explanations of occasional confusion of sexual, osmoregulatory, and stress response functions. Multiple arrows indicate multi-step chemical reactions while single arrows suggest single reactions.

becoming sexually mature during the stress of their downstream migration are less surprising when testosterone can be seen as part of the synthetic pathways of corticosteroid metabolism. Similarly, the combining of sexual maturation and the stress of upstream migration in salmon probably also enhances the hormone response compared to either stress taken alone; control of the corticosteroid system also probably operates in an overlapping fashion. ACTH from the pituitary appears to stimulate production of both corticosteroid hormones, both of which in turn probably exert overlapping negative feedback effects on the pituitary, although the details are far from clear.

Cortisol has the strongest influence on carbohydrate metabolism, although it is not without influence in osmoregulation. Increased levels of cortisol help to maintain blood glucose levels by depressing the rate of synthesis of fat from carbohydrates and by decreasing the utilization of blood glucose. Glycogen deposition in the liver increases at the same time as glucose production from muscle protein (gluconeogenesis) increases. As an example of the latter, adult salmon lose up to 60% of their skeletal muscle during upstream migration and spawning. Increased cortisol levels are also associated with diminished numbers of lymphocytes in circulation and a reduced inflammatory response. Increased cortisol level is a major component of a generalized stressed response (Chapter 12) whose results are not all necessarily beneficial. The reduced inflammatory response for example, leads eventually to increased susceptibility to infection, especially when combined with gluconeogenesis via increased breakdown of plasma antibodies. For unknown reasons, the plasma antibodies (a fraction of the beta globulins) decrease faster during stress than the other plasma proteins.

Corticosterone has its major effects on osmoregulation in fish (Butler, 1973). Administration of corticosterone to rainbow trout impaired the uptake and increased the excretion of Na^+ by the gills. The site of effect could have been on either the active transport system or on the epithelial permeability. Injections of corticosterone in rainbow trout decreased the kidney glomerular filtration rate by about 25%, a change similar to that seen in migratory trout (steelhead) as they adapt to seawater. On the other hand, cortisol also has osmoregulatory functions. Injections of cortisol in rainbow trout caused increased activity of the Na^+/K^+-activated ATPase which energizes the active transport system. Adrenalectomized eels survived in seawater with relatively normal Na^+ levels and urine flows when given daily injections of cortisol. Cortisol also enhanced the uptake of water from isolated segments of eel intestines, while aldosterone, deoxycorticosterone acetate, corticosterone and cortisone acetate had no demonstrable effect. In general, corticosterone and cortisone seem to play some role in seawater osmoregulation in most fish and a major role in a few fish, but cortisol seems to predominate in most teleosts.

In mammals, aldosterone plays a large role in osmoregulation, probably by being the final active agent into which other steroid hormones are eventually converted. Aldosterone injected into fish has moderate effects on osmoregulation, but studies which incubated corticosterone (the immediate precursor of aldosterone) *in vitro* in head kidney tissue failed to show any synthesis of aldosterone. Further, examination of fish blood showed no aldosterone present, except in lungfish. The aldosterone question was unresolved for fish at the time of writing, but evidence appeared mostly negative and demonstrated the problems of using mammalian hormone to investigate control of functions in fish.

3. Chromaffin Cells

Chromaffin cells reside in the head kidney of teleosts in the tissue immediately surrounding the posterior cardinal vein. The cells may be interdispersed with adrenocortical cells. Chromaffin cells secrete catecholamines into the blood passing through the posterior cardinal vein inside the kidney.

From what little is known about chromaffin cells of fish, it appears that they are largely comparable to the adrenal medulla of higher vertebrates. The adrenal medulla produces two catecholamines, epinephrine and norepinephrine, which differ slightly in their effects. Both increase heart rate and systolic blood pressure, the latter by different mechanisms—norepinephrine causes increased peripheral resistance while epinephrine increases cardiac output in spite of decreased resistance in muscles, liver and brain. Epinephrine strongly increases oxygen consumption and blood glucose, while norepinephrine does so only weakly. There are many such differential responses to the two hormones, and different species have different proportions of them. Some general effects of catecholamines in fish are listed in Table XII-1 as part of the discussion of stress responses.

Epinephrine and norepinephrine also are produced in other sites besides the chromaffin cells. The brain, particularly the hypothalamus, has areas rich in epinephrine. The sympathetic nerve endings tend to produce mostly norepinephrine. Catecholamines produced at nerve endings often do not appear in the blood stream since their action is very localized.

A few preliminary experiments (unpublished) with catecholamines in fish in my laboratory support the idea of fish being similar to higher vertebrates. Fish subjected to decompression stress (bends) showed increased levels of epinephrine but not norepinephrine (see more about stress response in Chapter XII). Placid rainbow trout used in aquaculture showed lower epinephrine levels both when resting and when swimming strongly than the wilder steelhead (migratory) rainbow trout. A suggestion from the latter work was that one criterion for cultured fish that would be exciting for sportsmen to catch should be the ability to produce large amounts of epinephrine.

H. JUXTAGLOMERULAR CELLS

These endrocrine cells are named for their location beside the glomerulus in the kidney. For a change, this system in mammals is not fully understood (Nishimura and Ogawa, 1973). The secretory cells are readily visible histologically and have been seen in teleost fish and all of the land vertebrates examined, but seem absent in elasmobranchs and cyclostomes, at least in the few species looked at. They may also occur in ratfish *(Hydrolagus colliei).* In man, the juxtaglomerular cells contain renin granules. Upon release renin acts as an enzyme to form angiotensin I from angiotensinogen (secreted in excess by the liver), which then becomes angiotensin II and acts to raise blood pressure and increase kidney function. There also probably are effects on ion transport and aldosterone secretion. This enzyme-hormone complex is often called the renin-angiotensin system (RAS).

The situation in fish is unclear. The renin granules in fish stain similarly as in mammals, but extracts of fish granules injected into mammals cause no change in blood pressure (no vasopressor effect), possibly because of species differences. Extracts from fish granules injected into fish produced a vasopressor effect for a few minutes. Renin granules occur abundantly in both glomerular and aglomerular fish but are not as localized as in mammals. Renin-like granules have been seen histologically in corpuscles of Stannius but constitute only about 1% of the total supply, so one wonders about their significance. The RAS may be active during adaptation to salt water. The RAS of fish probably differs chemically from that of land vertebrates and may have rather different functions than in mammals.

Chapter XI: REPRODUCTION

A. OVERVIEW

Persons interested primarily in reproduction sometimes seem to view the whole life of fish as relevant to reproduction. This is true to the extent that the existence of any species is not imaginable without reproduction. Further, the reproductive success of any individual clearly depends on its survival to reproductive age. Even then the individual still must obtain and store sufficient raw materials and chemical energy to produce the sex products required for reproduction. Thus reproduction involves efficient functioning of all of the organ systems. Much of the importance of studying bioenergetics (Chapter 5) arises from efficiency considerations. Appropriate behavior and coordination with seasonal temperature regimes and other environmental factors also play a role. However, this chapter is primarily concerned with reproduction from early stages of gamete development through early stages of development of the progeny.

Fish use a wide variety of strategies for allotting their energy (usually lipid) reserves between reproduction and other body functions. The winter flounder of the U.S. Atlantic coast uses a strategy in which individuals having little or no lipid reserve produce no gametes. Apparently they "decide" to postpone reproduction during poor years in favor of growing larger to maximize survival and gambling that the following year will be better. In another strategy, yellow perch in Lake Washington near Seattle showed an inverse relationship between visceral fat and ova. The female perch accumulated visceral fat for about half the year and then apparently used that fat for producing eggs. There may also be a direct relationship between the varying amounts of visceral fat from year to year and the size or number of eggs. In other words, reproductive effort may be proportional to the energy available.

Since they reproduce only once, Pacific salmon try to reproduce regardless of their energy reserves. This strategy is not quite comparable to the previous two examples because the problem of partitioning energy in Pacific salmon involves a strenuous, sometimes lengthy migration in addition to maturation of the gametes at the same time. Both processes occur without any food intake. Thus salmon migrating different distances necessarily have different quantities of lipid reserves. In addition to lipids, salmon also consume a considerable part of their body protein during migration and spawning. The manner of development of such an extreme reproductive strategy through natural selection seems obvious—any fish having insufficient stored energy to reach the spawning grounds does not contribute to the gene pool. No intermediate levels of reproduction are possible for a salmon, just all or none, though this is not to deny recognizable differences between individuals in egg quantity and quality based on lipid content and size of the eggs.

Thus, three conspicuous strategies for reproduction are: 1) to spawn only when sufficient lipid (energy) is available; 2) to spawn in proportion to the energy available; and 3) to spawn at the expense of all other functions, even if the individual subsequently dies. There may be other related strategies.

Another series of reproductive strategies concerns the size and number of eggs produced in comparison to the amount of parental care (effort = energy) devoted to the survival of the eggs and offspring. At the one extreme, fish having small eggs usually produce them in large numbers. One female cod *(Gadus)* may produce several million eggs, for example, which have a relatively low rate of survival. Small eggs hatch in a few days and resulting larvae require microscopic food immediately. At the other extreme, large eggs may be relatively few and require long incubation periods. The larvae are relatively large at hatching, require no food for several days, or even weeks afterwards and then can eat relatively large food. The successful culturing of salmonids rests to a considerable degree on the ease of incubating the relatively large eggs and the ease of feeding the larvae. Most salmonids produce eggs in the range of hundreds to a few thousands per female. Northern midshipmen *(Porichthys notatus)* attach a few dozen pea-sized eggs to the ceiling of a rocky burrow in the intertidal zone and the male guards them until they hatch. Sticklebacks *(Gasterosteus)* are famous for building nests in which the female lays a few eggs which the male fertilizes, guards and ventilates. These eggs are large only in relation to this rather small fish. Larger eggs usually have a better survival rate than small eggs. Perhaps the example of the largest and fewest eggs rests with the skates, which produce a large leathery case containing four eggs, each almost as large as a chicken egg.

When any fish produces fewer and fewer eggs, there must be better and better assurance that most of them will survive. Parental care of the eggs and sometimes of the fry after hatching is one such type of strategy. In addition to the sticklebacks mentioned above, other examples of parental care include the penpoint gunnel *(Apodichthys flavidus)* which curls its eel-like body around its egg mass, the large rubbery mass of eggs which is guarded and fanned by the male lingcod *(Ophiodon elongatus)* and several members of the genus *Tilapia* (family Cichlidae) in which males hold eggs in their mouths during incubation. There are many other examples of parental care of eggs and young among ornamental fish. And, of course, in the seahorse and its relative the pipefish the eggs develop in a specialized pouch in the *male!*

Perhaps the ultimate expression of parental care of eggs in fish is to keep them in the visceral cavity until they reach some advanced stage of development—i.e., the young are "born" alive. Livebearing occurs in several unrelated groups of fish, including guppies *(Poecilia reticulata)*, the rockfish (genus *Sebastes* of the family Scorpaenidae) and several members of the surfperch family (Embiotocidae) such as the shiner perch *(Cymatogaster)*, striped seaperch *(Embiotoca)* and pile perch *(Rhacochilus)*. Livebearing requires considerable anatomical modification compared to egg-laying since the male must have a copulatory organ for transferring sperm (sometimes packaged into spermatophores) into the female, and the female may store the sperm in a special chamber for up to six months. Then the female must accommodate the developing eggs, provide for their oxygen supply and waste elimination, and eventually have a liberation route for a relatively large (compared to passing unfertilized eggs) juvenile fish to be "born." Livebearing will be discussed in more detail in Section F of this chapter. In a few species, internal fertilization is followed by normal egg-laying.

There are some problems in describing the anatomy of teleost reproductive systems because of their wide range of variations on a few basic plans. Hoar (1969) indicated a generalized plan for the urogenital systems of teleosts with major variations for the more distantly-related groups (Fig. XI-1). One generalization is that eggs are commonly released into the peritoneal cavity and reach the outside through a funnel, while sperm almost always remain inside of ducts. When eggs are not released into the peritoneal cavity, they are shed into the cavity of an expanded ovary or ovarian duct. Salmon were first thought to confine their eggs this way, but current opinion (Henderson, 1967) suggests that salmon eggs are free and exit through a funnel and a very short oviduct in the posterior visceral cavity. The problem in making such determinations is that the funnel and duct resemble mesenteries, all of which are fragile and easily distorted during dissection. Further, hatcherymen normally take salmon eggs by killing the fish and slitting open the visceral cavity and so have no need to know about the normal membranes and ducts. In most livebearing fish, the chamber for incubation of the young is probably an expansion of the ovary and perhaps part of the oviduct. Some authors have called such structure a uterus.

There are further problems of naming which I will mention only briefly. The kidney goes through a series of embryological stages—pronephros, mesonephros and metanephros—which also reflect the probable evolutionary sequence of kidney development. Adult hagfish and lampreys have only a pronephros; Hoar's diagram (Fig. XI-1) shows teleosts as having a mesonephros; and higher vertebrates have a metanephros. The term opisthonephros has also been used for teleosts to indicate that they might not fit the general evolutionary pattern for the tetrapod vertebrates. This would not be surprising since there were only a few peculiar types of fish—the lobefins, of which the modern coelacanth is a distant relative—which evolved into the land vertebrates. The other 99.9% of the fish species continued for another 400 million years to evolve into more "effective" species

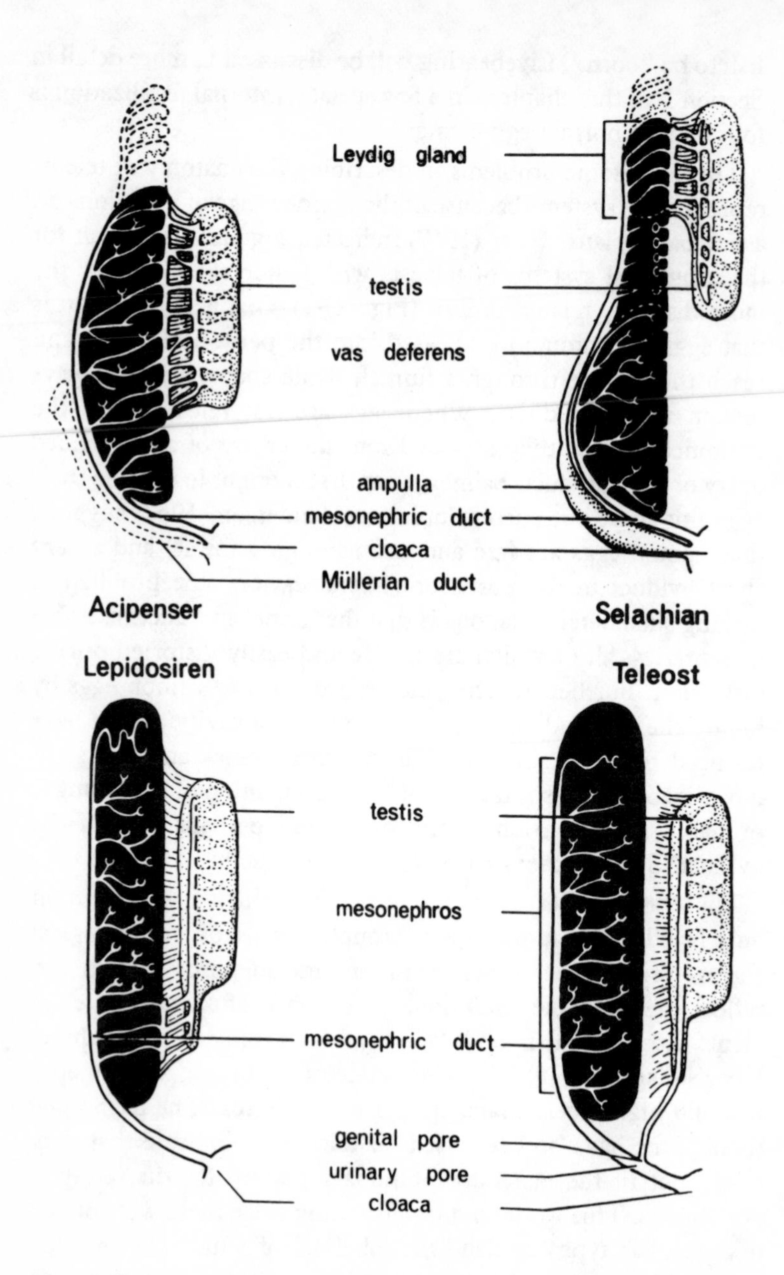
Leydig gland
testis
vas deferens
ampulla
mesonephric duct
cloaca
Müllerian duct
Acipenser
Selachian
Lepidosiren
Teleost
testis
mesonephros
mesonephric duct
genital pore
urinary pore
cloaca

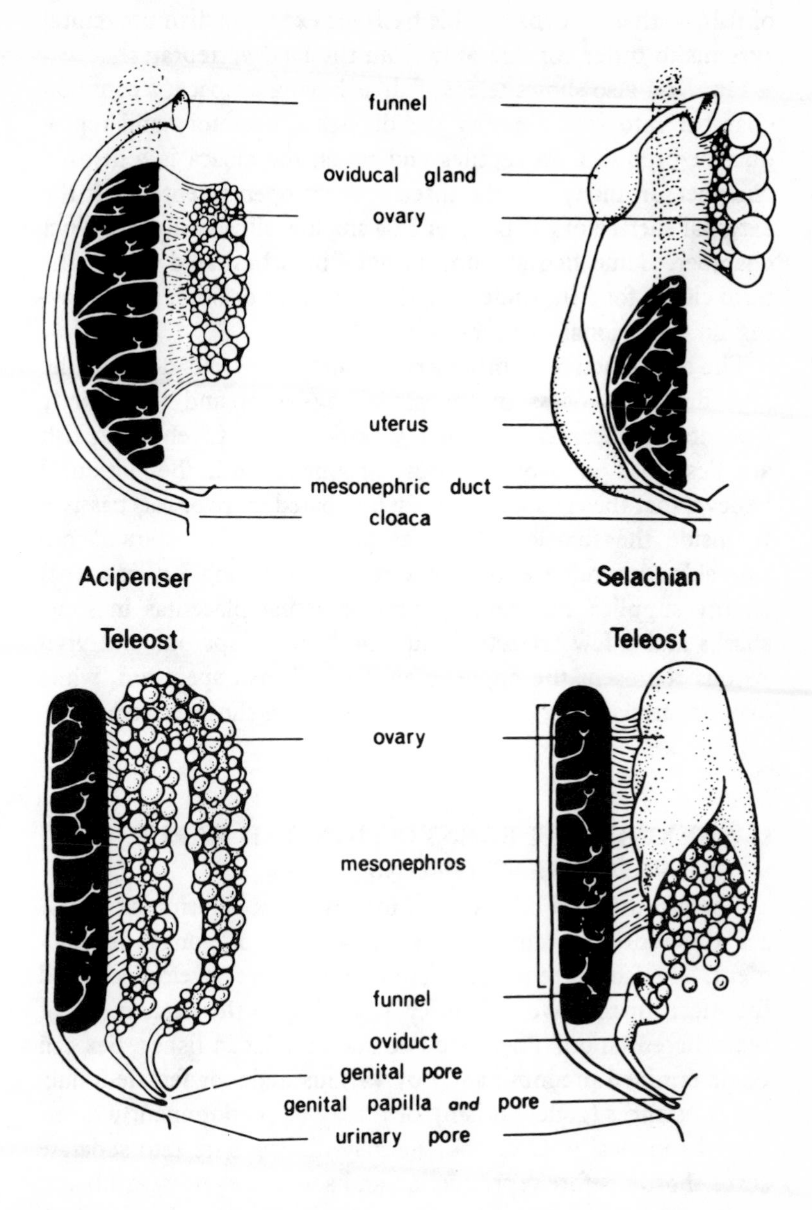

Fig. XI-1. Representative types of urogenital systems in male (opposite page) and female (above) fishes. (From Hoar, 1969, with permission of author and publisher.)

of fish, so there is considerable basis for expecting fish urogenital systems to differ considerably from the land vertebrates.

Fig. XI-1 also shows teleost fish as having a cloaca, a common chamber into which empty the digestive, excretory and reproductive products. In reptiles and birds, the cloaca is a distinct chamber. In many fish the three systems open essentially at the external surface of the body or in a shallow slit, hardly a distinct chamber. Some fish anatomists have thus hesitated to use the term cloaca for fish, preferring the term vent or other terms having no evolutionary connotations.

The reproductive terms oviparous and viviparous describe the reproductive process in the strict egg-layers and the strictly placental livebearers (mammals), respectively. Livebearing fish are described as ovoviviparous, originally with the notion, I believe, that the eggs were simply incubated more or less passively inside the female. More recent studies have shown considerable dependence of the developing young on maternal energy supplies and development of actual placentas in some sharks and a few teleosts. Thus the terms oviparous and viviparous represent the opposite ends of a broad spectrum, while ovoviviparous covers everything in between those extremes.

B. SEXUAL DIFFERENTIATION AND EARLY DEVELOPMENT OF GAMETES

The fact that fish have several modes of sex determination and sex differentiation complicates explanation of gonad development. The factors required to stimulate gamete development and the mechanisms involved differ according to the basic mode of sex differentiation. There are two such modes in fishes. Sex can be determined predominantly by various male- or female-inducing substances (gonochorism), or it may be predominantly determined genetically. Gonochorists may differentiate into separate sexes shortly before reproduction and sometimes possess a bisexual gonad temporarily. Examples include lampreys, hagfish, eels *(Anguilla anguilla)*, herring *(Clupea harengus)*, minnows *(Phoxinus laevis)* and paradise fish (*Macropodus* spp.). Or gonochorists may differentiate early in life and never show intersexuality. Examples of the latter include the platy *(Xiphophorus maculatus)*, the medaka *(Oryzias latipes)*, and the guppy and mollies (*Poecilia* spp., *Poeciliopsis* spp.), the latter of which can produce all-female populations. Genetically determined sexuality can be complicated by the occurrence of more numerous sex-related genes in regular (autosomal) chromosomes than in the sex chromosomes. Rainbow trout *(Salmo gairdneri)* may be intermediate between these two groups. Thus it is possible for a fish to have a mixture of male and female genes in which sex-related genes in the autosomes can override the influence of the sex chromosomes in the presence of the two sets of genes of opposite sexes. Further, administration of sex hormones may cause varying responses ranging from no effect to complete functional sex reversal, depending on the particular mixture of the above factors in the particular species of fish. The point of mentioning these complications now is to point out that gamete development may require different processes in different species of fish and that the comparatively straightforward mammalian type of sexual development cannot necessarily be used as a model for fish.

In spite of complexities of internal organization, many fish in temperate climates simply initiate gonadal development in response to seasonal changes in temperature and photoperiod. In the pumpkinseed sunfish *(Lepomis gibbosus)*, for example, gonadal development typically began in late May when pond water temperatures exceeded 12.5 °C and daylengths neared 15 hours. Laboratory experiments showed, however, that the true minimum daylength was between 12.0 and 13.5 hrs and that a minimal temperature of around 14 °C was required at the same time. Females may have had slightly higher temperature requirements than males. Neither warm temperature nor long photoperiod alone had much effect. After spawning in August there was no renewal of gonadal activity, even though daylengths and temperatures exceeded the minimums needed for development, suggesting that there is also a refractory period after spawning. Warm temperatures (17.5 °C) and short photoperiod (10.5 hrs) caused regression of gonadal development (Burns, 1976). Other fish having similar reproductive patterns include minnows *(Phoxinus laevis)*, medaka *(Oryzias latipes)*, stickleback

(Gasterosteus aculeatus) and sunfish *(Lepomis cyanellus).* In the tropics where the predominant seasons are simply wet or dry, spawning often occurs at the onset of the wet season. Such patterns of reproduction cued to seasonal changes coordinate the sexual development of both sexes and ensure favorable environmental conditions for early development of the young. The coordination is mediated through endocrine glands which influence behavior as well as physiology.

The early development of gametes—spermatogenesis or oogenesis—seems to resemble that of other vertebrates in most respects. Cells destined to become eggs proliferate and enlarge, each becoming surrounded with nutritive (follicular) cells. The granulosa layer of the follicle is recognized as providing the yolk material for the developing egg, but most other possible functions of the follicle are still speculative. In general, gametogenesis depends on pituitary gonadotropins. Although hypophysectomy does not entirely prevent gametogenesis in lower forms such as lampreys, all teleosts tested showed blocking or severe reduction in gametogenesis when the gonadotropins were chemically blocked. How many gonadotropins occur in the pituitary is unknown, although one experiment showed that ovine luteinizing hormone would overcome most effects of the chemical blocking agent suggesting that there may be only one. Some of the follicle cells probably produce estrogens under control of pituitary gonadotropin(s) and may continue to do so after the egg leaves the follicle. Estrogens such as estrone and estradiol most likely control the development of secondary sex characteristics in female fish while the hormone progesterone, which maintains uterine development during pregnancy in mammals, may be absent in fish. Male differentiation may be similarly induced by such corticosteroid androgens as testosterone, or androstenodione. Much remains to be done in clarifying the sex hormones and their roles in teleost fish.

C. COMPOSITION OF OVARIAN FLUID

Once released from the follicle, ova lose the supply of nutrients provided by the follicular cells and depend upon ovarian fluid to meet their requirements for nutrients and oxygen. The origin,

Component	Units	Amount
Ca^{++}	mg %	8.00±1.04[a]
Inorganic phosphate		11.49±4.41
Glucose		5.25±3.30
Blood urea nitrogen		0.93±0.68
Uric acid		0.18±0.07
Cholesterol		2.00±0.19
Total protein	g %	0.04±0.05
Albumen	"	0.06±0.02
Total bilirubin	mg %	0.05±0.05
Alkaline phosphatase	IU	7.00±3.89
Lactic dehydrogenase	"	5.50±2.05
SGOT/340	"	0.00
Na	Meq/L	5.50±3.11
K	"	1.25±0.19
Cl	"	8.00±0.16
Mg	"	1.48±1.05

[a]Mean ± SD.

Table XI-1. Chemical composition of ovarian fluid from rainbow trout (*Salmo gairdneri*). (From Satia, et al., 1974, with permission of author and publisher.)

quantity and turnover rate of ovarian fluid are unknown, although Satia, et al. (1974) examined its composition in rainbow trout (Table XI-1). Clearly, this ovarian fluid taken at the time of spawning was much more dilute than plasma. Unpublished work in my laboratory suggested that coelomic fluid from migrating (sexually immature as yet) adult coho salmon resembled an ultrafiltrate of plasma. If ovarian fluid is an ultrafiltrate, then the dilutions observed by Satia, et al. could have been produced through uptake of dissolved materials by the maturing eggs. The dynamics of ovarian fluid are further discussed in the section on viviparity (Ch. XI-F).

D. SPAWNING BEHAVIOR, RELEASE AND FERTILIZATION OF EGGS

Spawning behavior appears mostly under the control of hormones and environmental factors and often seems beyond the control of the individual fish. It ranges from that of the ocean sunfish, in which a group of fish of mixed sexes simply swim along together spewing out millions of eggs and sperm into the water, to the complex courtship, nest-building and territorial behavior of sticklebacks. A bizarre case is the hermaphroditic fish *Serranus subligarius,* in which pairs of fish each exhibit distinctive coloration while each releases the appropriate gametes. Then the pair exchanges coloration and sex products and releases the gametes from the other half of their bisexual gonad. In general, the teleosts show a greater range of reproductive behavior and mechanisms than all the land vertebrates put together.

Thus there is no simple relationship in fish between endocrine organs and sexual development and behavior. In mammals there are two pituitary gonadotropins, follicle stimulating hormone (FSH) and luteinizing hormone (LH). The ovary responds in turn by producing two hormones, estrogen and progesterone. In fish an LH-like hormone appears to regulate both maturation and release of ova (ovulation) from the ovary into the peritoneal cavity. An FSH-like activity is either absent or very weak or occurs in only certain species. Ovulation is usually followed soon afterward by spawning, so a spawning hormone which acts directly on the central nervous system has been postulated but not proved. In female sticklebacks *(Gasterosteus)* spawning was possible only after being for two or three days in the vicinity of a male stickleback engaged in nest-building. Other workers found that several species were stimulated to carry out spawning behavior most effectively when injected with a combination of fish pituitary extract and mammalian oxytocin, the latter serving to stimulate contraction of the smooth muscle in the gonoducts. In male fish, one of the few hormonal effects demonstrated on sperm was a thinning of the semen just before spawning. There is extensive involvement of several hormones in male spawning behavior, especially in species in which the males carry out nest-building and other parental activities.

The details of fertilization of the eggs vary widely, depending on the circumstances of spawning. Most teleost eggs have a tough outer coating, the chorion, which has only one opening, the micropyle, for the entrance of sperm. Usually the micropyle closes after entry of one sperm to prevent multiple fertilization.

After fertilization, the chorion swells by imbibition of water (water hardening) and provides the egg with a protective coating which is permeable to dissolved substances. In the absence of fertilization, eggs and sperm survive only a few seconds to a few minutes in the external environment. Aquaculturists have extended this survival period by keeping eggs and sperm cold (even successfully freezing sperm) and by putting them into balanced salt solutions resembling body fluids.

A major process occurring in the first few hours after being shed into water is an intake of water and "hardening" of the chorion. The molecular basis for water intake and swelling of the egg remains speculative, but the structural changes involved are well described by Hurley and Fisher (1966). Fig. XI-2 shows a nearly mature trout ovum in the ovary. Microvilli connect the follicle cells directly to the ovum and are thought to transport nutrients directly into the cytoplasm of the ovum through the vitelline (cell) membrane. The microvilli pass through a dense layer, the zona radiata (chorion), which becomes the outermost layer of the egg after it is freed from the ovary. The microvilli remain in the hardened egg as pore canals. During swelling the chorion (also called radiate membrane now) thickens and enlarges, producing a water-filled (perivitelline) space between the chorion and the cell (vitelline) membrane. The hardened chorion thus protects against mechanical damage and invasion by microorganisms but remains totally permeable to water, oxygen and dissolved materials.

A number of environmental factors influence the developmental processes going on inside the fertilized egg. Temperature plays a major role, especially in eggs with longer incubation periods such as those of salmonids. Operators of salmon hatcheries have long used the approximation that the product of the temperature and the number of days of incubation produces a constant value (TD = k). More recent work shows that this relationship remains linear over only narrow temperature ranges and becomes curvilinear over larger temperature ranges with stronger inflections of the curves near the lethal temperatures (Blaxter, 1969). Attempts to establish meaningful Q_{10} values as predictors of egg developement have been frustrated because of the lack of

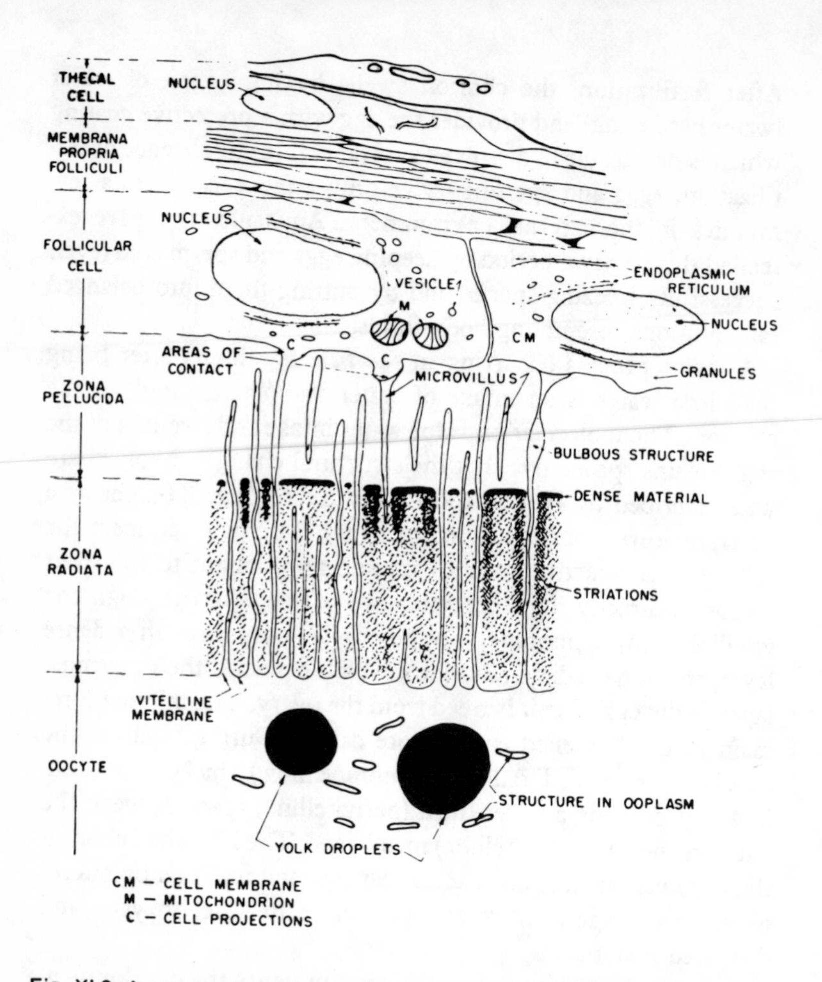

Fig. XI-2. A composite diagram of a portion of the peripheral region of a developing trout egg with its enveloping membranes, as seen with the electron microscope at approximately 5000X. (From Hurley and Fisher, 1966, with permission of the author and the National Research Council of Canada.)

linearity between temperature and development rates. The temperature range for successful development and hatching varies over a wide range depending on species. For example, smelt eggs develop optimally at 0.5 °C and tolerate no more than 6 °C, while *Fundulus* eggs develop over a range of from 12 °C to 27 °C with less than 2% abnormalities. Low salinities may retard or accelerate growth, depending on species. Low oxygen levels

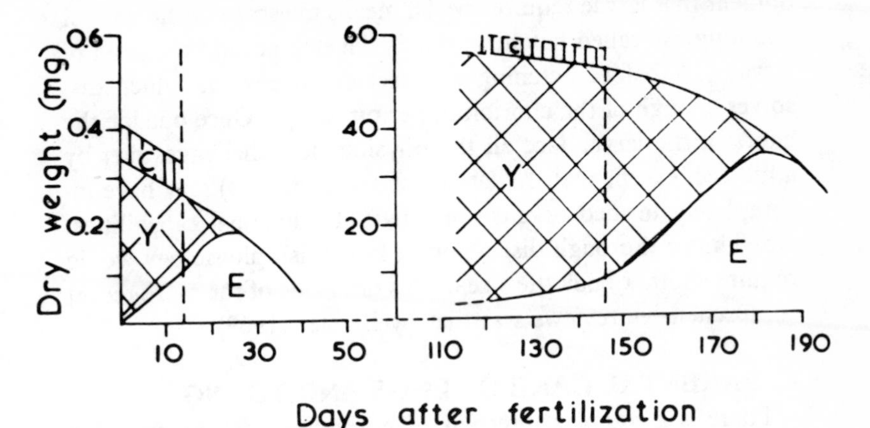

Fig. XI-3. The relative proportions of yolk (Y), embryo (E), and chorion (C) during the development of a small egg (herring) and a large egg (salmon). The vertical dashed line represents hatching. Note the difference in the time scale between herring and salmon. (From Blaxter, 1969, with permission of author and publisher.)

typically retard development. Bright light and ultraviolet light may accelerate development, but also produce increased abnormalities and mortality rates in salmonid eggs.

The amount and type of yolk material influence many aspects of egg and larval development. The type of yolk material often determines whether eggs float or sink. Eggs containing oil droplets (herring, flatfish) float; eggs containing solid yolk material (salmonids) sink. The general pattern of yolk utilization for either type of egg is similar, except that oil-filled eggs often have relatively short developmental periods (Fig. XI-3). Particularly in the case of oil-filled eggs, determining how much oil remains in the larvae can be difficult since the original single droplet can be broken up into smaller droplets and remain hidden in various body cavities. Further, some larval fish begin feeding before completely consuming the yolk oil and perhaps produce new oil which could be mistaken for yolk material. In the case of solid yolk material, the granules are surprisingly concentrated, as little as 1/3 of their mass being water compared to a 75-85% water content for more typical tissues.

The chorion is such a tough protective covering during development that larvae require special means to escape at the time of hatching. So-called hatching glands develop around the head of larvae just before hatching and secrete an enzyme which dissolves enough of the chorion to permit escape. Once outside the chorion, the larvae face all the physiological challenges met by adult fish except that the larvae (or fry or alevins) may have incompletely formed organs, only partial functional capabilities, etc. Thus a thorough discussion of larval fish physiology would require another book the size of this one. One of the more recent comprehensive reviews was that by Blaxter (1969).

E. PARENTAL CARE OF EGGS AND YOUNG

There is a complex interplay of hormones, each producing a mixture of physiological and behavioral responses, which controls parental behavior in fish. Parental behavior ranges from none to nest-building to guarding the developing eggs to various rearing chores after the eggs hatch. Thus parental care is a very large subject of which only one of the better known examples—sticklebacks—will be described here as representing an intermediate level of complexity (Fig. XI-4).

Reproduction in sticklebacks begins with a migration from seawater into freshwater, so the migration stress and osmoregulatory changes enlarge the scope of the hormonal changes involved. Unlike some fish which require several environmental cues, stickleback reproduction seems to be triggered by long photoperiods. Subsequent increases in prolactin and thyroxin stimulate preference for freshwater and migratory restlessness, respectively. Pituitary and then gonadal sex hormones bring on successive stages of territoriality, nest-building, courtship and spawning as shown in the diagram. Any stage of this behavior can be stopped and started experimentally by chemically blocking hormone action, removing endocrine glands or giving hormone injections. The parental behavior—fanning of the eggs in sticklebacks—presumably indicates the continued production of hormones by the gonad after ovulation and spawning. One could speculate that the evolutionary origin of parental behavior arose

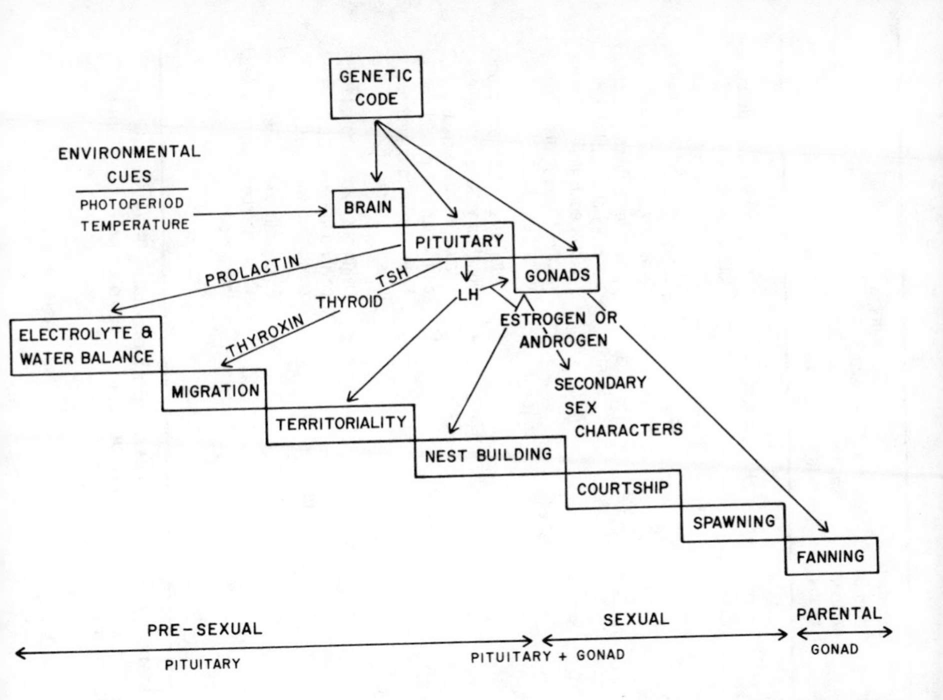

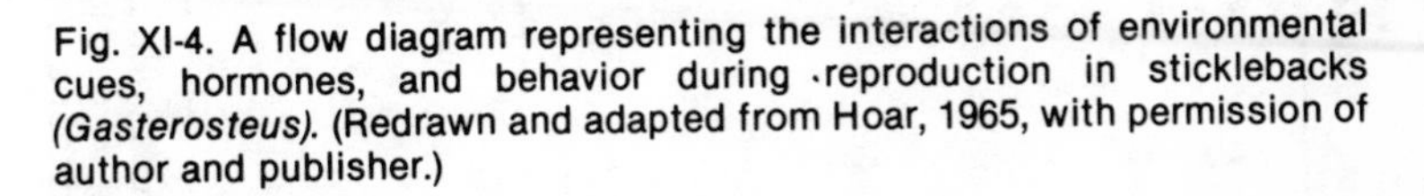
Fig. XI-4. A flow diagram representing the interactions of environmental cues, hormones, and behavior during reproduction in sticklebacks *(Gasterosteus)*. (Redrawn and adapted from Hoar, 1965, with permission of author and publisher.)

from the failure of the reproductive system to turn off production of hormones immediately following spawning. Then, increasingly extended care of eggs eventually continued long enough for at least some of the eggs to hatch, leading to parental care of fry. While the actual origins of parental care in fish remain unknown, physiologists know that some modern species have elaborate control systems which produce periods of nonfeeding in the adults both before and after spawning, presumably so they won't eat their offspring.

F. PHYSIOLOGY OF LIVEBEARING

There are several families of fishes which show some degree of livebearing; the Poeciliidae (guppies, etc.) and Anablepidae (foureye fish) are ovoviviparous, while the Jenynsidae, Goodeidae, Brotulidae (some) and Embiotocidae (surfperch) are

State of gestation	Area of ovigerous folds	Functional area of young	Thickness of ovarian epithelium	Capillary density of ovary surface	Volume of fluid in ovarian tissue	Volume of ovarian fluid	Secretory activity of ovarian tissue	Notes
1) Virgin ovary	+		+	+	+	+	+	
2) Inseminated ovary	+	+	+	+	++	+	++	Embryonic yolk sac develops following fertilization and is the first exchange surface elaborated. Regresses early.
3) Early gestation	++ to +++	+ to max.	++	++	+++	+	+++ to max.	Ovarian walls, followed by ovigerous folds, become more secretory and capillarized. Fluid in epithelium and connective tissue forms ''reservoirs'' or ''lakes.'' Embryonic gut becomes functional as yolk sac atrophies. Spatulate extensions formed on median and caudal fins. Scales begin to appear and gills to differentiate.
4) Late gestation	+++ to max.	+++	+++ to max.	+++ to max.	+++ to +	++ to max.	+++ to +	Ovigerous folds penetrate gill cavity and come into close association with gills. Embryonic spatulate extensions absorbed towards end of this period.
5) Post parturition	+		+	+	+	+	+	

Table XI-2. Summary of ovarian and embryonic changes during gestation in live-bearing perch (Embiotocidae). Five periods are distinguished: 1) No development; + = small; + + = developed; + + + = well developed; max. = maximally developed. (From Webb and Brett, 1972a, with permission of author and publisher.)

viviparous. Of the viviparous group, only the embiotocids are common and have been well studied physiologically (Webb and Brett, 1972 a, b).

Viviparity probably arose from ovoviviparity with increasing dependence of the young on the parent for all their needs. Webb and Brett (1972a) suggested that two or sometimes three critical problems must be solved to accomplish this. The first problem occurs when the yolk material is exhausted–the developing young must have some means to obtain nutrients from the mother. Second, the ability of the unspecialized ovary to supply oxygen to eggs must be greatly expanded to meet the respiratory needs of the advanced developmental stages. Finally, if solving the previous problem included the expansion of the respiratory surfaces of the young, as in the viviparous perch to be discussed below, then some rapid changes in these surfaces must take place at birth before the associated osmoregulatory problems become overwhelming. Some further details of this are shown in Table XI-2.

The increased respiratory needs of the developing young are emet in large part by the enlargement of both the ovarian and embryonic respiratory surface areas. As the young pile perch grow at a rate of nearly 5% of body weight per day, the ovary obviously must expand considerably to contain them. The volume of ovarian fluid surrounding the young increases from about 15 ml average to over 60 ml just before birth. This fluid is most likely produced and exchanged by the greatly increased vascularity of the ovary rather than being seawater taken in through the birth canal. In addition, the inner surface of the ovary enlarges greatly, producing vascularized folds, some of which enter the opercular cavity during late gestation and provide oxygen to the developing gills of the young.

The respiratory suraces of the young expand to even a greater extent during the developmental period than the ovary. At first the body surface has many capillaries but then it declines in importance as scales develop and oxygen diffusion rates through the skin decrease accordingly. Then the fins expand greatly, with spoon-shaped lobes extending beyond the fin rays (Fig. XI-5). The whole fin is highly vascular until birth, when most blood

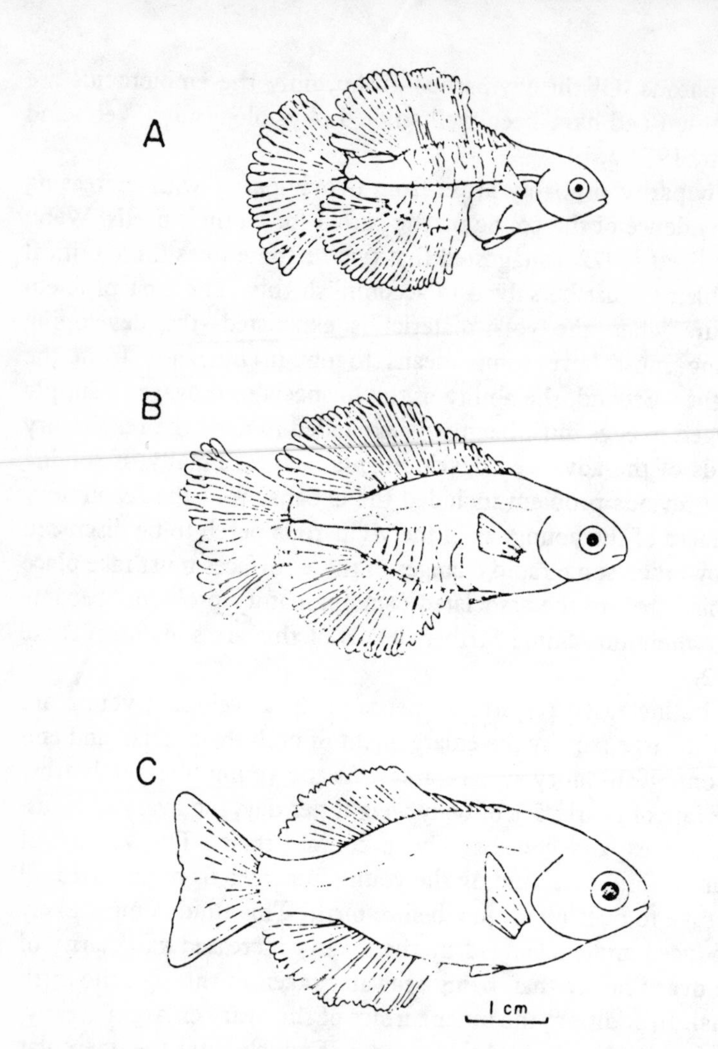

Fig. XI-5. Drawings taken from photographs illustrating structural changes of the young during gestation. A. Striped seaperch *(Embiotoca lateralis,* 0.64g) in the middle gestation period, showing the extensive venous system particularly of the fins and spatulate extensions. The arterial system and fin rays are omitted. The gut, shown stippled, still protrudes at this stage. B. Late gestation pile perch *(Rhacochilus vacca,* 1.43g). The dotted line crossing the body indicates the limit of scale differentiation rendering the anterior portion of the body opaque. The veins in the fins are still prominent and the spatulate extensions are still present. The fin arteries and rays are omitted. C. Pile perch *(R. vacca,* 4.41g) young just after birth. The body is completely covered by scales, and vessels in the fins greatly reduced. The fin rays can be more clearly seen as shown in the drawing. The spatulate extensions of the median and caudal fins have largely been absorbed.

circulation to the fins and the excess fin tissue rapidly decreases to prevent osmoregulatory problems. The gills develop last but of course do not carry the entire respiratory load until after birth. These changes in surface areas are summarized in Fig. XI-6. The sum total of these areas appears in Fig. XI-7 and indicates a general decrease on a weight-specific basis of the effective exchange area of the developing young. This situation leads to the hypothesis that birth may occur when the available respiratory surfaces can no longer meet the steadily increasing oxygen demand produced by the continued growth of the developing young (see also Fig. XI-10 and the associated text).

The oxygen consumption rates for developing young and for pregnant adults are shown in Fig. XI-8. While the oxygen consumption rate of the brood increases steadily throughout gestation with only a slight leveling-off just before birth, the brood affects the oxygen consumption of the adult noticeably only during the last two-thirds of gestation.

Fig. XI-6. Interrelations between the various surface areas of young striped seaperch and their average wet weight. Development of scales begins at a weight of 1.0 g and is complete at 2.1 g, eliminating body surface as an exchange site. At birth, the blood supply to the fins largely stops; therefore the only exchange surface remaining is the gills. The total effective area is the sum of the gill area, fin area, and effective body area. (From Webb and Brett, 1972a, with permission of author and publisher.)

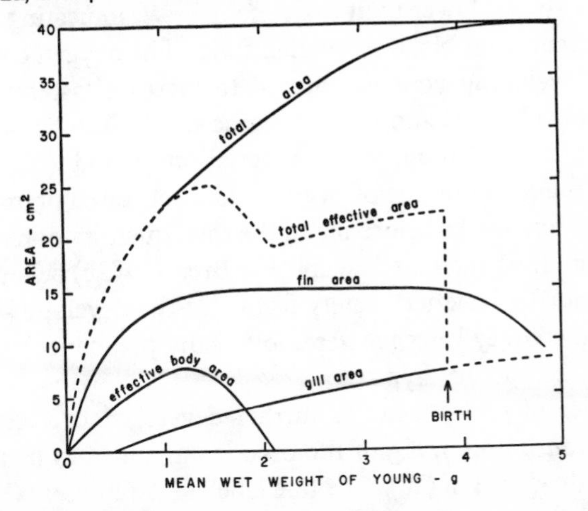

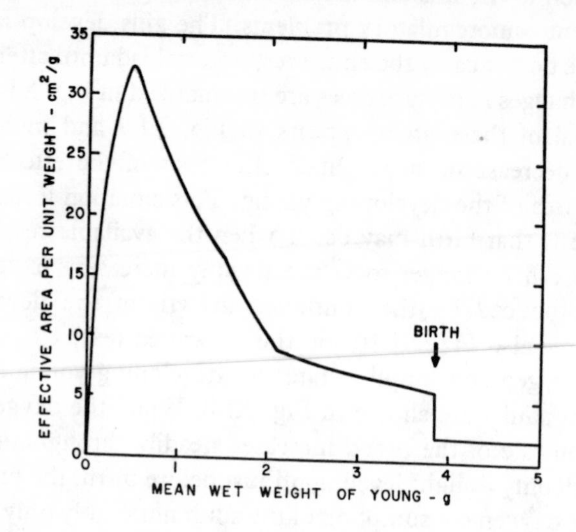

Fig. XI-7. Relation between effective area per unit weight of the young striped seaperch. At birth, decreased blood supply to the fins produces rapid loss of exchange surface, leaving only the gills after that time. (From Webb and Brett, 1972a with permission of author and publisher.)

It appeared that the oxygen supply to the developing young was limited by the ovarian blood flow. The $P\text{-}O_2$ of the ovarian fluid was always lower than the venous $P\text{-}O_2$, indicating a diffusion gradient from blood to ovarian fluid. The oxygen consumption of developing young appeared to increase faster than the ovarian blood could supply it, because the $P\text{-}O_2$ of the ovarian fluid decreased throughout the gestation period (Fig. XI-9). Because the total volume of ovarian fluid increased dramatically in the last few weeks before birth, the total quantity of oxygen in the ovarian fluid increased. Webb and Brett (1972b) believed that this provided a beneficial safety factor for the developing young as an "emergency" oxygen reservoir if the pregnant female experienced temporary hypoxia.

The idea of safety factor is further developed in Fig. XI-10. Oxygen consumption (Q) of the brood is assumed to be proportional to their weight (W). The line labelled A represents the surface area of the brood available for uptake of oxygen and shows a peak early in the gestation period. The values for L come from the diffusion distance between ovarian blood and ovarian fluid. Thus 1/L and A/L represent transfer factors for the ovary and for the ovary and brood system as a whole, respectively. The shaded area is an estimate of the excess capability of the ovary to meet the oxygen demands of the brood if the demand of the brood increases or if the external availability of oxygen decreases—i.e., the safety factor. At least in the terms of this graph, birth occurs when the ability of the ovary to deliver oxygen cannot meet the needs of the brood. This idea seems strengthened by observa-

Fig. XI-8. Rate of oxygen consumption of unborn pile perch *Rhacochilus vacca)* and of pregnant adult striped seaperch *(Embiotoca lateralis)* at ambient temperatures increasing from 10° to 18°C during gestation. Solid line shows oxygen consumption of young as a function of their weight. Solid circles indicate values for pile perch; open circle for striped seaperch. Open squares and broken line show oxygen consumption of pregnant adults as a function of time. The ringed point is for minimum oxygen consumption of a female fish without young added to the expected rate of oxygen consumption of a typical brood. This substantially agrees with the observed data from pregnant female fish. (From Webb and Brett, 1972b, with permission of author and publisher.)

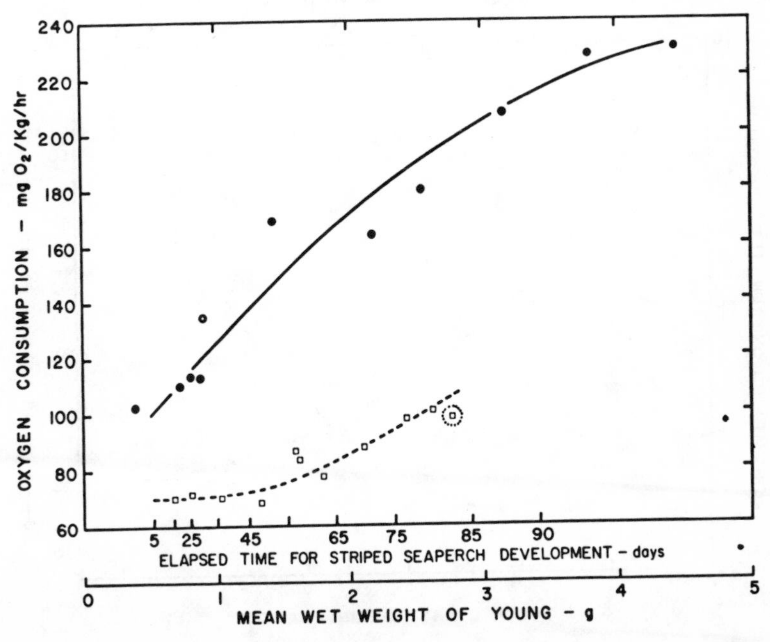

tions that the young may weigh as little as two grams (are about half-grown) at birth and still survive. This suggests that some pregnant females may actually experience emergency situations in which oxygen consumption exceeds the safety factor as it diminishes later in the gestation period and that premature birth follows.

After birth, exposure to seawater of the enlarged fins and the associated blood supply would be expected to cause dehydration and an influx of ions far greater than the fish's osmoregulatory system could cope with. Thus the newly-born fish shunts its blood away from its fins almost immediately after birth; the fins

Fig. XI-9. Relation between the intra-ovarian oxygen environment of pile perch *(Rhacochilus vacca)* and the increasing biomass and oxygen demand in the ovary (brood weight: ovary weight) during gestation. Oxygen tension in the ovarian fluid is shown by solid circles and solid line. Oxygen content of the ovarian fluid is shown by the dotted line and increases during gestation as the volume of the ovarian fluid increases, even though the oxygen tension decreases. (From Webb and Brett, 1972b, with permission of author and publisher.)

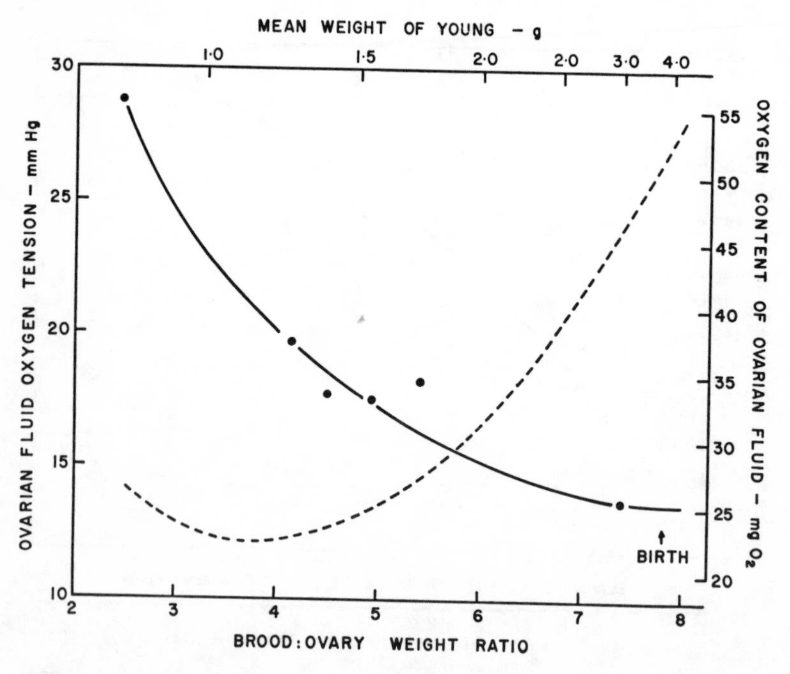

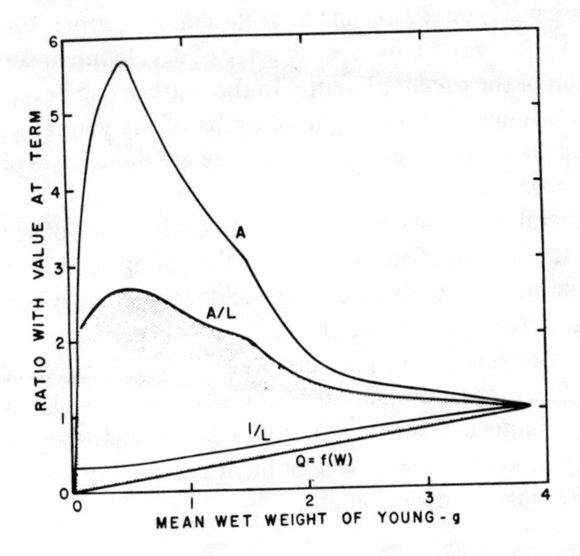

Fig. XI-10. Interrelations between structural features of the brood/ovary exchange system and the demand of the young, according to their average wet weight. The demands of the young, Q, are a function of their weight, f (W). A is the effective surface area of the young per unit weight and L is the diffusion distance between ovarian fluid and blood in the young. All values are shown as proportions of their values at birth, which was assigned a value of 1. The stippled area represents the calculated safety factor for the young in relation to the minimum that would be required to meet the requirements of the young. Successful premature births can occur at about 2 g or larger when poor environmental conditions prevent the adult female from meeting even the minimal requirements for the young. (From Webb and Brett, 1972a, with permission of author and publisher.)

decrease in size to approximately adult proportions within two to three days, and gills become the only respiratory surface. The free-swimming young are thus soon functioning independently with their own respiratory and osmoregulatory controls.

A somewhat different situation occurs in the Anablepidae (foureye fish) and Poeciliidae (mollies and guppies) because gestation is intrafollicular (inside the egg sac) rather than intraovarian. Early stages are much like other livebearing fish, but after yolk absorption the tissues surrounding the yolk continue to expand, coming into close contact with the follicular walls. This is called a follicular pseudoplacenta and serves as the major exchange surface between the two blood systems. In both families this system regresses later in the gestation. In guppies the young are born at a comparatively early stage of development soon after reduction of the pseudoplacenta. In the foureye fish development continues longer, with the gut and gills of the young being the major exchange surfaces. In neither case are there fin expansions to be reabsorbed at birth.

In general it appears that the selective advantage of viviparity rests in the improved survival rate of the young produced by the larger size and more advanced stage of development at birth. The limiting factor seems to be the capacity of the ovary to deliver sufficient oxygen, and perhaps nutrients, to the developing young. The highest development of viviparity in fish is thus seen in the Embiotocidae and the Anablepidae because they have the longest period of prenatal development and are born at the most adult-like stage of development.

Chapter XII: APPLIED PHYSIOLOGY

A. OVERVIEW

In the preceding chapters, each organ system was described as a largely separate entity. However, no fish in the real world operates only one organ system at a time, but all of them together, all of the time. One author studying the effects of heat shock came to the conclusion that the cause of death in this case was not from failure of any particular organ or organ system, but from failure of the integration of the systems. Perhaps this was a way of saying that the central nervous system failed or perhaps only said that we really didn't know what failed. In any event, this chapter will explore some of the interactions among organ systems and will describe some of the ways in which whole fish respond to a variety of externally-imposed situations.

Most organ systems operate under compromise rules dictated by other organ systems, by the external environment and by inescapable physical laws. For example, adjustment of respiratory functions for maximal ease of gas exchange would place a severe or impossible load on the osmoregulatory organs. Thus the two systems each have conflicting requirements which dictate that their own functioning is always a compromise with that of the other system. This is true to varying degrees with all of the organ systems, and the ultimate compromise among all of the systems is probably the one which uses the least energy under ordinary circumstances. This optimum compromise is sufficiently recognizable by the fish so that most adjustments to organ system functions following any kind of disturbance to the optimum balance serve to return the fish to optimum. This tendency to seek an optimum balance of functions and to maintain it is called homeostasis.

Life is rarely so ideal and unchanging that any one set of compromises remains optimum for long periods of time. There are temperature changes, salinity changes and dietary changes, or it is time to migrate, to spawn or to stop feeding for the winter. There may be disease organisms to resist, predators (including man) to escape, pollutants to endure and environmental modifications such as dams to survive. Responses to these kinds of changes may be temporary if the changes are only temporary, may be tolerated for long periods of time if sublethal or may lead to an adjustment of functions and adaptation of a new optimum balance of compromises. There are limits to any fish's abilities to change and to the length of time which some changes can be tolerated, of course. The borderline between sublethal and lethal levels of any change is typically fuzzy and variable from individual to individual, as well as being different for different species of fish.

B. GENERALIZED STRESS RESPONSE

Any factor which causes fish to depart from a state of optimum balance of organ function is a stressor, and the fish's response to this stressor is a stress response. The idea that this response has a fairly consistent pattern originated in mammalian research with Selye in 1950. The idea is also applicable to fish, although one would obviously expect differences between mammals and fish because of one being a warm-blooded air-breather and the other a cold-blooded aquatic animal. Study of stress responses in fish began in the 1960's with Snieszko, Wedemeyer and others, but understanding of the responses in different circumstances and species of fish is still rather incomplete compared to mammals. Physiological data on stress responses is sufficiently fragmentary yet that one cannot say whether there are several different stress responses or variations of a single general stress response to different stressors. At present there is only a group of responses which seem related to the mammalian stress model, while the role and significance of other physiological changes is unclear. Let's look at the clearly identifiable responses first.

The concept of a generalized stress response which was developed with data from experiments on mammals seems to have applicability to fish (Wedemeyer, 1970). The three stages of mammalian stress responses are: 1) resistance, in which the animal attempts to regain its original level of homeostasis; 2) adaptation, in which a new level of homeostasis develops under the influence of a chronic stressor; and 3) exhaustion, when the animal exhausts its reserves of energy, hormone precursors or other factors which were needed to maintain the new homeostasis. Fish under stress, even at a fairly early stage of responding, may show decreased resistance to disease. Thus when fish disease specialists are investigating a disease outbreak at a hatchery, one of the things they ask about is whether anything stressful happened in the previous two weeks or so. The exhaustion stage of the stress response leaves a fish in poor physiological condition—for instance, spawned-out Pacific salmon die in a state of what might be termed total physiological exhaustion. In general, when chronic stress exceeds the sustainable response level of the fish, debilitation follows, as often evidenced by outbreak of disease.

There are two major components in a stress response which operate in two different, but overlapping time frames. The catecholamines, epinephrine and norepinephrine, mediate changes beginning in less than a second and lasting minutes to hours, presumably under control of the sympathetic nervous system. Cortisol mediates changes beginning in less than an hour and lasting weeks or even months. Catecholamine production by the sympathetic nervous system is probably the most rapid and transitory, while the chromaffin cells in the head kidney may produce catecholamines on a long term basis. At least in mammals, the adrenal medulla produces them as long as the stress continues. Cortisol is produced in fish by the interrenal bodies in the head kidney under the influence of ACTH from the pituitary gland. Cortisol production has been seen to begin in as little as fifteen minutes, but maximal production rates require as much as twenty-four hours to achieve.

The effects of catecholamines can be characterized collectively as preparation for emergency action. They lead to enhanced alertness (restlessness, responsiveness), mobilization of energy reserves, increased respiratory capacity and other related functions. Specific effects of increased catecholamines are listed in Table XII-1. There are also effects of catecholamines known

Effect	Comment
1. Increased blood glucose	1. Normal levels of 50-150 mg/100 ml increased to 200-300 mg/100 ml.
2. Increased heart rate	2. Injection of epinephrine can nearly double the heart rate in salmonids.
3. Increased cardiac output	3. Can occur in salmonids mostly through increased stroke volume and little increase in heart rate.
4. Increased respiratory ventilation	4. Usually accompanied by decreased proportion of the available oxygen being removed from the water.
5. Vasodilation and vasoconstriction of arterioles	5. Contraction of some arterioles and relaxation of others causes most of the increased cardiac output to go to the head, heart, and skeletal muscles and not to the viscera in mammals; probably similiar in fish.
6. Increased numbers of thrombocytes in blood	6. Probably results from contraction of the spleen.
7. Decreased peristalsis or total stasis in the gut	7. Not directly observed in stressed fish, but is seen in mammals. Growing number of observations suggest that it happens in fish.
8. Conversion of glycogen to glucose in liver and muscle; release of stored lipids	8. This is the source of energy for the increased levels of activities listed above.

A number of additional, significant effects are known in mammals, but have not been investigated in fish, including:

Increases in:
Gluconeogenesis in liver
Production of steroid hormones
Epithelium permeability to water and ions
Calcium reabsorption in bone
HCl secretion in the stomach
Insulin release by the pancreas
Thyroxin release

Decreases in:
Protein synthesis inliver
Glycogen synthesis
Smooth muscle tension

Table XII-1. Effects of increased catecholamines in fish.

from mammals which have not been seen in fish. One of those effects is the shunting of blood flow away from the visceral organs and increasing the blood flow into muscles. This effect was looked for once in rainbow trout and not found, which is not to say yet that blood shunting does not occur in fish.

The quantities of catecholamines accompanying these effects are rarely known because chemical assay of epinephrine and norepinephrine is difficult and requires large samples of plasma because of their low concentration. In adult salmon which had already returned to the hatchery, resting levels of catecholamines were about 1 ug/ml and about 20 µg/ml after 25 minutes of stress (fish held out of water). In resting rainbow trout, total concentrations of epinephrine rose from about .007 µg/ml to a maximum of about 0.160 µg/ml an hour after being disturbed for 10 minutes. The only agreement between these two sets of data is that epinephrine increased during stress. The numbers are so different that they are difficult to correlate with mammalian data or to generalize upon. Epinephrine and norepinephrine also occurred in varying proportions in fish. In mammals, the catecholamines are about 70% norepinephrine, and this also appears to be the case in carp. In coho salmon, the adrenalin-like substances are about 70% epinephrine, and in the few other cases studied there were some 50-50 mixtures. No pattern has been established.

The effects of cortisol are widespread among many organ systems, and the biological significance of the effects is only partially understood. In contrast to the effects of catecholamines where there appears to be considerable survival benefit, the effects of cortisol, especially during long-term stress, frequently seem deleterious. Primary effects seem to be at sites of ion regulation—gill membranes, kidney tubules, urinary bladder and gut epithelium (in fish in seawater). At these sites, cortisol seems to be antagonistic to prolactin—that is, permeability of the membranes to ions is increased by cortisol. Additional specific effects of cortisol are listed in Table XII-2.

As can be seen, many of the effects are deleterious at their extreme or if continued for extended periods of time. A classic case of a fatal stress response was seen in *Lophius* by Homer Smith

Target Organ(s) or Functions	Effects
General	
Protein Metabolism	Increased protein synthesis via enhanced activity of messenger RNA; the proteins synthesized seem to be mostly proteolytic.
	Inhibition of growth.
	Mobilization of protein from thymus, spleen, and liver.
Carbohydrate Metabolism	Reduced utilization of carbohydrate.
	Increased glucose production from tissue protein.
	Deposition of glycogen in the liver.
Circulatory System	Mobilization of leucocytes.
	Reduced inflammatory response.
In Freshwater	
Gills	Increased retention of ions.
Kidney	Increased reabsorption of NA+ by kidney tabules.
Testes	Precocious sexual maturity in downstream migrant smolts (males only).
In Seawater	
Gills, Intestine	Increased production and activity of NA-K-ATPase, giving enhanced salt absorption in gut and enhanced salt excretion by the gills.

Table XII-2. Effects of increased corticosteroids (especially cortisol) in fish and similar vertebrates.

and his students who studied kidney functions in this fish during the 1930's and 1940's. Although they described the effects as "laboratory diuresis" rather than as a stress response and they did not measure cortisol levels, the symptoms fit what we now know about cortisol. Upon capture in seawater and confinement in laboratory aquaria, the normally low level of urine production increased dramatically and the water content of the urine increased. Especially if injured even slightly (usually scale loss), the fish would die of dehydration in as little as 24 hours. In my opinion, this was a stress (cortisol) response involving large increases in the permeability and reduced resorptive abilities of the kidney tubules, with similar changes also occurring in the gills. Since the changes resulted in the death of the fish (and great frustration for researchers for over 20 years), it hardly seems to have been an adaptive response. Although early stages of the response may have been useful, the later stages certainly are not.

At one time there was a belief that there might be a single, generalized response by fish to any kind of stress. As more information becomes available for comparison of different species and different stressors, the idea can be seen as too simple. While there are similarities in responses to different stressors—fish have a limited number of ways to respond to stress—the responses are by no means identical. A comparison among several kinds of stress follows this point even though there are no data on comparable physiological functions in all cases (see summary in Table XII-3). The available data suggest four degrees of response: (1) a full stress response involving increased production of catecholamines and cortisol; (2) a partial stress response emphasizing either catecholamines or (3) cortisol, but not both; and (4) responses to disruptions of homeostasis which the fish does not recognize as stressful situations and which involve neither catecholamines nor cortisol. Examples of each of these will be discussed below and include, respective to the order above: (1) salmonid migration (Section C); (2) short-term hypoxia (Section D); (3) gas bubble disease (Section G); and (4) IHN virus disease (Section H). In general, however, the following examples of stress responses are presented because there is physiological information available about them and not because they particularly fit the stress response categories.

Since most data on stress in fish do not include measurements of catecholamines, I will use decreases in blood chloride as being indicative of increased cortisol and increases in blood glucose as indicative of increased catecholamines. These are, at best, only general approximations because both indicators can be affected by factors other than the two hormones.

C. MIGRATION AS STRESS IN SALMON

1. Smolt (downstream) Migration

At some time after hatching,which is variable for each species of salmon, juvenile salmon become restless and change their orientation from swimming against the river current (or staying in a

Plasma or Serum Factor	Cold Shock[1]	Heat Shock[1]	Hypoxia[2]	Gas Bubble Disease[3]	IHN Disease[4]	Smolt Migration[5]	Vibriosis[6]
Na+		↓	NC	NC		↑	↓
Cl−	Severe ↓	↓		NC	NC or ↓	↑	↓
K+		↓	NC	↑		↑	↑
Mg++						↑	↑
Ca++				↓	NC or ↓	↓	↓
PO_4−−−				↑	NC or ↓		
HCO_3−	↑	↓			↓		
CO_2	↑	↓					
Osmolarity					↓		↓
Total Protein				↓	NC		↓
Albumen				↓	NC		
Cholesterol				↓			
Glucose				NC	NC		↑
pH	↓ or ↑	↓			↑		
RBC Count					↓		↓
Leucocyte Count					NC		↓
Hematocrit					↓	↑	
Hemoglobin					↓		
LDH Enzyme				NC	↑		↑
Clotting Time				↓ then ↑			
Tissue Water	NC	Slight ↓					↑
O_2 Consumption			NC			↑	
Heart Rate		↑ then ↓	↓ (stroke vol. ↑)			↑	
Urine			pH ↓				

↑ = increase, ↓ = decrease, NC = no change, blanks = no data

1 from Houston, 1973
2 from Shelton, 1970
3 from Newcomb, 1974
4 from Amend and Smith, 1974
5 from Miles and Smith, 1968
6 from Harbell, Steve, U. Wash. MS Thesis, 1976. Unpublished.

Table XII-3. Directions of change for a variety of physiological functions under the influence of different stressors. Data adapted from several sources: 1 - Houston, 1973; 2 - Shelton, 1970; 3 - Newcomb, 1974; 4 - Amend and Smith 1974; 5 - Miles and Smith, 1968; 6 Harbell, Steve. U. Wash. MS Thesis, 1976. (Unpublished). Arrows (↓) indicate decrease in rate of function or decreased concentration, (↑) indicate increase, NC indicates no change.

lake) to swimming determinedly downstream. The physiological preparations for the trip are influenced by photoperiod—accelerated by long days or delayed by short days—but not postponed indefinitely. If confined in freshwater and prevented from migrating, however, most salmonid species eventually revert to their freshwater way of life and abandon their state of preparedness to enter seawater. Natural occurrences of this reversion have resulted in land-locked varieties of sockeye salmon—kokanee—in which the entire life history occurs in freshwater. Chinook and coho salmon introduced into the Great Lakes also carry out their life cycle in freshwater but spawn in streams and migrate into the lakes for rearing. Other migratory salmonids include steelhead and cutthroat trout, dolly varden and the ayu (a small Japanese salmonid). Table XII-4 lists the typical migratory characteristics for the common North American salmonids. Other migratory groups include the anguilid eels, the smelts, shad and lampreys.

Much of the study of the salmon smolt migration took place before the development of the present concept of a stress response which we discussed above. Therefore the picture is fragmentary.

The ion content of the plasma of Atlantic salmon is shown in Table XII-5A & B, and changes of plasma ions during adaptation to seawater are shown for coho salmon in Fig. XII-1. The general direction of change seems to be in anticipation of levels found in seawater-adapted fish. This phenomenon has been called preadaptation. Associated changes include the disappearance of the parr marks (vertical black [melanophore] bars on the side of the body) and increased silveriness resulting from a steady deposition of guanine on the underside of the scales and in the skin. Basal metabolism increased under the influence of increased thyroxin levels and was presumed to be responsible for the migratory restlessness observed by many investigators. Involvement of other hormones was suggested by the occasional occurrence of sexually mature male chinook smolts among the downstream migrants. In mammals, corticosteroid hormones can have sex-stimulating side effects, so perhaps this is also true in fish. In Atlantic salmon smolts, cortisol increased four- to five-fold during migration over the premigratory level and presumably in-

Name	Freshwater Residence	Marine Residence	Spawning
Oncorhynchus nerka Sockeye (Red) Salmon	1-3 years in lakes, may become land-locked and mature in lakes (called Kokanee)	2-3 years	In streams adjacent to lakes, usually in early fall, then die.
O. kisutch Coho (Silver) Salmon	½-1½ yrs. in streams (sometimes lakes, can mature in f.w.)	most for 3 yrs.	In streams, in late autumn, then die.
O. tshawytsha Chinook (King) Salmon	A few weeks up to 1 year	3-8 years (most 4-5)	In larger streams, near mouth or far upstream, during most of the year.
O. gorbuscha Pink (Humpy) Salmon	Migrate downstream soon after emergence from the gravel	2 years	In streams, often close to seawater in early autumn, then die.
O. keta Chum (Dog) Salmon	Migrate downstream soon after emergence from the gravel	most for 4 yrs.	In streams close to the sea, sometimes inter-tidally in late autumn, then die.
Salmo gairdneri Steelhead Trout	2-3 years	2-4 years	In streams, in late winter, may spawn 2 or 3 times. Summer-run steelhead enter stream in early sum-mer, spawn in winter.
S. clarki clarki Coastal Cutthroat Trout	May migrate down-stream when quite small, live in estuaries for a year or more	—	In smaller streams in the spring.
S. salar Atlantic Salmon	1-3 years in streams may become landlock-ed (Sebago)	1-2 years	In streams in autumn.
Salvelinus malma Dolly Varden	Most spend 3 years in freshwater	2-3 years	In streams in autumn.

Table XII-4. Migration, spawning, and rearing charateristics of migratory salmonids.

A

Salmo salar	Na^+	mM/kg K^+	Ca^{++}	Cl^-	Medium
Parr	117.0±17.6	2.19±.85	2.33	129.8±21.5	f.w.
Smolts	131.2±52.0	3.03±.91	2.00±1.4	182.7±10.5	f.w.
Post-smolts	155.8±44.0	3.28±.51	-	132.8±45.3	f.w.
Smolts	159.3±25.3	3.62±1.6	-	166.3±56.7	s.w. (2 wks.)
Adults	211.9±106.2	3.15±2.1	3.43±1.4	156.7±72.2	s.w.

B

Physiological characteristic	**Level in smolt, compared with that in parr**
Thyroid section	Increases
Gill microsome, NA^+-K^+ ATPase enzyme activity	Increases
Body silvering	Increases
Salinity tolerance and preference	Increases
Hyposmotic regulatory capability	Increases
Ability to grow well in full-strength sea water (salinity 35%)	Increases
Oxygen consumption	Increases
Ammonia production	Increases
Blood glucose	Increases
Buoyancy (swim bladder, Atlantic salmon)	Increases
Migratory behavior	Increases
Weight per unit length (condition factor)	Decreases
Body total lipid content	Decreases
Liver glycogen	Decreases

Table XII-5. Physiological changes in salmonids during downstream migration (smolting). A. Plasma ions in Atlantic salmon at various stages of the life cycle. Values are mean ± SE. (Adapted from Parry, 1961). B. General description of physiological changes occurring during the parr-smolt transformation of Pacific salmon *(Oncorhynchus)* and Atlantic salmon *(Salmo salar)*. All or most of those changes must be evident to ensure acceptable health, adequate osmoregulatory function, and survival. (From Wedemeyer and Yasutake, 1977, with permission of author and publisher.)

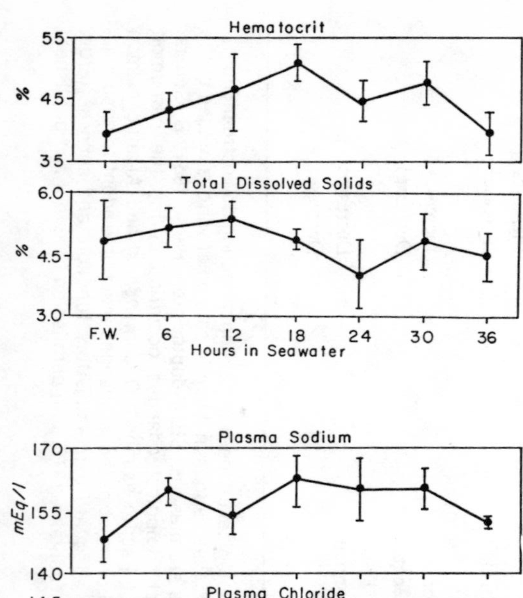

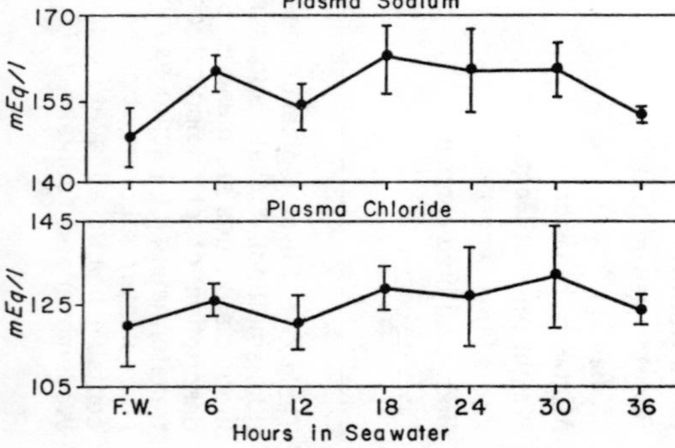

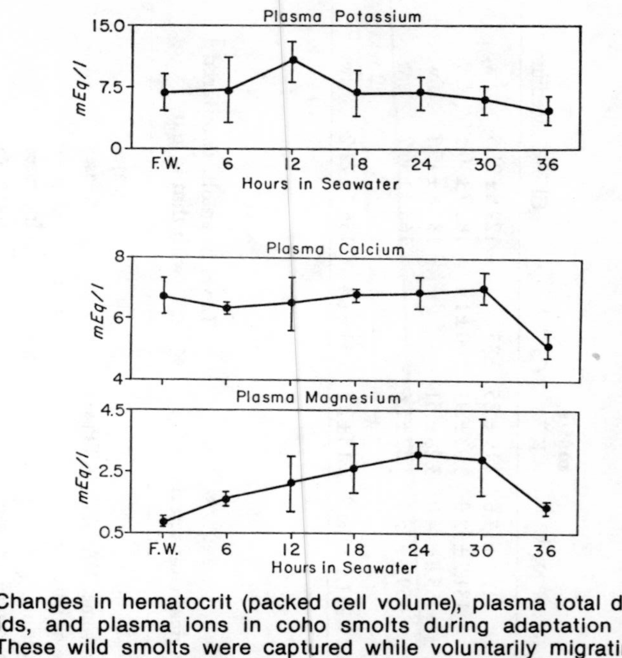

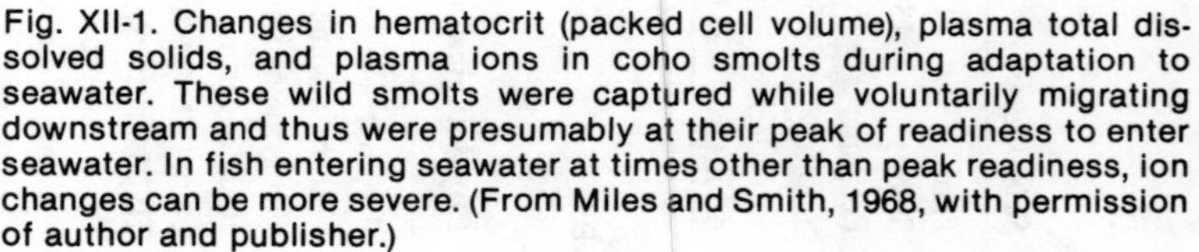

Fig. XII-1. Changes in hematocrit (packed cell volume), plasma total dissolved solids, and plasma ions in coho smolts during adaptation to seawater. These wild smolts were captured while voluntarily migrating downstream and thus were presumably at their peak of readiness to enter seawater. In fish entering seawater at times other than peak readiness, ion changes can be more severe. (From Miles and Smith, 1968, with permission of author and publisher.)

creases similarly in other migratory salmonids. Smolting changes in general are summarized in Table XII-5B.

A major difference between smolting and spawning migrations is that the spawning migration can be terminal for individuals of many species, but the downstream migration must be survived by most individuals. The downstream stress response cannot be so great as to compromise the survival of the fish, otherwise there would be no spawning migration. Because rainbow trout require three or four days of gradually increasing salinity before they are able to survive in full-strength seawater, it was thought that a transitional period was necessary for migratory salmonids. This does not appear to occur in nature, except where the estuary is long and a gradual salinity change is the product of the geography to be transversed. It is not a physiological requirement. Abrupt changes in salinity cause no mortality in fish which are ready for such changes. Further, a behavioral preference for seawater (produced by thyroxin) locks fish into the seawater mode of osmoregulation and prevents reversion to freshwater when available in the estuary. The importance of the seawater preference was seen in our laboratory, where the freshwater tolerance of chinook smolts in seawater took nearly a month to disappear after they entered seawater. They still survived a return to full freshwater after 8-10 days in seawater. They were fully capable of surviving in freshwater, had they the desire to do so. In another case in our lab, coho smolts spent nearly a month in seawater and then all survived a transfer back to freshwater. In contrast, an adult coho which had been in freshwater no more than three days was transferred back to seawater and died overnight, apparently of dehydration. Thus salmon smolts appear much more versatile and much less stressed during their migration than adults.

This is not to say that the smolt migration is not without its stresses. Entry into saltwater has been observed to rekindle a latent (carrier state) furunculosis infection which then killed most of one year's coho at one hatchery release on Hood Canal (Washington). We recovered marked chinook smolts in an industrialized estuary only three to four days after their release from the hatchery 30 miles upstream. Their stomachs were empty except for wood chips, fir needles and other exotic particles of ingestible size. Apparently they were either having problems finding suitable food or were having difficulties in learning which natural particles (after having only hatchery pellets) were edible. In chum salmon in a spawning channel, the advantage of going to sea in spite of the problems was clearly demonstrated. Fish which stayed in the spawning channel increasingly fell behind in size compared to their cohorts in the estuary and never did catch up in size even after later entry into seawater. In spite of the problems, migration thus appears preferable in the long run to staying in freshwater.

Readiness to migrate is indicated in several ways. With coho smolts, a minimum size (regardless of age or season) of 68-70 mm fork length is required for survival in seawater. Coho can smolt in the same spring as they hatch, in the autumn or the following spring, depending on how fast they grow to the minimum size. Chum salmon smolt at a much smaller size and can migrate almost as soon after emergence as they can swim. Chum salmon eggs spawned intertidally, however, such as at the mouth of a stream, show increasing mortality in proportion to the exposure to seawater. At the other extreme, sockeye salmon in northern oligotrophic lakes may grow very slowly and remain in freshwater for three years before they migrate. An increase in the activity of an ATPase enzyme which is activated by Na^+ or K^+ has also been identified in coho, chinook and steelhead as part of the pre-adaptation to seawater (Zaug and McLain, 1970). This increased enzyme activity presumably provides the energy to operate the chloride "pump" at the increased rates needed in seawater. Cessation of feeding and metabolism of lipid reserves is also characteristic of the downstream migration. Although Table XII-3 shows no information for glucose and shows increased Cl^- due to entry into seawater, I believe that both cortisol (involved in seawater osmoregulation) and catecholamines probably increased.

If migration does not occur, most pre-adaptive changes regress to freshwater status. The migratory restlessness and higher metabolic rate return to normal. The ATPase activity decreases from its intermediate (between low freshwater and high seawater) level to freshwater levels. Presumably cortisol and catecholamine levels also normalize. These migratory changes occur to some degree even in landlocked strains of salmonids.

2. Spawning Migration

The obvious importance and spectacular nature of the upstream migration of the anadromous salmonids has led to numerous studies of the physiology and behavior involved. As an athletic feat, to swim hundreds of miles at velocities averaging 2.8 body lengths per second (calculated for the Frazer River sockeye migrating about 800 miles) is impressive. To do so entirely on stored energy with no food intake and to finish the development and maturation of the eggs and sperm during migration is extraordinary. To see large (by freshwater standards) fish arrive in hordes on shallow upstream spawning grounds in full spawning regalia—males with hooked noses, both sexes often brightly colored (sockeye with bright green heads and bright red bodies, for example)—go through their spawning rituals and then die is spectacular. On the other hand, other reasons for studying these same migrations include man's impact on salmon: water pollution (including air supersaturation of water by dams); the impact of conflicting land and water usage on migration routes and spawning and rearing areas; and overfishing. Major Pacific salmon runs in the western U.S., British Columbia and Alaska have been declining for years, some since the early 1900s. Atlantic salmon and shad migrations on the Atlantic coast of the U.S. have been gone about that long. One pessimistic author noted an excellent relationship between the decline of the salmon runs in the Columbia River and the increased number of biologists studying the runs, as though firing all of the biologists would bring the runs back.

The spawning migration of Pacific salmon probably was first studied as a stress response by O.H. Robertson and co-workers in the Sacramento River chinook in the late 1950s and early 1960s because of its striking resemblance to Cushing's syndrome in man (Table XII-6). When they caught salmon offshore from San Francisco, they found low levels of cortisol which rose four- to five-fold after 24 hours of confinement in tanks aboard the

Changes	Cushing Syndrome	Spawning salmon
A *General*		
Weight	Gain or no change	Loss starvation
Muscle tissue	Wasted	Wasted
Weakness	Present	Present
Osteoporosis	Present	Absent [?]
Increased susceptibility to infection	Present	Present
Increased sensitivity to stress	Present	Present
Blood pressure	Hypertension	
B *Blood*		
17-OHCS conc.	High	High
Glucose	Hyperglycemia	Hyperglycemia
Sodium	Normal or increased	Decreased
Potassium	Normal or decreased	Decreased
Total serum protein	Decreased	Decreased
Gamma globulin	Decreased	Variable
Cholesterol cone	Increased	Decreased
Blood urea nitrogen	Normal	Normal
Protein-bound iodine	Normal or decreased	Decreased
Polycythemia	Present	Absent
C. *Histology*		
Adrenal	Hyperplasia	Hyperplasia
Pituitary	Hyaline degeneration of basophils	Degeneration of both basophils and acidophils
Kidney	Nephrosclerosis	Nephrosclerosis
Pancreas variable	Islet hypertrophy constant	Islet hypertrophy
Lymphoid tissue	Depletion	Depletion
Liver	Degeneration variable	Degeneration constant
Cardiovascular system	Arteriosclerosis	Degenerative changes beginning arteriosclerosis

Table XII-6. Comparison of biological, biochemical, anatomical changes in Cushing's Syndrome in humans with those in spawning salmon. (Adapted from Robertson, et al., 1961.)

research vessel. They found increasing baseline levels of cortisol at increasing distances upstream until baseline cortisol levels peaked on the spawning grounds at about the same level as high-seas fish confined in tanks. Fish along the migration route attained similar peak cortisol levels if held after capture. The conclusion was that salmon could produce only a maximum of a five-fold increase in cortisol levels and that the maximal level was fixed regardless of the starting level.

The stimulation of cortisol production (via ACTH from the pituitary) is only partly understood. The muscular exertion and physical battering endured during the upstream swimming past or over various obstacles is an obvious stress, although this is probably not the whole story. In the Fraser River system recently, about 100,000 early run sockeye arrived in late summer on their autumn spawning grounds in the Horsefly River, but most of them died from a warm-water disease (Columnaris) before spawning. Since there was a dam upstream from the spawning area, officials decided for the following year to discharge water from below the reservoir's thermocline and provide the spawning grounds with cool water. Arriving on the spawning grounds the following year, many sockeye then died from a bacterial gill disease before spawning. Upon measuring cortisol levels in salmon in this run, plasma cortisol was already elevated in fish caught in seawater 60 miles south from the mouth of the Fraser River. Although possibly of a genetic nature, the reason for the early stress response in these fish is unknown.

This anecdote about Fraser River sockeye illustrates a major effect of long-term elevated cortisol levels. Increased protein catabolism stimulated by cortisol eventually includes the plasma proteins which contain the antibody proteins. Decreased level of antibodies eventually produces increased susceptibility to disease. In spawning Pacific salmon, a common disease organism is *Saprolegnia,* a fungus which grows in tan, spongy masses on fish and erodes away the skin and flesh. In contrast to the loss of antibodies from the plasma proteins, however, the heart rate and blood pressure of spawned-out, dying pink salmon were nearly normal—the heart and blood vessels were among the last organs to fail.

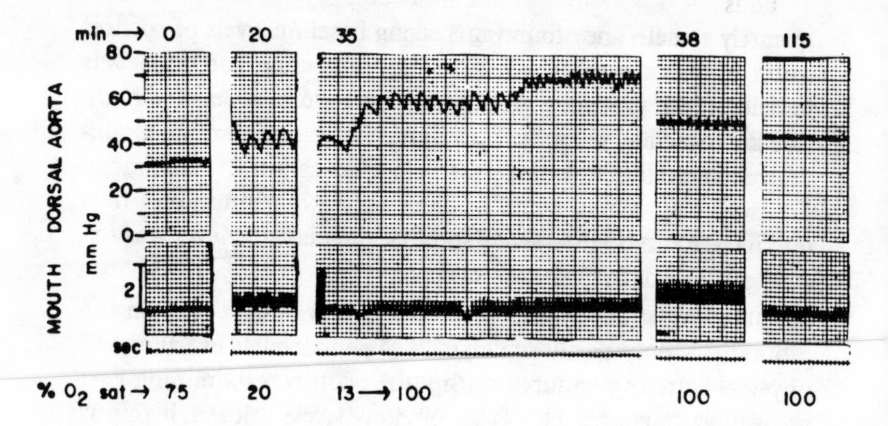

Fig. XII-2. Acute hypoxia and recovery in a 1.5 kg adult sockeye salmon. The tunnel respirometer was closed at zero time, then fresh water circulation begun again after 35 min. (From Randall and Smith, 1967, with permission of author and publisher.)

It is not obligatory that Pacific salmon die after spawning, and most other salmonids, of course, spawn several times. When sockeye salmon were spawned in artificial holding facilities and kept in disease-free freshwater afterwards, they survived for up to eleven months. Feeding began spontaneously about two months after spawning, and they reverted to their sea green and silver coloring. The hooked nose, however, did not regress, and their ability to osmoregulate in seawater was not tested. There are stories by commercial fishermen, however, of catching spawned-out chinook salmon in gill nets set in seawater near the mouth of streams in southeastern Alaska, so perhaps the readjustment to seawater is possible for adult salmon after spawning. Steelhead trout have been shown to develop fatty deposits (arteriosclerosis, an early stage of hardening of the arteries) in their coronary arteries during the spawning migration. The arteries become normal again when the fish return to seawater. Although about two-thirds of the spawning steelhead die (including fishing mortality), it appears that migration stress in steelhead is generally less severe than in Pacific salmon—less tissue depletion, little development of the hooked nose, etc. Atlantic salmon appear to respond to migration stress about the same as steelhead.

Thus the spawning migration of salmonids is highly stressful (usually lethal!) and the role of cortisol is well established. The roles of glucose and catecholamines appear not to have been studied, but probably would be complicated anyway by the severe muscular exertion and tissue deterioration due to prolonged starvation.

D. ACUTE AND CHRONIC HYPOXIA

The immediate response of salmonids to decreased environmental oxygen is to increase the ventilation rate and cardiac output. Opercular movements increase although heart rate may slow down. Blood pressure increases only slightly. The increased expenditure of energy for increased gill ventilation and perfusion apparently meets the oxygen demand until the oxygen saturation of the water (at 15°C) falls to about 25 percent. If returned to oxygen-saturated water at that point, blood pressure rises, heart rate increases and ventilation continues at a rate above normal for some period of time until the accumulated oxygen debt is paid off (Fig. XII-2). If the oxygen level continues to decline (Fig. XII-3), then a point is reached at which the increased effort and increased oxygen demand from that effort do not produce the needed oxygen. Fish often turn belly up at this point (and

Fig. XII-3. Terminal hypoxia in a 1.7 kg adult sockeye salmon. The tunnel respirometer was closed at zero time and the fish's responses recorded as it used up the oxygen in approximately 28 liters of water at 15°C. (Note: Zero pressure level in the pressure record for the mouth changed slightly.) (From Smith and Randall, 1967, with permission of author and publisher.)

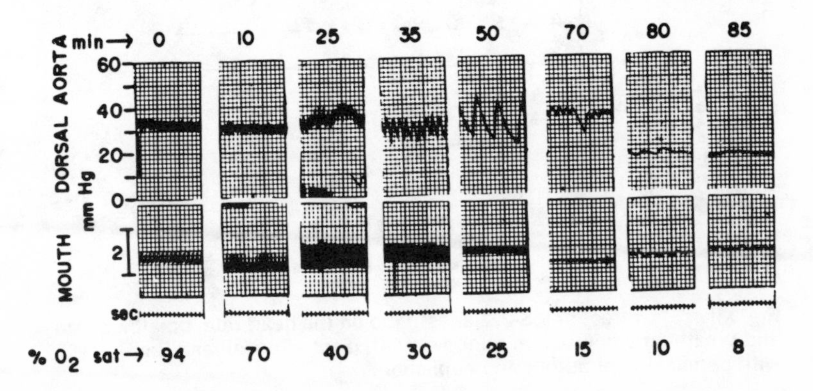

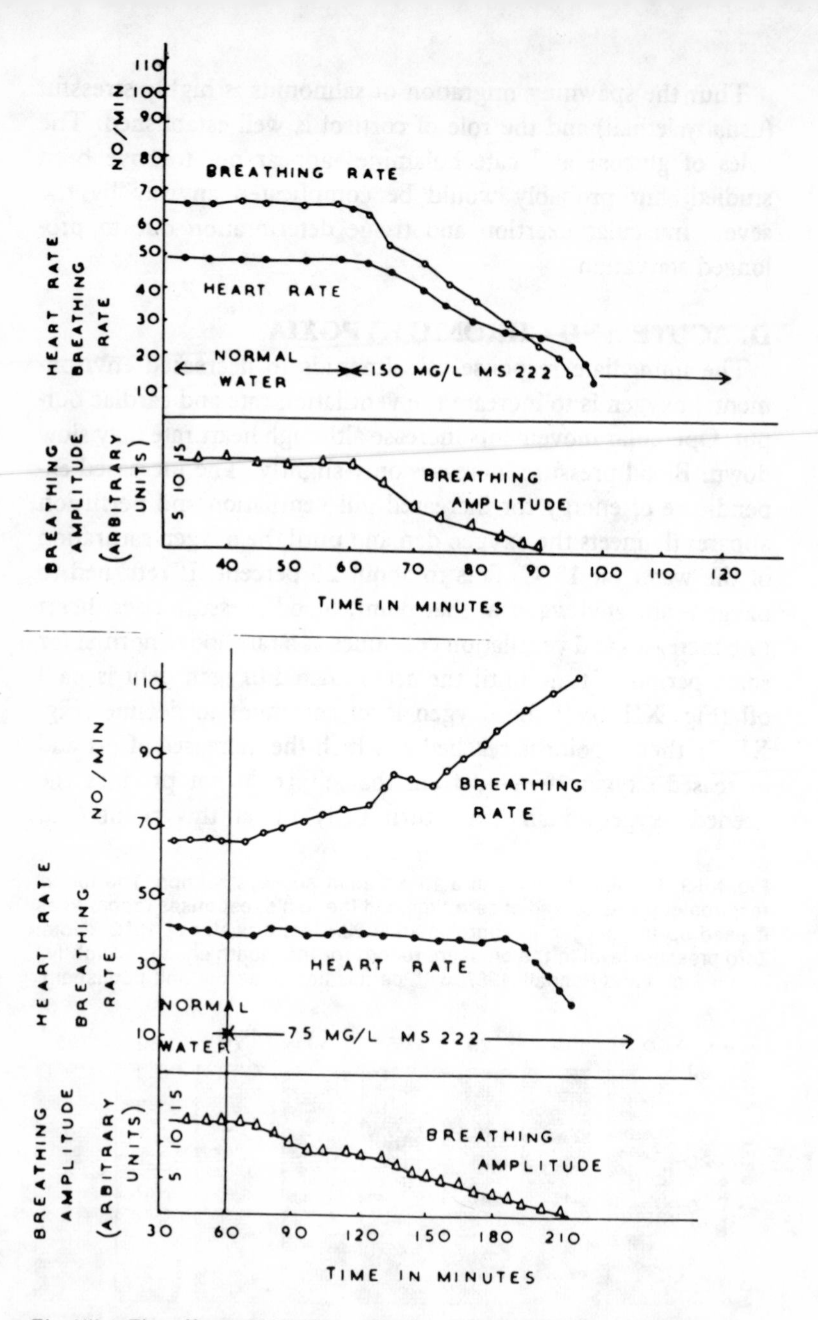

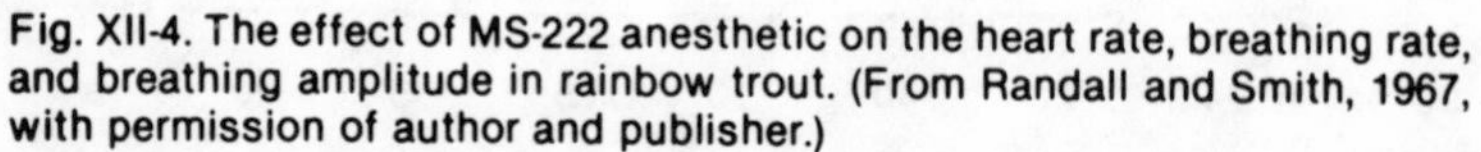
Fig. XII-4. The effect of MS-222 anesthetic on the heart rate, breathing rate, and breathing amplitude in rainbow trout. (From Randall and Smith, 1967, with permission of author and publisher.)

presumably lose consciousness), while both heart rate and breathing slow down markedly. If this situation continues without respite, the fish dies. Prolonged anesthesia produces similar hypoxic effects (Fig. XII-4).

Until 1970 when discharge of raw sewage into the Duwamish River estuary (Seattle) stopped, the oxygen-consuming action of the sewage produced a hypoxic "plug" in the upper estuary which did not flush out with tidal action. Tests simulating the low oxygen problem were run in a swimming chamber to see whether adult coho salmon could pass through this area or not to reach their upstream spawning areas in the river. Fish swam steadily at 56 cm/sec (1¼ miles/hr) for an hour at moderate levels of dissolved oxygen (DO) and then experienced a decreasing DO. Changes in their respiration are shown in Fig. IV-7. As ventilation volume increased, the effectiveness (extraction coefficient = percent of available oxygen removed from water passing over the gills) decreased. Some adjustment and recovery occurred toward the end of the second hour, even though the DO continued to decline. Most of the fish became fatigued during the second hour of the swimming test; a few did not and might have made it through the area of low DO into the upper river.

The occurrence of fatigue was correlated with a sudden increase in blood lactate (Fig. XII-5). Resting levels of lactate remained stable or increased only slowly at DO level down to about 5 mg O_2/liter, but then began to increase rapidly as the DO continued downward. In the swimming chamber, the fish had no choice but to continue swimming. In the real world, the fish might have reduced its swimming velocity or have avoided the situation and changed the physiological limits slightly. Under the given circumstances, any sizeable body of water having 4.5 mg O_2/liter or less along the migration route would constitute a real barrier to fish passage.

Here are some examples of chronic hypoxia that occur naturally. In some respects, warm-water fish have relatively little oxygen available because of its decreased solubility at higher temperatures and because the fish have a relatively high rate of basal metabolism. In eutrophic waters with dense algal populations, the DO level is high during daylight and falls to low levels

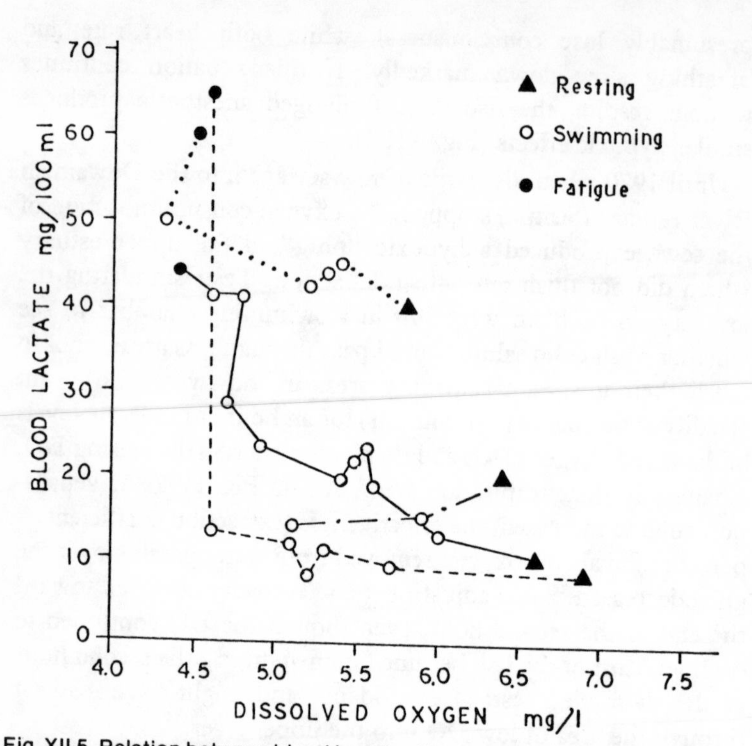

Fig. XII-5. Relation between blood lactate, dissolved oxygen and duration of swimming in four adult coho salmon (3 to 4 kg) swimming in seawater at constant velocity until they were unable to swim because of fatigue. Blood samples were taken at 15 min intervals from a catheter implanted in the dorsal aorta before, during and at the end of the period of swimming. Each line thus proceeds from a resting (preswimming) value on the right to a final fatigue value on the left. Three of the four fish changed from aerobic (low lactate) to increasingly anerobic metabolism (lactate accumulation) near 4.5 mg O_2/liter. (From Smith, et al., 1971.)

during the night as the algae respire but do not produce oxygen. This might be considered an example of chronic hypoxia to the extent that there are repeated daily pulses of low DO. Fish living at high altitude experience reduced partial pressures of oxygen. Fish swimming in dense schools have reduced DO when they pass through water already exhaled by fish ahead of them. Some pollutants, especially heavy metals such as lead and zinc, precipitate mucus on the gills and restrict the amount of oxygen which can enter the fish.

The responses to these various situations are known in

fragmentary fashion. Trout in mountain lakes up to 10,000 feet high have the same level of O_2 in their tissues (about 25 mm Hg) as lowland trout. Goldfish acclimated to 15 percent of air saturation swam as well as fish acclimated to 100 percent air saturation, suggesting that swimming was not directly linked to respiration. Largemouth bass, on the other hand, showed decreased food consumption and growth rates at DO levels below 4 mg/liter, as well as at levels higher than normal air saturation.

Adjustments to chronic hypoxia favor the continued effiency of the oxygen-transport system under the new conditions. After about a week of hypoxic conditions, there were changes in the characteristics of the hemoglobin which increased the oxygen-carrying capacity of the blood and decreased the partial pressure of oxygen required to saturate the hemoglobin to any given level—i.e., the blood saturation curve moved upward and to the left of normal. These changes allowed the hemoglobin to continue to load and unload oxygen in the most favorable part of the dissociation curve. Also, ventilation rates may increase during hypoxia and heart beat is more likely to be in phase with the opercular movements than normal, at least in salmonids.

Although not shown in Table XII-3 under acute hypoxia, epinephrine responds to short-term chasing, capture, struggling and any other circumstances which interfere with normal respiration (see also Chapter X-G, Table X-1; also Table XII-1). Blood glucose should rise, although I found no data to confirm this. Cortisol responds more slowly than the catecholamines, requiring up to 24 hours to reach maximal levels. Since acute hypoxic responses and recovery take only an hour or two, cortisol should play a very small role, if any, in hypoxic stress.

E. ACUTE HEAT AND COLD AS STRESSORS

After oxygen, temperature is the next most pervasive factor of aquatic life. Except for a few specialized fish with heat exchange systems, the internal temperature of fish must be essentially the same as that of their environement.

Some general concepts about temperature limits for long-term exposure are illustrated in Fig. XII-6 in the form of a performance polygon for sockeye salmon. The slope to the top and bot-

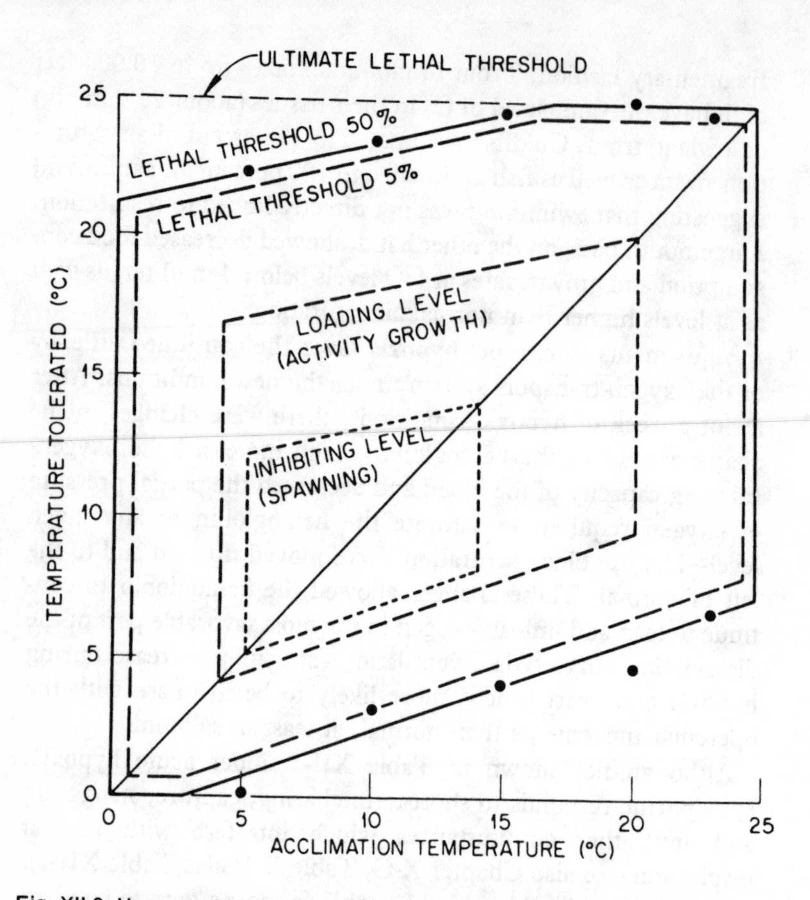

Fig. XII-6. Upper and lower lethal temperatures for young sockeye salmon plotted to show the zone of tolerance and, within this zone, more restrictive zones for activity, growth or spawning. The term threshold is equivalent to "incipient lethal temperature" in most scientific literature. (From Coutant, 1972, with permission of the author and publisher.)

tom lines shows that the higher the temperature to which the fish is accustomed (acclimated), the higher are both the upper and lower lethal limits. That the five percent lethal threshold and the 50 percent lethal threshold (incipient lethal) temperatures are so close together can be taken as an indication of considerable ability of these fish to operate effectively over a large portion of their temperature range. On the other hand, the smaller inner polygons indicate that operating at temperatures near the extremes is not without its costs—growth is reduced outside of the middle polygon, and spawning is inhibited outside of the inner polygon. This general concept is applicable to most fish, although the temperature ranges on either or both the vertical and horizontal axes would have to be adjusted appropriately for species in different climates.

Some limits on short term heat stress are shown in Fig. XII-7, which is one form of dose-response curve with adaptations to show the effects of various acclimation temperatures. The slanting line A-B relates to the rising incipient lethal temperature with increasing acclimation temperature seen in the previous figure. To the left of line A-B, the lines slope upward as in-

Fig. XII-7. Median resistance times to high temperatures among young chinook salmon acclimated to the temperature indicated. Line A-B denotes rising incipient lethal temperatures with increasing acclimation temperature. This rise eventually ceases at the ultimate incipient lethal temperature (line B-C). (From Coutant, 1972, with permission of author and publisher.)

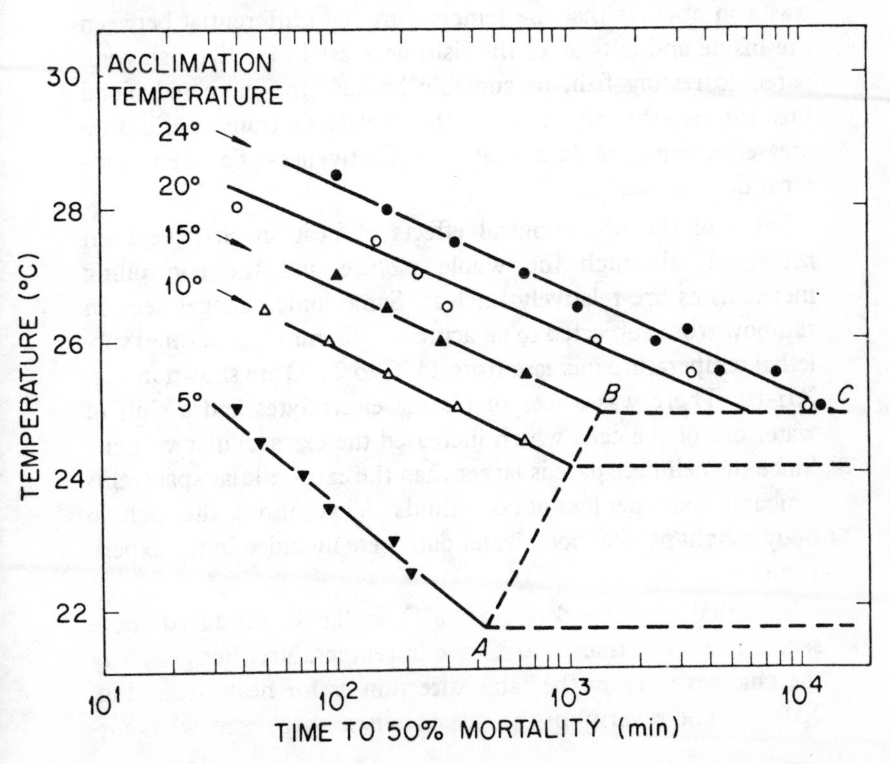

creased temperatures produce mortality more and more rapidly. Another way to put this is that mortality can be produced by a more or less constant "dose" of heat or that heat times exposure time is constant—the greater the rate of heat application, the less time needed for a given dose. This dose-response relationship is typical of many pollutants or other deleterious agents.

There can be considerable differences in the effect of a temperature change on fish, depending on how it is applied. Fig. XII-8, for example shows that a slow rise in temperature and rapid return to normal was more lethal than a sudden rise and slow return. A slow rise and slow return was nearly as lethal. One reason for this is that fish do not follow a change in environmental temperature instantaneously, but lag behind. While gills are excellent heat exchangers, the temperature lag for a rapid rise in temperature prevents the fish from reaching as high a temperature internally as when the rise is gradual (Fig. XII-9 B and A, respectively). In some of my own unpublished experiments, it was also obvious that the temperature lag (differential between the inside and outside of the fish) decreased in active fish compared to resting fish, presumably because the increased blood flow through the gills increased the rate of heat transfer. Thus increased activity should increase the effectiveness of a given short-term dose of heat.

Some of the physiological effects of heat shock have been measured, although the whole picture and the controlling mechanisms are relatively unclear. Some ionic changes seen in rainbow trout subjected to an acute (15-20 minute rise time) sublethal temperature increase from 11 °C to 21 °C are shown in Fig. XII-10. There was a loss of plasma electrolytes and a shift of water out of the cells which increased the extracellular volume. Since the cellular space is larger than the extracellular space, this probably was a net loss of body fluids (dehydration), although no body weight or total body water data were included in the experiment.

Sublethal cold shock from 11 °C to 0.5 °C produced more severe responses than a 10 °C rise in temperature, but generally the changes were in the same direction as for heat stress (Fig. XII-11). The severity of the response is perhaps somewhat sur-

Type of Exposure	Number Exposed	Number Losing Equilibrium	Mortalities	% Mortality
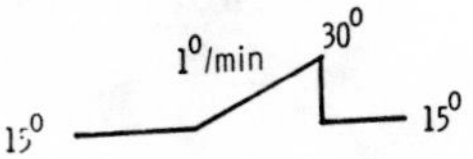 15^0 — 1^0/min — 30^0 — 15^0	10	10	10	100
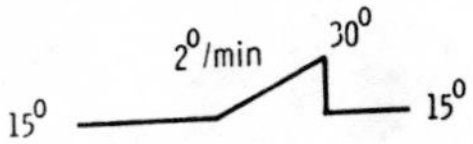 15^0 — 2^0/min — 30^0 — 15^0	20	7	4	20
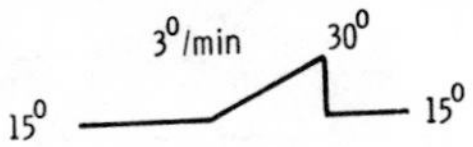 15^0 — 3^0/min — 30^0 — 15^0	8	4	0	0
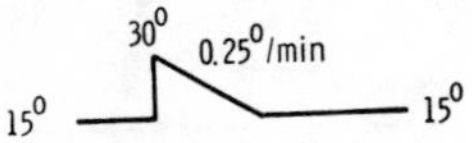 15^0 — 30^0 — 0.25^0/min — 15^0	10	9	7	70
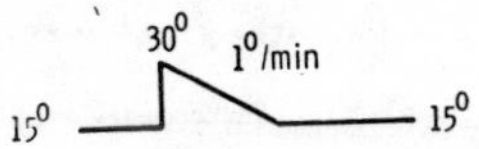 15^0 — 30^0 — 1^0/min — 15^0	5	2	1	20
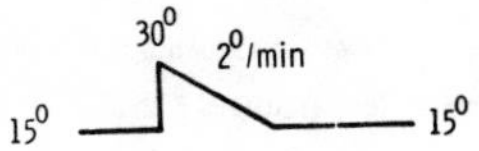 15^0 — 30^0 — 2^0/min — 15^0	12	--	1	8
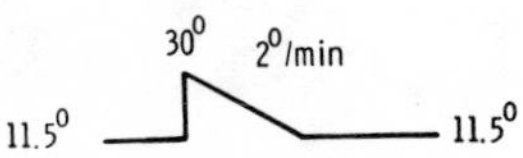 11.5^0 — 30^0 — 2^0/min — 11.5^0	5	5	3	60
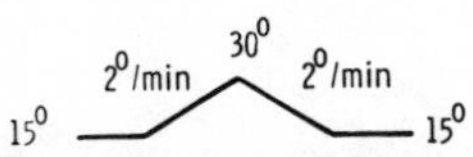 15^0 — 2^0/min — 30^0 — 2^0/min — 15^0	5	5	5	100
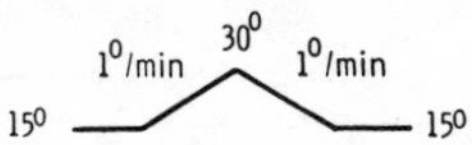 15^0 — 1^0/min — 30^0 — 1^0/min — 15^0	25	25	24	96

Fig. XII-8. Effects of sudden and slow rates of temperature change on coho salmon. The interpretation of the differing results appears in Fig. XII-9. (From Dean, 1973, with permission of the author.)

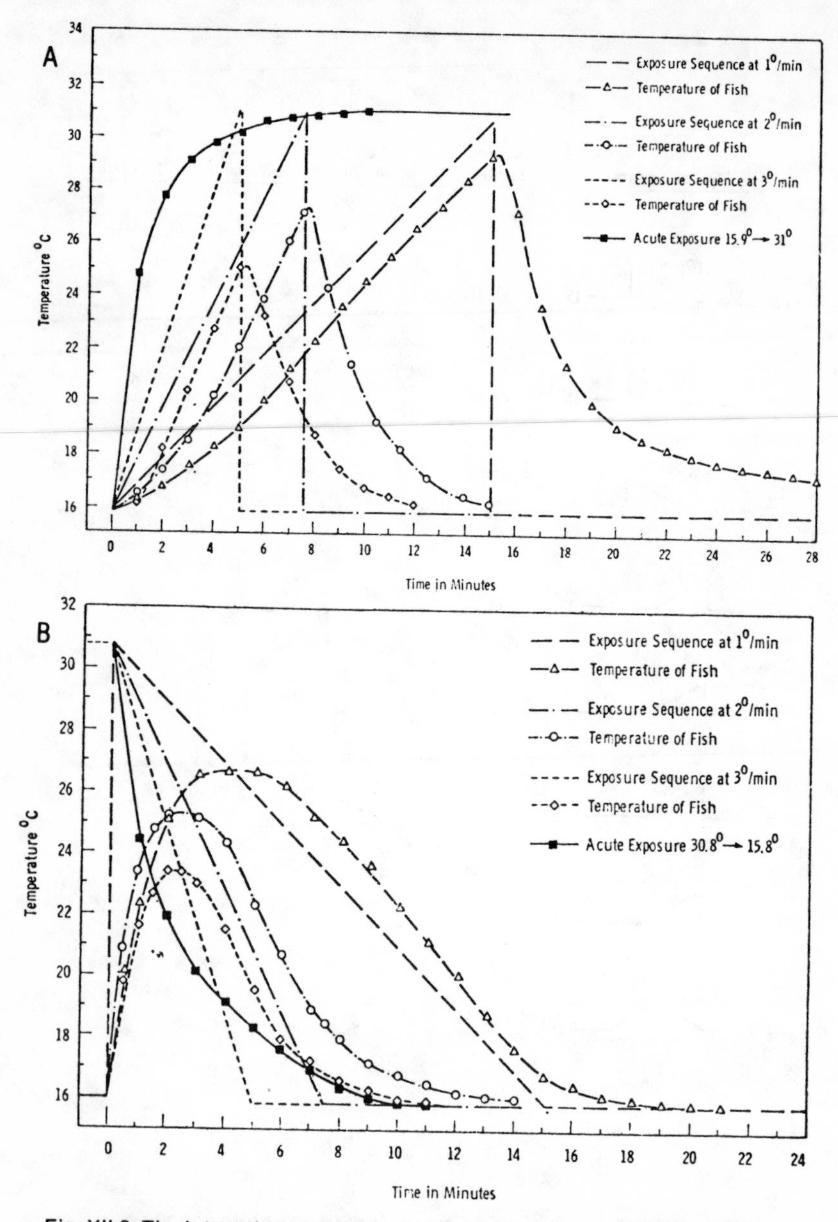

Fig. XII-9. The internal temperatures in fish exposed to heating and cooling at the different rates shown in Fig. XII-8. A. Temperatures of water and fish exposed to a 15°C increase in water temperature. B. Temperature of water and fish exposed to a sudden increase of 15°C and cooled at different rates. (From Dean, 1973, with permission of the author and Dr. Walter Chavin, Ed.)

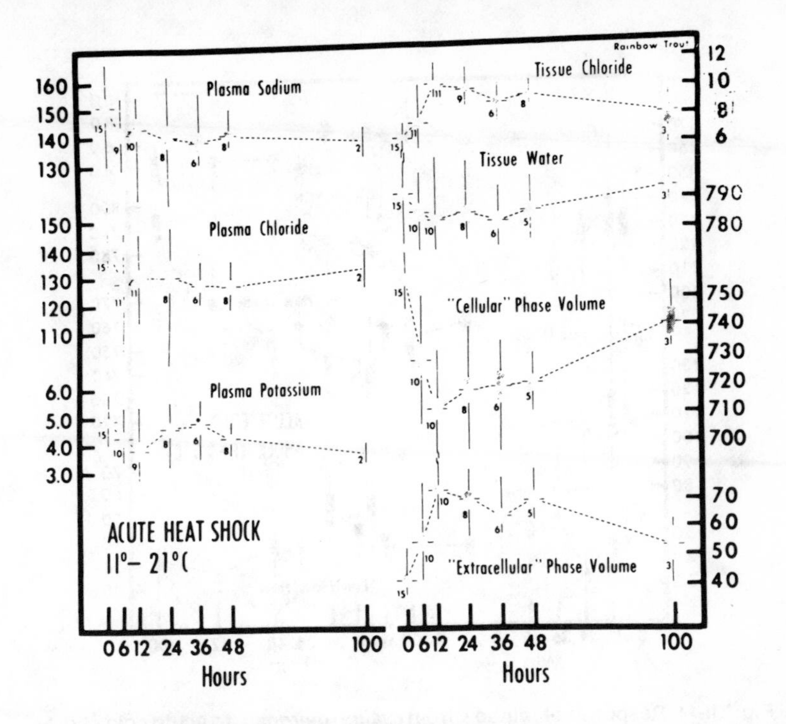

Fig. XII-10. Plasma and tissue electrolyte levels and tissue water content and distribution in rainbow trout, *Salmo gairdneri,* subjected to acute (11° to 21°C), sublethal heat shock. Cross bar, sample mean; vertical line, sample range; vertical bar, standard error of the mean; numerals, sample size. (From Houston, 1973, with permission of the author and Dr. Walter Chavin, Ed.)

prising ·since such a temperature change can be used as an anesthetic and gives the impression only of slowing down normal functions rather then disrupting them. However, it is clear that osmoregulation in salmonids is quite sensitive to temperature change, although the mechanisms for causing these osmoregulatory changes are unclear.

Species which are more eurythermal than salmonids probably respond differently than salmonids. Goldfish given a lethal temperature shock of being transferred from 22°C to 2-3°C showed no major fluid shifts, only a severe loss of plasma chloride (Fig. XII-12). Similar experiments were done with tench, but the results were difficult to compare with goldfish or trout because

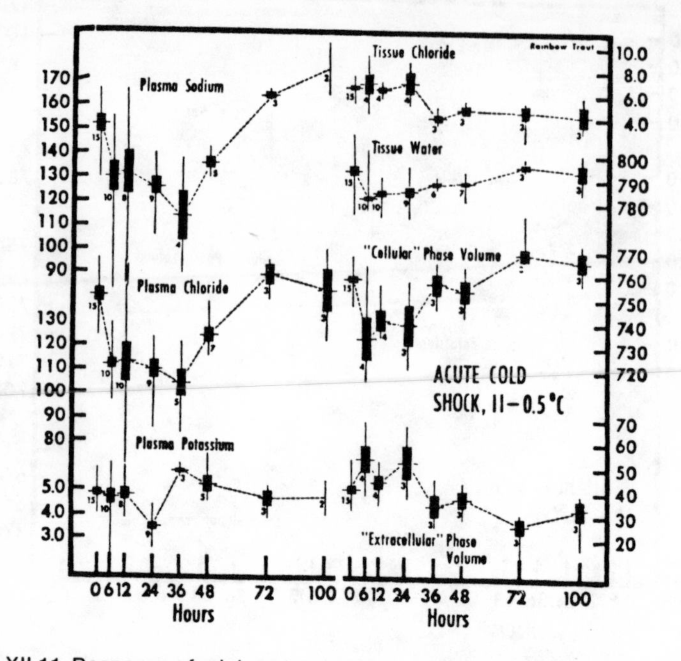

Fig. XII-11. Response of rainbow trout *(Salmo gairdneri)* to acute cold shock (11°-0.5°C). Data presentation same as in Figure XII-10. (From Reaves, et al., 1968, with permission of the publisher.)

of major differences in the rates of temperature changes used. The physiological effects of major temperature changes on marine fish are largely unknown. A possible generalization about the effects of acute temperature change is that more rapid changes in temperature produce more severe responses than slow changes. Houston (in Chavin, 1973) discussed many further hypotheses and speculated about other possible generalizations.

F. EFFECTS OF ANESTHESIA, HANDLING AND SCALE LOSS

While there are many substances that have been used as anesthetics for fish, their use has now been mostly limited by a recent restriction by the U.S. Food and Drug Administration (FDA). If any anesthetic is to be used on fish which could be actually or

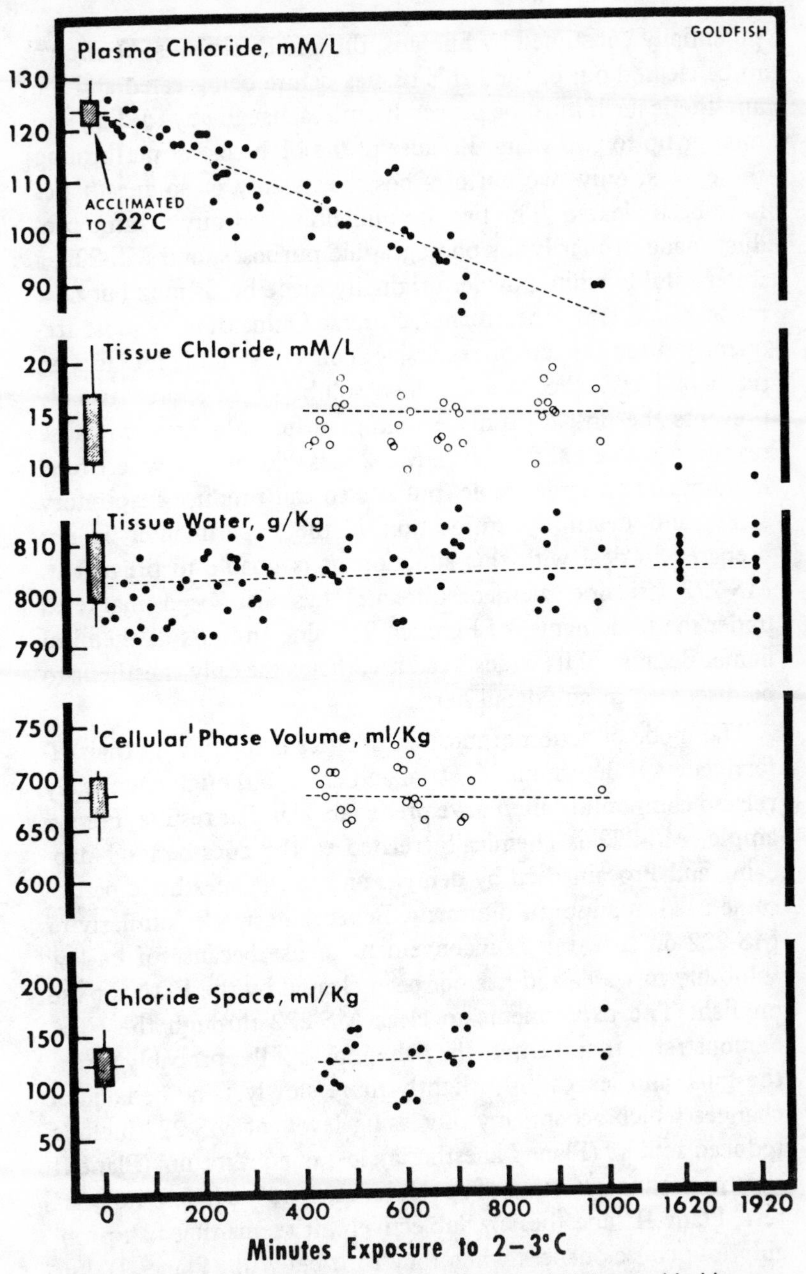

Fig. XII-12. Effects of lethal cold shock on plasma and tissue chloride concentrations and tissue water content and distribution in goldfish, *Carassius auratus*. (From Houston, 1973, with permission of the author.)

potentially consumed by humans, the anesthetic must be shown to be cleared out of the fish's tissues before being eaten and the anesthetic itself must be proven harmless, usually by feeding it to rats for up to two years. Because of the high cost of performing these tests, only two out of a possible twenty or so anesthetics have been cleared. The two are quinaldine, an oily organic product made primarily for photographic purposes, and MS-222, a water-soluble white powder originally made by Sandoz but now made by several other manufacturers. Quinaldine is most frequently used as an inexpensive collecting tool for use in tidepools, etc., because its low solubility in water usually prevents the dosage from becoming sufficiently great to cause respiratory arrest and kill fish. MS-222 is widely used where total immobilization is desirable, but it also can produce respiratory arrest and death by suffocation if the experimenter is not prepared to deal with this situation or is unable to prevent it. MS-222 (tricaine methanesulfonate) has also been marketed under the trade names of Finquel, Tricaine and its full chemical name. Because of its widespread use, this is the only anesthetic to be discussed in any detail here.

The mode of action of most anesthetics is unknown—their effectiveness is determined by trial and error, although chemically related compounds often have predictably similar results. For example, MS-222 is chemically related to the anesthetics Novocaine and Procain used by dentists and to the anesthetic benzocaine used in sunburn ointment. Benzocaine works similarly to MS-222 on fish, but is inconvenient to use because of its low solubility in water and has not been cleared by the FDA for use on fish. The experiments to clear MS-222 through the FDA demonstrated that it enters the fish quite rapidly, probably across the gills, and leaves only slightly more slowly. The behavioral changes which accompany increasing levels of MS-222 include reduced activity (Plane I anesthesia), loss of equilibrium (Plane II anesthesia) and cessation of opercular movements (respiratory arrest, Plane III anesthesia). Subjectively, it seems that fish probably lose consciousness when they turn belly up (Plane II), but there is no objective evidence such as changes in brain wave activity to substantiate this interpretation.

Anesthesia can be stressful to fish in at least two ways. First, either cessation of breathing or keeping the fish out of water for tagging, etc., produces hypoxia, lactate accumulation and the other associated effects discussed earlier in this chapter (Fig. XII-4). Second, tricaine is acidic, the pH depending on the concentration. The usual stock solution of 100 mg tricaine/ml water has a pH of about 3, and a full anesthetic dose of 100 mg tricaine/liter water produces a pH of about 6-6.5, but a tranquilizing dose of 10-20 mg/liter has only a minimal effect on the pH of the ambient water. Continued exposure to tricaine for a maximum of 12 minutes produced a typical cortisol/adrenalin-type stress response which did not occur with the same exposure to benzocaine or to neutralized tricaine. Effects of tricaine included increased blood glucose, cholesterol and blood urea nitrogen and decreased interrenal ascorbate (precursor to ACTH). Cortisol levels did not increase, but the exposure time was judged to be too short to allow such a response. Because of little response to neutralized tricaine, the stress response was judged to be primarily a problem of pH regulation rather than a response to the anesthetic. Problems of pH should be reduced in areas having hard water, since this research was done in very soft water.

Handling stress is primarily a problem of hypoxia which is aggravated by the strenuous activity associated with being chased, thrashing about in a net, etc. If there is scale loss added to the handling stress, the results can be lethal. When capturing juvenile salmon for tagging during their high-seas migration, a research team found that loss of more than about 30 percent of their scales caused death in 24 hours or less. Descaled fish became less and less active and died quietly in a state of flaccid paralysis. The major cause of death appeared to be a drastic increase in plasma Mg^{++} which was thought to be sufficiently high to block the transmission of nerve impulses across the myoneural junction. The fish also suffered a loss of up to 20 percent of their body weight, although control fish (no descaling or tagging) held out of water for the same length of time survived the same degree of weight loss. Associated with this severe weight loss was up to a five percent decrease in body (nose to tail notch) length. These same fish developed and survived blood lactate levels in excess of

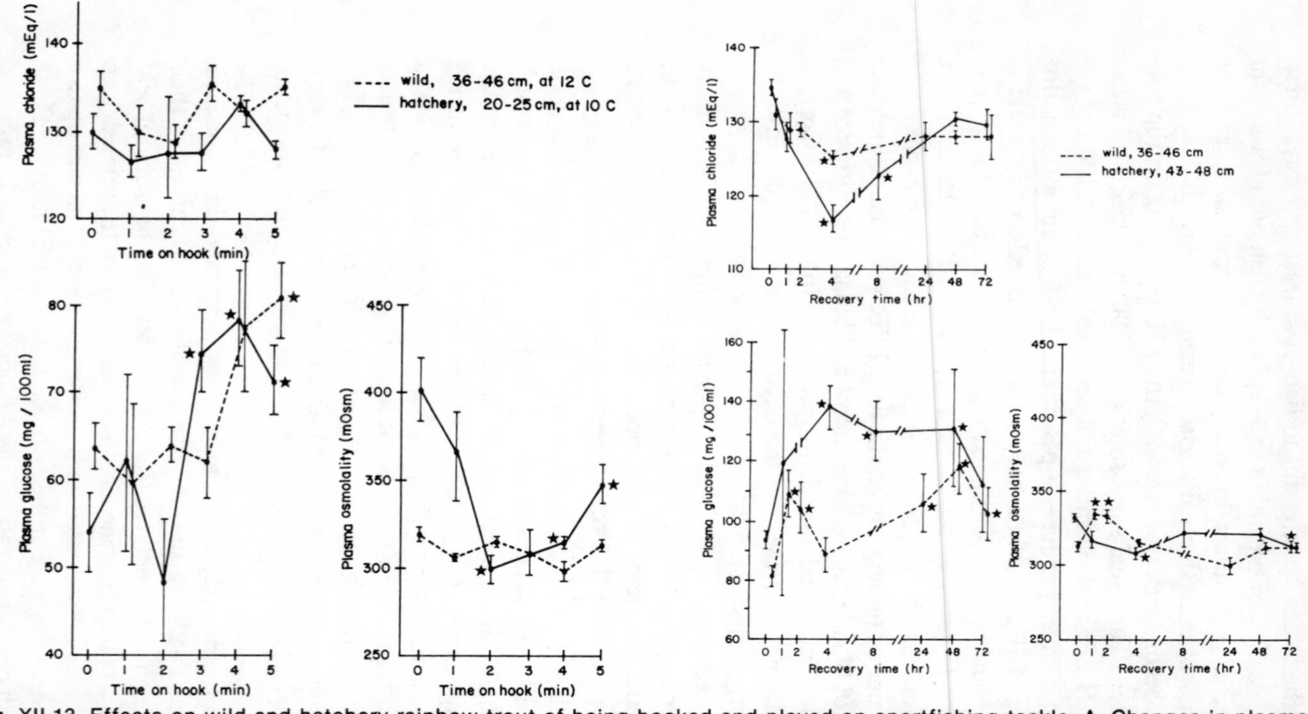

Fig. XII-13. Effects on wild and hatchery rainbow trout of being hooked and played on sportfishing tackle. A. Changes in plasma chloride, glucose, and osmolarity after 0-5 min of hooking stress at 10°C. B. Changes in the same variables during recovery from 5 min of hooking stress at 12°C. Data points are ± standard errors of groups of 5 to 10 fish. Stars indicate significant difference from the initial level, $P < 0.05$ (From Wydoski, et al., 1976, with permission of the publisher.)

500 mg percent, while levels of only 150-200 mg percent are lethal to adult salmon.

Scale loss in freshwater can also be lethal, but the reasons are not clear. For equivalent sizes of fish and degrees of scale loss, the mortality rates were about the same. Their weight increased in comparison to unfed control fish which lost weight at a moderate rate. There was no source of Mg^{++} in freshwater to cause muscular paralysis, although blood Mg^{++} was not measured. The physiological cause of death from scale loss in freshwater invites further work.

Compared to the stress of scale loss, the stress of capture by hook and line was surprisingly small. Wydowski, et al. (1976) found increased blood glucose and decreased blood chloride levels in proportion to the length of time that rainbow trout were played on sport fishing gear. The responses continued to increase for up to four hours after release, but had returned to normal by 72 hours later (Fig. XII 13 A & B). The response was slightly greater at higher temperatures (20°C vs 10-12°C) and in larger fish (43-48 cm vs 20-25 cm). Since the values for neither factor ever exceeded the normal ranges described for rainbow trout, the authors of this paper concluded that the stress was within the physiological ability of the fish to cope with the situation.

The fact that wild rainbows in the study above showed less change in plasma chloride and blood glucose than the hatchery fish leads to interesting speculation when compared to some related work in my laboratory (unpublished). We caught about ten fish from this same stock of wild rainbow trout and held them under hatchery conditions. They ate little or nothing and steadily deteriorated while we waited for their recovery from the stress of capture and transport; all died within a month. We obtained swimming performance measurements on a few of these fish before they died. Even though they were severely stressed, they swam at much higher velocities before becoming fatigued and swam more efficiently (lower tailbeat frequency for a given velocity relative to size) than the very placid hatchery fish with which we compared them. Sportsmen describe such wild fish as being very exciting to catch, while the placid hatchery fish fight about excitingly as hooking into a rubber boot. In another related study, the same placid rainbows had lower catecholamine levels, both resting and swimming, than hatchery steelhead (migratory rainbows somewhat comparable in vigor to the wild rainbows). One interpretation of these results is that hatchery rainbows have had much of their ability to produce catecholamines (= vigor?) and respond to stress bred out of them. Instead the hatchery fish had high rates of food intake and maximal growth rates. One might suspect that this would be a necessary result of high-density rearing practices. Fish which produced high levels of catecholamines when crowded and confined would not be expected to survive and grow well in a hatchery because so much of their metabolic energy would be siphoned off by the stress response and made unavailable for growth.

If one set out to develop a stress-resistant fish, it is unclear whether one should seek a fish which produces a maximal or minimal stress response. Too much response would be detrimental for any long-term stressor such as confinement or crowding, but too little response could be inappropriate for short-term stress such as capture or escape from a predator. This could suggest either different degrees or different kinds of responses to different stressors. It also suggests that there will probably never be a single variety of fish which is ideal for all circumstances. Different strains with different physiological characteristics will be developed for different purposes, much as has been done with domestic animals.

G. RESPONSE TO AIR SUPERSATURATION OF WATER

Marsh and Gorham published in 1898 the results of a study which demonstrated that dissolving air into water under pressure kills fish. Now, 83 years later, we are not greatly advanced beyond that point. The solubility of air (mostly nitrogen and oxygen) in water is proportional to the hydrostatic or gas pressure. Gas bubbles carried down underwater at the foot of the dam or waterfall can partially or completely dissolve into the surrounding water. Similar situations occur when there is an air leak on the intake (suction) side of a pump. Well water coming up from underground is sometimes supersaturated. The gas content in a fish's blood equilibrates to that of the water within an hour or two. Whenever either the fish or the water comes near enough to the surface so that the hydrostatic pressure is less than needed to keep the excess gas in solution, bubbles eventually form. If these bubbles block a blood vessel for a vital organ, the fish dies. This is basically like the bends experienced by divers, except that bubbles in fish may take over a month of continued exposure to low-level supersaturation before they form. Some of the significant physiological changes which accompany exposure to supersaturated water appear in Fig. XII-14. In addition to the changes seen, it is also significant that there were no changes in blood glucose and neither alkaline phosphatase (APase) or glutamic oxalacetic transaminase (SGOT), both of which commonly increase after tissue damage. With changes occurring in blood ions and proteins, it would appear that cortisol was involved in the response, but with no change in blood glucose, it would appear that catecholamines did not increase—i.e., that only a partial stress response occurred.

H. PHYSIOLOGICAL CHANGES ASSOCIATED WITH DISEASE

Infectious hematopoietic necrosis (IHN) is a rhabdoviral disease of rainbow trout and sockeye or chinook salmon. The most conspicuous effect of the disease is extensive necrosis of the hematopoietic tissue of the kidney. Secondary necrosis occurs in the liver, pancreas and gut. The disease is most active at 10 °C and does not occur naturally above 15 °C. Fish which were infected with IHN virus at 10 °C and then moved to 18 °C survived the infection and developed antibodies against IHN after about two months. Further, their serum injected into other fish provided passive protection against IHN. It was not known whether the survivors of the IHN infection carried subclinical levels of the infection or were actually free of the disease. The major means of controlling the disease at the time of writing was to destroy the infected fish, an issue which was hotly debated by trout growers since destroying fish is costly to them and there was no clear evidence of the disease's mode of transmission—i.e., no evidence

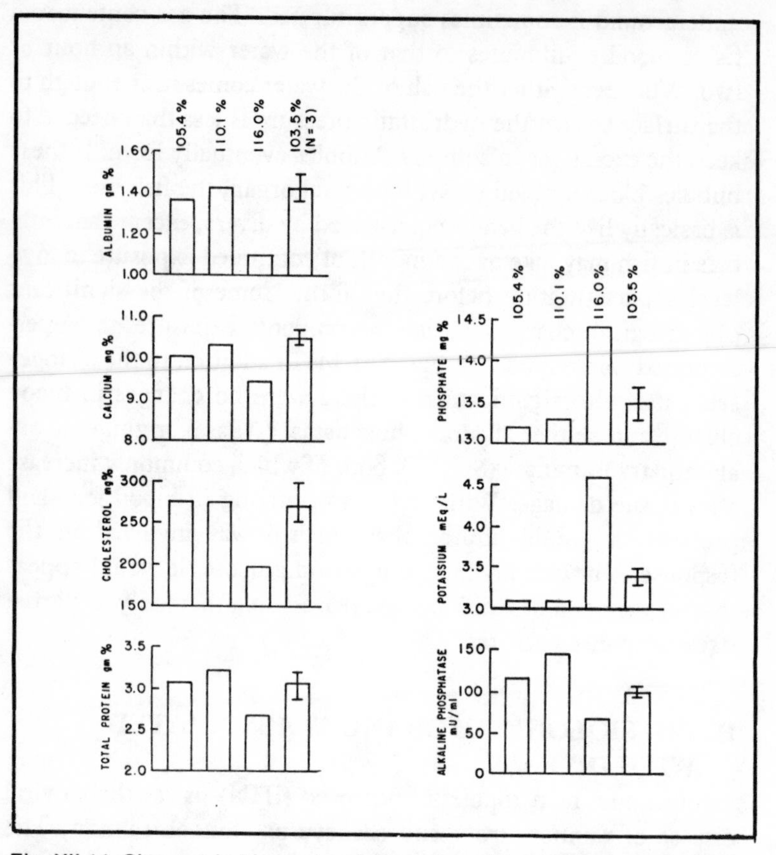

Fig. XII-14. Changes in blood chemistry of juvenile steelhead trout, *Salmo gairdneri,* due to chronic exposure to several levels of nitrogen plus argon supersaturation. Control values at 103.5% N_2 + Ar are mean and range. Steelhead at 116.0% saturation showed a 46% incidence of external signs of gas bubble disease (bubbles in the lateral line and in fins) while no signs occurred at lower supersaturations. There were no mortalities at any of these levels of supersaturation. (From Newcomb, 1974, with permission of author and publisher.)

that destroying infected fish really prevented further infection.

The physiological changes associated with a developing IHN infection are shown in Fig. XII-15. The decrease in the blood cell parameters (PCV, Hb, RCB) was expected, since IHN virus attacks the blood cell-forming tissues, but the decrease in HCO_3^- was probably more significant physiologically. The acid-base

tend to confirm this—increases in non-protein nitrogen, urea, uric acid, creatinine and ammonia all fit into the picture of muscle breakdown. The decrease in sialic acid is thought to be the result of the pathogen cleaving certain glycoproteins on the surface of muscle cells before the pathogen enters the cells. There were major decreases in serum lipids which could have been explained by any of several mechanisms, none of which was clearly identifiable as responsible for causing death. Severe hypoglycemia occurred in carp infected with *A. salmonicida,* but there was hardly any change at all in blood glucose in the brook trout described in Table XII-8. This may be a species difference or the result of a different route of entry of the pathogen into the fish. With little change in blood glucose or blood ascorbate (not quite the same as measuring ascorbate reserves in the head kidney, but probably similar), it would seem that neither catecholamines nor ACTH and cortisol were much involved in the response to the disease.

I. PHYSIOLOGICAL EFFECTS OF SOME TOXICANTS

There is extensive information on the concentrations of many pollutants or toxicants which kill fish and other organisms. There is much less information about the mechanisms by which these toxicants act. There are several reasons for this. Government regulatory agencies don't need to know how a substance acts before they can set limits on its discharge into a body of water, only what the maximum allowable concentration should be. Regulation has usually been accomplished by setting the allowable limit as some small percentage of the concentration which kills 50 percent of the test population in 96 hours (96 hr LD 50 bioassay). Finding the manner in which the toxicant interferes with physiologial functions is much more difficult and time-consuming than 96-hour bioassays, although there is increasing need to do this with the increasing emphasis on sublethal effects of pollutants. Further, the short time available for the first pollutant regulations based on bioassay testing for a large number of pollutants also minimized most work on physiological mechanisms by the regulatory agencies.

Component	Sample	Control		Infected		Level of significance
		X	Sm	X	Sm	
Total protein (g/100 ml)	Serum	2.2	0.4	1.5	0.3	S
Hemoglobin (g/100 ml)	Blood	9.11	1.8	6.02	1.2	S
Urea (mg/100 ml)	Blood	1.5	0.3	3.2	0.5	S
Uric acid (mg/100 ml)	Blood	1.0	0.1	1.2	0.1	S
Total creatinine (mg/100 ml)	Blood	1.6	0.2	2.2	0.3	S
Amino acid-N (mg/100 ml)	Blood	19	2.1	28	3.5	S
Ammonia (mg/100 ml)	Serum	0.1	0.01	0.15	0.01	S
Glucose (mg/100 ml)	Blood	55.8	9.5	83.9	16	S
Pyruvic acid (mg/100 ml)	Blood	0.25	0.02	0.21	0.02	N.S.
Lactic acid (mg/100 ml)	Blood	5.37	0.6	6.61	0.8	N.S.
Ascorbic acid (mg/100 ml)	Blood	0.51	0.04	0.59	0.04	N.S.
Sialic acid (mg/100 ml)	Serum	96.5	19	52.5	11	S
Total lipids (mg/100 ml)	Serum	1104	220	768	150	S
Total fatty acids (mg/100 ml)	Serum	381	61	203	38	S
Free fatty acids (mg/100 ml)	Serum	63	7.1	37	4.5	S
Triglycerides (mg/100 ml)	Serum	290	55	230	35	S
Total cholesterol (mg/100 ml)	Serum	240	31	118	19	S
Free cholesterol (mg/100 ml)	Serum	80	11	32	6	S
Inorganic—P (mg/100 ml)	Serum	18.8	3.1	16.4	2.8	N.S.
Acid soluble—P (mg/100 ml)	Serum	23.2	4.1	20	3.5	S
Lipid—P (mg/100 ml)	Serum	19.2	3.7	12	1.6	S

A few pollutants seem to have fairly direct effects during acute exposure. Ammonia (NH_3—see also Table II-6A concerning ionization of ammonia) affects salmonids at concentrations beginning about 2 parts per million (ppm) and is killing some fish by the time it reaches 15 ppm. The most obvious effect in trout is an increase in total permeability, because urine flow increases dramatically and tissues such as skin and gills swell up with water. Since fish excrete ammonia, the prevention of ammonia poisoning is of obvious importance in any kind of water re-use system, be it a tropical fish aquarium or a huge hatchery. It is also an important factor in assessing the impact of discharges from fish hatcheries and from sewage treatment plants with incomplete oxidation of ammonia or other nitrogenous compounds.

Chlorine is a chemical widely used to treat a variety of problems involving incomplete oxidation of effluents and thereby also sterilizing the water by killing any pathogens present. Many municipal sewage plants chlorinate their effluents just before discharge, and most people are familiar with chlorine in swimming pools. Persons keeping fish as pets also quickly become aware of the necessity of removing or neutralizing the chlorine in drinking water to keep from killing their fish. Study of the problem is very recent and still ongoing, but a few effects were known at the time of writing. First, the chlorine itself is not very toxic but reacts with ammonia by replacing one or sometimes two of the hydrogen atoms with chlorine to give chloramines. Chloramines are toxic at concentrations of a few parts per billion (ppb). Two of the known effects of chloramines are the gradual destruction of blood cells and the oxidation of the iron in hemoglobin to a stable state (methemoglobin) so that it will not transport oxygen. The

Opposite page: Table XII-8. Biochemical comparison of blood from brook trout *(Salvelinus fontinalis)* infected with *Aeromonas salmonicida* (furunculosis) and from uninfected controls. X: mean; Sm: Standard deviation; S: Significant; N.S.: Not significant; Test t: Significant difference at P = 0.05; (From Smith and McClean, 1976, with permission of author and publisher.)

latter effect is similar to that of carbon monoxide on people. These are probably not the only effects of chloramine because, although the reduction of oxygen transport capability at the time of death was severe, it did not seem severe enough to kill the fish by suffocation.

There was great hysteria in the early 1970's about mercury in fish and in the environment, as well as other heavy metals such as lead and zinc. The extreme concern was probably unwarranted, although research continues on the complex roles of heavy metals in the aquatic environment at a more deliberate pace. Some of the severe industrial abuses have been corrected, and metabolic studies on humans showed that more of the heavy metals pass through the body (rather than being deposited) than was first believed. The presence of heavy metals in the aquatic environment should be cause for concern, but need not always be cause for alarm.

Some of the effects of heavy metals on fish are quite direct. In England, the water runoff from mine tailings became somewhat acid and picked up a considerable load of dissolved zinc (Zn^{++}). This seeemed both to stimulate the production of mucus and to precipitate it. This process eventually covered the gill membranes sufficiently thickly that the fish suffocated. In another 96-hour study relating to pulpmill effluents in Canada, zinc reduced the numbers of small lymphocytes present in circulating blood in coho salmon but did not affect the number of erythrocytes.

Lead is also a toxic substance, and chronic exposure in trout causes such things as blackened tails, spinal curvature, caudal fin erosion and finally lethargy and paralysis. Eggs and larvae are more sensitive to lead than adults. The problem with lead, however, is to determine the concentrations at which these effects occur. Where there is hard water, lead interacts in a variety of ways with the other materials dissolved in water to form carbonates, sulfates, chlorides and hydroxides of lead which are variously soluble, colloidal or precipitated according to the pH of the water. Thus analysis of the amount of lead actually reaching a functional site inside the fish requires very sophisticated chemical techniques. The point at which lead interferes with metabolic functions is not known, but since the effects of lead intoxication partly resemble those for dietary deficiencies of vitamin C and the amino acid tryptophan, lead may block either or both of those metabolic pathways.

An even more complex chemical situation occurs with industrial effluents. In the case of kraft (sulfate) pulp mills, the interactions between the great variety of natural compounds in wood with the pulping chemicals produce a tremendous array of different compounds, of which a reputed 3500 have been identified so far. The physiological effects are similarly diverse (Table XII-9). A significant finding from this research was that bleached kraft mill effluent (BKME) was not particularly toxic if the pH was adjusted to neutral before the fish were exposed to it. In one rather startling experiment, fingerling salmon grew faster in dilute BKME than in clean water, although the fish in the BKME showed some distress—elevated blood glucose and changes in blood cell counts. The dark color of the BKME may have reduced the territorial aggressiveness of the fish by making fish less visible to each other in the 30-gallon tanks and reduced the effects of crowding—anxiety, chasing, etc.—thereby allowing more energy to be used for growth.

Short-term studies of swimming performance in dilute BKME showed relatively minor effects on carbohydrate metabolism (Fig. XII-16). Muscle glycogen decreased and plasma lactate increased during swimming much the same in control fish as in those swimming in BKME. Liver glycogen decreased and plasma glucose increased and did not recover in the BKME fish until after the end of the swimming period. The significance of the lack of recovery of plasma glucose and liver lactate was not determined in these experiments but might relate to the fish's capabilities to repeat the swimming effort.

Another approach for studying BKME was to use one of its constituents, dehydroabietic acid (DHAA) in purified form as a test substance. During 48 hours of continuous swimming at low, intermediate and high levels, several blood parameters changed during exposure to DHAA. The number of leucocytes decreased and whole blood clotting times increased compared to non-DHAA controls in the low and intermediate, but not the high,

FUNCTION OR SYSTEM AFFECTED	EFFECT OF KME	SPECIES	SIZE	TEMP. °C	CONC. TESTED	APPROX. THRESHOLD	COMMENTS
RESPIRATORY	"Coughing" response elevated	Rainbow	8-10 in.	11 ± 1	—	1.1% of full strength KME	
	"Coughing" response elevated	Sockeye	207-321 g	10.5 ± 0.5	—	0.1-0.2 LC 50	Possible adaptation
	Ventiliation volume increased	Sockeye	207-321 g	10.5 ± 0.5	—	0.2 LC 50	Possible adaptation
	Oxygen uptake increased	Sockeye	207-321 g	10.5 ± 0.5	—	0.33 LC 50	Possible adaptation
CIRCULATORY	Arterial oxygen tension reduced	Sockeye	To 2 Kg	10.5 ± 0.5	0.36 LC 50	—	No adaptation
	Arterial oxygen tension reduced	Rainbow	150 g	10	0.33 LC 50	—	No adaptation
	White blood cell/ thrombocyte counts reduced	Coho	Juveniles	11 ± 1	—	0.1 TL_M96	After 21 days exposure
	Blood neutrophil count elevated	Coho	Juveniles	11 ± 1	—	0.25TL_M96	After 200 days exposure
METABOLISM	Plasma glucose elevated	Coho	Juveniles	11 ± 1	—	0.10 TL_M96	After 200 days exposure
	Blood & muscle lactate elevated	Coho	Juveniles	11 ± 1	—	0.25TL_M96	After 200 days exposure
	Swimming ability reduced	Coho	Juveniles	13 ± .5	—	0.20TL_M96	—
	Muscle glycogen depressed	Coho	Juveniles	11 ± 1	—	0.10TL_M96	After 200 days exposure

Opposite page and above: Table XII-9. Sublethal effects of bleach kraft mill effluent (BKME) on various physiological functions of salmonids. (From Davis, 1968, MSc. Thesis, University of British Columbia, unpublished, with permission of Dr. John C. Davis.)

FUNCTION OR SYSTEM AFFECTED	EFFECT OF KME	SPECIES	SIZE	TEMP. °C	CONC. TESTED	APPROX. THRESHOLD	COMMENTS
GROWTH	Growth rate decreased	Sockeye	2.4-2.8 g	15	—	10-25% full strength KME	Exposed over about 8 weeks
	Growth rate decreased	Sockeye	Fingerlings	8	—	0.05-0.1 LC 50	Bleach waste
	Food conversion efficiency reduced	Sockeye	2.4-2.8 g	15	—	10-25% full strength KME	8 week exposure
	Growth rate enhanced	Coho	4-10 g	10-13	—	0.1-0.2 LC 50	Several wks. exposure
	Growth rate enhanced	Coho	Juveniles	11±1	—	0.1-0.25 TL_M96	70 day exposure
BEHAVIOR	Feeding behavior affected	Coho	4-10 g	10-13	—	0.1-0.2 LC 50	Response lasted for 2 weeks then disappeared
	Fish slow & "unresponsive"	Coho	4-10 g	12-13	—	0.15 LC 50	—
	"alarm response" slowed	Sockey	Fingerlings	8	—	0.4 LC 50	Bleach waste
	Orientation to water current affected	Sockeye	Fingerlings	8	—	0.8 LC 50	Bleach waste
	Fish avoid effluent	Sockeye	3 CM	10	—	0.2 LC 50	Within 1 hour
	Fish avoid effluent	Atlantic Salmon	7.7-14.8 CM	17±0.2	—	10 ppm (approx. .006 LC 50)	"Vague" response
	Fish avoid effluent	Atlantic Salmon	7.7-14.8 CM	17±0.2	—	3.77XLC 50	Strong response
REPRODUCTION	Egg mortality	Coho eggs	—	6-6.5	—	0.36 LC 50 for underyearling coho	—
HISTOLOGY	Histological & cytochemical change of liver, kidney, intestine	Sparus macrocephalus	25 CM	—	—	Field exposure in cages off mill in seawater	12-24 hr exposure (Sublethal)
ENDOCRINOLOGY	Plasma cortisol up	Sockeye	168-200 g	12.5	0.50 LC 50	—	After 2 hr.
	Mucous production increased	"Salmonids"	Juveniles	—	—	2-5% full strength KME	—
IMMUNOLOGICAL	Disease resistance reduced	Coho	10 g	12.0	0.40 LC 50	—	Methemoglobinemia resistance

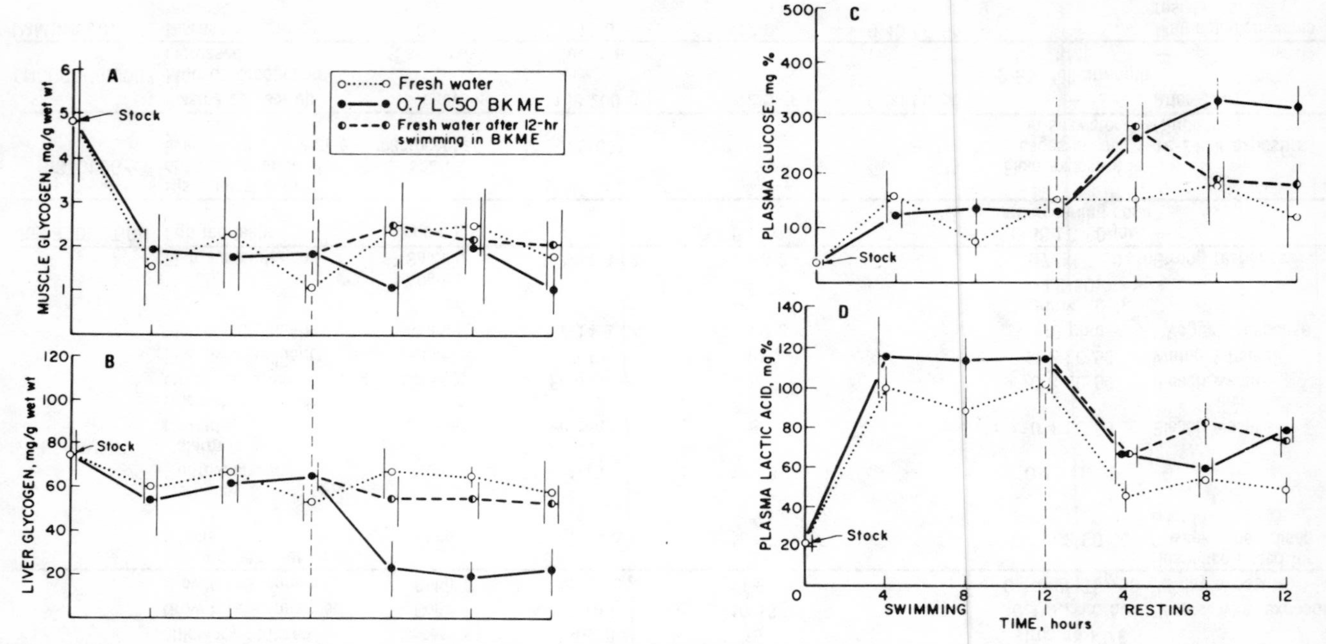

Fig. XII-16. Short term effects of bleached kraft mill effluent (BKME) on carbohydrate metabolism of juvenile coho salmon. Fish were exposed for up to 12 hr. at moderate levels of exercise in BKME (concentration = 70% of the LC50 value) and then allowed to rest either in the same water or were changed to fresh water. Data are shown for muscle glycogen (A), liver glycogen (B), plasma glucose (C) and plasma lactate acid (D).

(From McLeay and Brown, 1975, with permission of author and publisher.)

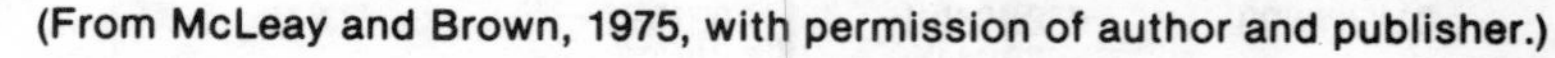

exercise levels. Hematocrits, erythrocyte sedimentation rates and red blood cell counts were not different from controls at any exercise level. The protective value of high levels of exercise may seem strange at first but could relate to the experiments described in the previous paragraph. Therefore, swimming delayed changes in plasma glucose and liver glycogen until after the swimming period. The BKME was thought to decrease respiratory capabilities in some way, and thus the role of exercise in optimizing gas transfer may also have compensated for early respiratory effects of DHAA. The fact that exercise also delayed the onset of bradycardia during decreasing environmental oxygen levels is probably a related phenomenon. Both the BKME and DHAA studies illustrated the need to perform studies of pollutants under a variety of physiologically realistic situations.

I am suggesting that the BKME effects were a combined cortisol-catecholamine response, while DHAA appeared to produce only a cortisol response, although in the latter case no measurements of catecholamine-related factors were made.

J. INDICATORS OF FISH HEALTH

For many years the working definition of fish health was the absence of disease. This represented the state of the art of those days in recognizing when fish were having problems and then trying to save the fish. The results were often unsatisfactory because one of the first symptoms of a problem was that the fish stopped feeding. The major route of administering medication to treat the problem—usually a disease—when it involved millions of fish in large hatcheries was to put antibiotics in their food. This meant that sick fish which had ceased feeding therefore received no medication and usually died. This lack of understanding of fish health was not surprising. Fish show very few external signs of ill health until the problems reach major proportions, and they have fewer means of communicating their discomforts to us than do domestic animals. In contrast, consider how much a medical doctor depends on what you can *tell* him about your complaints.

There is a growing body of knowledge about fish health, beginning in the late 1960's, concerning the physiological changes to

Clinical test	Possible significance if:	
	Too low	Too high
1. Ammonia (in water, un-ionized)	No recognized significance	Gill hyperplasia, predisposition to bacterial gill disease
2. Blood cell counts		
a. Erythrocytes (red blood cells, RBC)	Anemias, hemodilution due to impaired osmoregulation, gill damage	Stress polycythemia, dehydration. hemoconcentration
b. Leukocytes (white blood cells, WBC)	Leukopenia due to acute stress	Leukocytosis due to bacterial infection
c. Thrombocytes	Abnormal blood clotting time	Thrombocytosis due to acute or chronic stress
3. Chloride (plasma)	Gill chloride cell damage, compromised osmoregulation	Hemoconcentration, compromised osmoregulation
4. Cholesterol (plasma)	Impaired lipid metabolism	Fish under chronic stress, dietary lipid imbalance
5. Clotting time (blood)	Fish under acute stree, thrombocytopenia	Sulfonamides or antibiotic disease treatments, affecting the intestinal microflora
6. Cortisol (plasma)	Interrenal exhaustion from severe stress	Fish under chronic or acute stress
7. Glucose (plasma)	Inanition	Acute or chronic stress
8. Glycogen (liver or muscle)	Chronic stress, inanition	Liver damage due to excessive vacuolation. Diet too high in carbohydrate
9. Hematocrit (blood)	Anemias, hemodilution due to gill damage	Hemoconcentration, dehydration, stress polycythemia

Clinical test	Possible significance if:	
	Too low	Too high
10. Hemoglobin (blood)	Anemias, hemodilution due to gill damage, nutritional disease	Hemoconcentration, dehydration, stress polycythemia
11. Lactic acid (blood)	No recognized significance	Acute or chronic stress, swimming fatigue
12. Methemoglobin (blood)	No recognized significance	Excessive No_2^- in water or use of O_2 instead of air in fish-hauling trucks
13. Nitrite (water)	No recognized significance	Methemoglobinemia in fish population
14. Osmolality (plasma)	External parasite infection, heavy metal exposure, hemodilution	Dehydration, salinity increases in excess of osmoregulatory capacity, stress-induced diuresis, lactic acidosis
15. Total protein (plasma)	Infectious disease, kidney damage, nutritional imbalance, inanition	Hemoconcentration, impaired water balance

Opposite page and above: Table XII-10. Outline interpretation of clinical tests used to assess the effects of environmental stress on fish health. The examples are based on salmonids, but are applicable, with caution, to other fishes. (From Wedemeyer and Yasutake, 1977, with permission of author and publisher.)

Hematological component	Estimated range
Bilirubin (mg/100 ml)	1.4-1.7
Blood urea nitrogen (mg/100 ml)	0.9-4.5
Chloride (mEq/l)	84-132
Cholesterol (mg/100 ml)	161-365
Clotting time(s)	
Aorta cannula	150-250
Cardiac or caudal vessel puncture	50-150
Caudal peduncle severed	20-60
Erythrocytes (10^6/mm^3)	0.77-1.58
Glucose (mg/100 ml)	41-151
Hemoglobin (g/100 ml)	5.4-9.3
Hematocrit (%)	24-43
Leukocytes (10^3/mm^3)	7.8-20.9
Differential count	
Lymphocytes (%)	89-98
Neutrophils (%)	1-9
Thrombocytes (%)	1-6
Total protein (g/100 ml)	2-6
Osmolality (mOsm/1)	288-339

Table XII-11. Hematological and blood chemistry ranges which can be expected in clinically healthy juvenile rainbow trout in soft water (100 ppm $CaCO_3$ or less) at 10°C. (From Wedemeyer and Yasutake, 1977, with permission of author and publisher.)

be expected during early stages of infections or other problems. Normal values of physiological characteristics have been sufficiently established so that departures from normal can be used as indicators of sublethal problems. In addition, nutritional deficiencies have been identified as causing some of the problems, and a number of environmental impacts on fish health were seen, examples of which have been already discussed in this chapter. Other environmental effects are now known to eventually show up as outbreaks of disease (Wedemeyer, Meyer and Smith, 1976). None of these realizations has come as a spectacular breakthrough, but rather as an accumulation of piecemeal insights into

Fish type and species	Hematological characteristic			
	Cl^- (meq/l)	Glucose (mg/100 ml)	pH	Hematocrit (%)
COLD WATER				
Brown trout	108.6	52.0	7.51	33.0
(Salmo trutta)	(6.9)	(6.5)	(0.08)	(4)
Brook trout	108.6	59.2	7.71	31.0
(Salvelinus fontinalis)	(1.6)	(4.3)	(0.06)	(5)
COOL WATER				
Northern pike	104.8	53.4	7.64	30.0
(Esox lucius)	(4.3)	(16.2)	(0.18)	(5)
Walleye	62.0	152.5	7.85	46.0
(Stizostedion vitreum)	(10.5)	(78.3)	(0.08)	(4)
WARM WATER				
Channel catfish	114.0	29.1	7.55	32.1
(Ictalurus punctatus)	(5.0)	(17.7)	(0.05)	(4.1)

Table XII-12. Mean blood chemistry values (± S.D.) for selected cold-, cool-, and warm-water fishes of interest in environmental monitoring. Normal ranges are not available, but the mean values ± 1.96 S.D. can be used as a guide. (From Wedemeyer and Yasutake, 1977, with permission of author and publisher.)

the ways in which fish, being typical vertebrates in many respects, have special adaptations to living in an aquatic environment rather than a terrestrial one like most other vertebrates.

Wedemeyer and Yasutake (1977) listed a number of environmental and physiological variables which show relationships to fish health (Table XII-10). The criteria for putting variables in this table included the availability of data for determining normal values and the ease of making the measurement in a modestly equipped fishery laboratory, as well as the theoretical appropriateness of the variables as indicators of important physiological functions. Considering the wide ranges given for some of the normal values (Table XII-11) for such variables as hematocrit or glucose, one may well wonder whether the normal variability is so high as to obscure sublethal deviations from normal. In the case of hematocrit, this is sometimes true—fish may be nearly dead before values go outside this normal range. On the other hand, people often have past experience with given populations and commonly know that the typical hematocrit at a particular hatchery normally ranges only from 29% to 32% and that a reading of 25% or 35% may signify the onset of a problem for that particular population of fish. Blood glucose, on the other hand, commonly exceeds 200 mg/100 ml as a result of even moderate, relatively short-term stress. Thus hematocrit readings may tell you very little unless you have prior information about the fish in question, while high blood glucose is a useful indicator of fish health in spite of its wide range of normal values. On the other hand, hematocrit readings are often available whenever some other analytic process requires blood to be centrifuged, so the hematocrit reading is often available with little or no extra effort and on this basis is worthwhile.

Once you leave the realm of rainbow trout and salmon (coho, chinook, Atlantic), the amount of data available on normal values and stress indicators becomes scarce indeed. Table XII-12 shows normal values for a few other fish. Similar data for wild fish is largely nonexistent. Much remains to be done before fish medicine comes even close to the status of veterinary medicine.

Bibliography

Alexander, R. McNeil. 1967. *Functional design in fishes.* Hutchinson University Library, London. 160 pp.

Alexander, R.M. 1969. The orientation of muscle fibers in the myomeres of fishes. J. Mar. Biol. Assoc. U.K. 49:263-290.

Amend, D.F. and L. Smith. 1974. Pathophysiology of infectious hematopoietic necrosis virus disease in rainbow trout *(Salmo gairdneri):* early changes in blood and aspects of immune response after injection of IHN virus. *J. Fish. Res. Bd. Canada* 31:1371-1378.

Bennett, M.V.L. 1971. Electric organs. pp 347-491 *in* Hoar, W.S. and D.J. Randall (Eds.), *Fish Physiology,* Vol. V. Academic Press, N.Y.

Bern, Howard A. 1967. Hormones and endocrine glands of fishes. *Science* 158:455-462.

Black, E.C. et al. 1962. Changes in glycogen, glucose and lactate in rainbow and kamloops trout, *Salmo gairdneri,* during and following muscular activity. *J. Fish. Res. Bd. Canada* 19 (3): 409-436.

Blaxter, J.H.S. 1969. Development: eggs and larvae. pp. 177-252 *in* Hoar, W.S. and D.G. Randall (Eds.), *Fish Physiology,* Vol. III. Academic Press, N.Y.

Blaxter, J.H.S. and M. Pattie Jones. 1967. The development of the retina and retinomotor responses in the herring. *J. Mar. Biol. Assoc. U.K.* 47:677-697.

Bodznick, David. 1975. The relationship of the olfactory EEG evoked by naturally-occurring stream waters to the homing behavior of sockeye salmon *(Oncorhynchus nerka* Walbaum). *Comp. Biochem. Physiol.* 52A: 487-495.

Brett, J.R. 1964. The respiratory metabolism and swimming performance of young sockeye salmon. *J. Fish. Res. Bd. Canada* 21 (5):1183-1226.

Brett, J.R. 1970. Fish - the energy cost of living. pp. 37-52 in McNeil, W.J. (Ed.) *Marine Agriculture.* Oregon State University Press.

Brett, J.R. 1971. Energetic responses of salmon to temperature. A study of some thermal relations in the physiology and freshwater ecology of sockeye salmon *(Oncorhynchus nerka). Am. Zool.* 11:99-113.

Brett, J.R. 1972. The metabolic demand for oxygen in fish, particularly salmonids, and a comparison with other vertebrates. *Resp. Physiol.* 14:151-170.

Brett, J.R. and N.R. Glass. 1973. Metabolic rates and critical swimming speeds of sockeye salmon (*Oncorhynchus nerka*) in relation to size and temperature. *J. Fish. Res. Bd. Canada* 30:379-387.

Brett, J.R. and D.A. Higgs. 1970. Effect of temperature on the rate of gastric digestion in fingerling sockeye salmon *(Oncorhynchus nerka). J. Fish. Res. Bd. Canada* 27:1767-1779.

Brett, J.R., J.E. Shelbourn and C.T. Shoop. 1969. Growth rate and body composition of fingerling sockeye salmon *(Oncorhynchus nerka),* in relation to temperature and ration size. *J. Fish. Res. Bd. Canada* 26:2363-2394.

Brett, J.R. and C.A. Zala. 1975. Daily pattern of nitrogen excretion and oxygen consumption of sockeye salmon *(Oncorhynchus nerka)* under controlled conditions. *J. Fish. Res. Bd. Canada* 32 (12):2479-2486.

Butler, D.G. 1973. Structure and function of the adrenal gland of fishes. *Am. Zool.* 13 (3):839-879.

Campbell, Graeme. 1970. Autonomic nervous systems. pp. 109-132 *in* Hoar, W.S. and D.J. Randall (Eds.), *Fish Physiology,* Vol. IV. Academic Press, N.Y.

Campbell, Graeme. 1975. Inhibitory vagal innervation of the stomach in fish. *Comp. Biochem. Physiol.* 50C:169-170.

Casillas, E., et al. 1975. Changes in hemostatic parameters in fish following rapid decompression. *Undersea Biomedical Research* 2 (4):267-276.

Casillas, E. and L.S. Smith. 1977. Effect of stress on blood coagulation and haematology in rainbow trout *(Salmo gaird-*

neri). *J. Fish. Biol.* 10:481-491.

Chan, D.K.O. 1977. Comparative physiology of the vasomotor effects of neurohypophysial peptides of the vertebrates. *Am. Zool.* 17 (4): 751-761.

Chiason, Robert B. 1966. *Laboratory anatomy of the perch.* 53 pp. Wm. C. Brown Co., Dubuque, Iowa.

Cooper, J.C., et al. 1976. Experimental confirmation of the olfactory hypothesis with homing, artificially imprinted coho salmon *(Oncorhynchus kisutch). J. Fish. Res. Bd. Canada* 33 (4):703-710.

Coutant, C.C. 1972. Biological aspects of thermal pollution II. Scientific basis for water temperature standards at power plants. *CRC Critical Reviews in Environmental Control* 3 (1): 1-24.

Davis, John C. 1966. The influence of temperature and activity on certain cardiovascular and respiratory parameters in adult sockeye salmon. MS Thesis, University of British Columbia, 1966.

Davis, J.C. and D.J. Randall. 1973. Gill irrigation and pressure relationships in rainbow trout *(Salmo gairdneri). J. Fish. Bd. Canada* 30:99-104.

Dean, John M. 1973. The response of fish to a modified thermal environment. pp. 33-63 *in* Chavin, W. (Ed.) *Responses of Fish to Environmental Changes.* C.C. Thomas, Springfield, Ill.

Enger, Per S. 1967. Hearing in herring. *Comp. Biochem. Physiol.* 22:527-538.

Flock, Ake. 1971. The lateral line mechanoreceptors. pp 241-263 *in* Hoar, W.S. and D.J. Randall (Eds.), *Fish Physiology,* Vol. V. Academic Press, N.Y.

Fry, F.E.J. 1971. The effects of environmental factors on the physiology of fish. pp 1-98 *in* Hoar, W.S. and D.J. Randall (Eds.), *Fish Physiology,* Vol. VI. Academic Press.

Fujiya, Masaru and J.E. Bardach. 1966. A comparison between the external taste sense of marine and freshwater fishes. *Bull. Jap. Soc. Sci. Fish.* 32:45-56.

Gainer, H., and J.E. Klancher. 1965. Neuromuscular junctions in a fast-contracting fish muscle. *Comp. Biochem. Physiol.* 15:159-165.

Gainer, H., K. Kusano and R.F. Mathewson. 1965. Electrophysiological and mechanical properties of squirrel fish sound-producing muscle. *Comp. Biochem. Physiol.* 14:661-671.

Greenwald, Lewis, L.B. Kirschner and Martin Sanders. 1974. Sodium efflux and potential differences across the irrigated gill of seawater-adapted rainbow trout *(Salmo gairdneri). J. Gen. Physio.* 64:135-147.

Greer-Walker, M. 1971. Effect of starvation and exercise on the skeletal muscle fibers of the cod (*Gadus morhua* L.) and the coalfish (*Gadus virens* L.) respectively. *J. Cons. Int. Explor. Mer.* 33 (3):421-426.

Hafeez, M.A. and P. Ford. 1967. Histology and histochemistry of the pineal organ in the sockeye salmon (*Oncorhynchus nerka* Walbaum). *Can. J. Zool.* 45:117-126.

Hagiwara, S., T. Szabo and P.S. Enger. 1965. Physiological properties of the electroreceptors in the electric eel, *Electrophorus electricus. J. Neurophysiol.* 28:775-783.

Halver, J.E. et al. 1973. Nutrient requirements of domestic animals No. 11. *Nutrient requirements of trout, salmon and catfish.* National Academy of Sciences, Wash., D.C. 57 pp.

Hammel, H.T., S.B. Stromme and K. Myhre. 1969. Forebrain temperature activates behavioral thermoregulatory response in arctic sculpins. *Science* 165:83-84.

Hammond, B.R. and C.P. Hickman Jr. 1966. The effect of physical conditioning on the metabolism of lactate, phosphate and glucose in rainbow trout *(Salmo gairdneri). J. Fish. Res. Bd. Canada* 23 (1):65-83.

Hara, T.J. 1974. Is morpholine an effective olfactory stimulant in fish? *J. Fish. Res. Bd. Canada* 31:1547-1550.

Hara, T.J., Kazuo Uedo and Aubrey Garbman. 1965. Electroencephalographic studies of homing salmon. *Science* 149:884-885.

Henderson, N.E. 1967. The urinary and genital systems of trout. *J. Fish. Res. Bd. Canada* 24 (2):447-449.

Hickman, C.P., Jr. 1968a. Glomerular filtration and urine flow in the euryhaline southern flounder, *Paralichthys lethostigma,* in seawater. *Canadian J. Zool.* 46:427-437.

Hickman, C.P., Jr. 1968b. Urine composition and kidney tubular function in the southern flounder, *Paralichthys lethostigma*, in seawater. *Canadian J. Zool.* 46:439-455.

Hickman, C.P., Jr. 1968c. Ingestion, intestinal absorption and elimination of seawater and salts in the southern flounder, *Paralichthys lethostigma. Canadian J. Zool.* 46:457-466.

Hickman, C.P., Jr. and B.F. Trump. 1969. The kidney. pp91-240 *in* Hoar, W.S. and D.J. Randall (Eds.), *Fish Physiology,* Vol. I. Academic Press.

Hirano, T., M. Satou and S. Utida. 1972. Central nervous system control of osmoregulation in the eel *(Anguilla japonica). Comp. Biochem. Physiol.* 43A:537-544.

Hoar, W.S. 1965. The endocrine system as a chemical link between the organism and its environment. *Trans. Roy. Soc. Canada,* Sec. III, Ser. IV, Vol. 3:175-200.

Hoar, W.S. 1969. Reproduction. pp1-72 *in* Hoar, W.S. and D.G. Randall (Eds.), *Fish Physiology,* Vol. III. Academic Press, N.Y.

Hochachka, P.W. and G.N. Somero. 1971. Biochemical adaptation to the environment. pp. 100-156 *in* Hoar, W.S. and D.J. Randall (Eds.), *Fish Physiology,* Vol. VI. Academic Press.

Holeton, G.F. 1974. Metabolic cold adaptation of polar fish: fact or artifact? *Physiol. Zool.* 47 (3):137-152.

Hopkins, C.D. 1974. Electric communication in fish. *Am. Scientist* 62 (4):426-437.

Houston, A.H. 1973. Environmental temperature and the body fluid system of the teleost. pp. 87-162 *in* Chavin, W. (Ed.) *Responses of Fish to Environmental Changes.* C.C. Thomas, Springfield, Ill.

Hudson, R.C.L. 1969. Polyneuronal innervation of the fast muscles of the marine teleost, *Cottus scorpius* L. *J. Exp. Biol.* 50:47-67.

Hunn, Joseph B. 1969. Chemical composition of rainbow trout urine following acute hypoxic stress. *Trans. Am. Fish. Soc.* 98 (1):20-22.

Hunn, J.B. 1976. Inorganic composition of gallbladder bile from freshwater fishes. *Copeia* 1976 (3):602-605.

Hurley, Donal A. and K.C. Fisher. 1966. The structure and development of the external membrane in young eggs of the brook trout, *Salvelinus fontinalis* (Mitchill). *Canadian J. Zool.* 44:173-190.

Hyodo, Y. 1964. Effect of X-irradiation on the intestinal epithelium of the goldfish, *Carassius auratus.* I. Histological changes in the intestine of irradiated fish. *Annotationes Zoologicae Japonenses* 37 (2):104-111.

Iwai, Tamotsu, 1967. The comparative study of the digestive tract of teleost larvae. II. Ciliated cells of the gut epithelium in pond smelt larvae. *Bull. Japanese Soc. Sci. Fish.* 33 (12): 1116-1119.

Johansen, Kjell. 1968. Air-breathing fishes. *Scientific American* 219 (4):102-111.

Johansen, Kjell. 1970. Air-breathing in fishes. pp. 361-411 *in* Hoar, W.S. and D.J. Randall (Eds.), *Fish Physiology,* Vol. IV. Academic Press.

Johnson, Donald W. 1973. Endocrine control of hydromineral balance in teleosts. *Amer. Zool.* 13:799-818.

Kaplan, Harriett and L.R. Aronson. 1967. Effect of forebrain ablation on the performance of a conditioned avoidance response in the teleost fish, *Tilapia h. macrocephala. Animal Behav.* 15 (4):438-448.

Kaplan, Harriett and L.R. Aronson. 1969. Function of forebrain and cerebellum in learning in the teleost, *Tilapia heudolotii macrocepaphela. Bull. Am. Mus. Nat. Hist.* 142:141-208.

Kapoor, B.G., H. Smit and I.A. Verighina. 1975. The alimentary canal and digestion in teleosts. pp. 109-239 *in* Russell, F.S. and M. Yonge (Eds.), *Advances in Marine Biology,* Vol. 13. Academic Press, N.Y.

Knudsen, E.I. 1975. Spatial aspects of the electric fields generated by weakly electric fish. *J. Comp. Physiol.* 99:103-118.

Konishi, J. and I. Hidaka. 1967. Stimulation of the chemoreceptors of the sea catfish by dilute electrolyte solutions. *Jap. J. Physiol.* 17 (6):726-737.

Kutty, M.N., et al. 1971. Maros-Schulek technique for measurement of carbon dioxide production in fish and respiratory quotient in *Tilapia mossambica. J. Fish. Res. Bd. Canada* 28 (9):1342-1344.

LaPointe, Joe. 1977. Comparative physiology of neurohypophysial hormone action on the vertebrate oviduct-uterus. *Am. Zool.* 17 (4):763-773.

Laurent, Pierre and Suzanne Dunel. 1976. Functional organization of the teleost gill. I. Blood Pathways. *Acta Zool.* (Stockh.) 57:189-209.

Maetz, J. 1971. Fish gills: mechanisms of salt transfer in freshwater and sea water. *Phil. Trans. Roy. Soc. Lond.* 262:209-249.

Markert, J.R., et al. 1977. Influence of bovine growth hormone on growth rate, appetite, and food conversion of yearling coho salmon *(Oncorhynchus kisutch)* fed two diets of different composition. *Can. J. Zool.* 55 1 :74-83.

McLeay, D.J. and D.A. Brown, 1975. Effects of acute exposure to bleached Kraft pulpmill effluent on carbohydrate metabolism of juvenile coho salmon *(Oncorhynchus kisutch)* during rest and exercise. *J. Fish. Res. Bd. Canada* 32:753-760.

Mearns, A.J. 1971. Lactic acid regulation in salmonid fishes. Ph. D. Thesis, University of Washington, Seattle. 135 pp.

Miles, H.M. 1971. Renal function in migrating adult coho salmon. *Comp. Biochem. Physiol.* 38A:787-826.

Miles, H.M. and L.S. Smith. 1968. Ionic regulating in migrating juvenile coho salmon, *Oncorhynchus kisutch. Comp. Biochem. Physiol.* 38A:787-826.

Muir, B.S. and J.I. Kendall. 1968. Structural modifications in the gills of tunas and some other oceanic fishes. *Copeia* 1968 (2):388-398.

Newcomb, T.W. 1974. Changes in blood chemistry of juvenile steelhead trout. *Salmo gairdneri,* following sublethal exposure to nitrogen supersaturation. *J. Fish. Res. Canada* 31:1953-1957.

Nishimura, Hiroko, and M. Ogawa. 1973. The renin-angiotensin system in fishes. *Am. Zool.* 13 (3):823-838.

Oduleye, S.O. 1975a. The effects of calcium on water balance of the brown trout, *Salmo trutta. J. Exp. Biol.* 63:343-356.

Oduleye, S.O. 1975b. The effect of hypophysectomy and prolactin therapy on water balance of the brown trout, *Salmo trutta. J. Exp. Biol.* 63:357-170.

Ohnesorge, F.K. and R. Rauch. 1968. Pharmacological investigations on the peristaltic movements of the gut of the tench *(Tinca vulg.). Zeit vergl. Physiol.* 58:153-170.

Ohnesorge, F.K.and G. Schmitz. 1968. The influence of bath temperature and adaptation temperature on the contractions of the isolated gut of the tench *(Tinca vulg.). Zeit vergl. Physiol.* 58:171-184.

Pang, P.K.T. 1977. Osmoregulatory functions of neurohypophysial hormones in fishes and amphibians. *Am. Zool.* 17 (4): 739-749.

Parry, Gwyneth. 1961. Osmotic and ionic changes in the blood and muscle of migrating salmonids. *J. Exp. Biol.* 38:411-427.

Parry, G. and F.G.T. Holliday. 1960. An experimental analysis of the function of the pseudobranch in teleosts. *Nature, Lond.* 183:1248-49

Pritchard, A.W., et al. 1971. The relation between exercise and biochemical changes in red and white muscle and liver in the jack mackerel, *Trachurus symmetricus. Fish. Bull.* 69 (2): 379-386.

Rahn, Hermann. 1966. Aquatic gas exchange: theory. *Resp. Physiol.* 1:1-12.

Randall, D.J. 1968. Functional morphology of the heart in fishes. *Am. Zool.* 8:179-189.

Randall, D.J. 1970. Gas exchange in fish. pp. 253-292 *in* Hoar, W.S. and D.J. Randall (Eds.), *Fish Physiology,* Vol. IV. Academic Press, N.Y.

Randall, D.J. 1974. The regulation of H^+ concentration in body fluids. pp. 89-94 *in* M.D.B. (Ed.) *Proceedings of the Canadian Society of Zoologists Annual Meeting,* June 2-5, 1974.

Randall, D.J. 1975. Carbon dioxide excretion and blood pH regulation in fish. pp 405-418 *in Chemistry and Physics of Aqueous Gas Solutions.*

Randall, D.J. and L.S. Smith. 1967. The effect of environmental factors on circulation and respiration in teleost fish. *Hydrobiologia* 29:113-124.

Reaves, R.S., A.H. Houston and J.A. Madden. 1968. Environmental temperature and the body fluid system of the freshwater teleost - II. Ionic regulation in rainbow trout, *Salmo gairdneri,* following abrupt thermal shock. *Comp. Biochem. Physiol.* 25:849-860.

Robertson, O.H., et al. 1961. Physiological changes occurring in the blood of the Pacific salmon *(Oncorhynchus tshawytscha)* accompanying sexual maturation and spawning. *Endocrinology* 68 (5):733-746.

Rommel, S.A., Jr. and J.D. McCleave. 1971. An electromagnetic system for studying the responses of aquatic organisms to weak electric and magnetic fields. *Bio-medical Engineering* 18 (6):421-424.

Rosen, M. William. 1959. Water flow about a swimming fish. *NOTS Technical Publication,* 2298. (MS Degree, UCLA).

Royce, W.F., et al. 1968. Models of oceanic migrations of Pacific salmon and comments on guidance mechanisms. *Fish. Bull.* 66 (3):441-462.

Satia, B.P., L.R. Donaldson, L.S. Smith and J.N. Nightingale. 1974. Composition of ovarian fluids and eggs of the University of Washington strain of rainbow trout *(Salmo gairdneri). J. Fish. Res. Bd. Canada* 31:1796-1799.

Schneider, P.W., Jr. and L.J. Weber. 1975. Neuromuscular function and acetylcholinesterase in the pectoral abductor muscle of largemouth bass *(Micropterus salmoides)* as evidenced by the effects of diisopropylfluorophosphate. *J. Fish. Res. Bd. Canada* 32 (11):2153-2161.

Schreck, C.B., et al. 1976. Physiological response of rainbow trout *(Salmo gairdneri)* to electroshock. *J. Fish Res. Bd.* Canada 33 (1):76-84.

Schreibman, M.P., J.F. Leatherland and B.A. McKeown. 1973. Functional morphology of the teleost pituitary gland. *Am. Zool.* 13 (3):719-742.

Schwartz, Erich. 1974. Lateral-line mechanoreceptors in fishes and amphibians. pp 257-278 *in* Fessard, A. (Ed.), *Handbook of Sensory Physiology,* Vol. III (No. 3): *Electroreceptors in lower vertebrates.* Springer-Verlag, Berlin.

Shehadah, Z.H. and M.S. Gordon. 1969. The role of the intestine in salinity adaptation of the rainbow trout, *Salmo gairdneri. Comp. Biochem. Physiol.* 30 (3):397-418.

Shelton, G. 1979. The regulation of breathing. pp 293-359 *in* Hoar, W.S. and Randall, D.J. (Eds.), *Fish Physiology,* Vol. IV. Academic Press, N.Y.

Shieh, H.S. and J.R. McClean. 1976. Blood changes in brook trout induced by infection with *Aeromonas salmonicida. J. Wildlife Diseases* 2:77-82.

Skadhauge, Eric. 1969. The mechanism of salt and water absorption in the intestine of the eel *(Anguilla anguilla)* adapted to waters of various salinities. *J. Physiol.* 204:135-158.

Smith, L.S. 1966. Blood volumes of three salmonids. *J. Fish. Res. Bd. Canada* 23:1439-1446.

Smith, L.S. and G.R. Bell. 1975. *A practical guide to the anatomy and physiology of Pacific salmon.* Misc. Special Publ. No. 27, Dept. of the Environment, Fisheries and Marine Service. Ottawa, Canada. 11 pp.

Smith, L.S., et al. 1967. Cardiovascular dynamics in swimming adult sockeye salmon. *J. Fish. Res. Bd. Canada* 24 (8): 1775-1790.

Smith, L.S., et al. 1971a. *Responses of teleost fish to environmental stress.* 114 pp. Water Pollution Control Research Series, No. 18050EBK02/71. Environmental Protection Agency, Water Quality Office, U.S. Govt. Printing Office.

Smith, L.S., et al. 1971b. Physiological changes experienced by Pacific salmon migrating through a polluted urban estuary. pp 1-7 in report of: *FAO Technical Conference on Marine Pollution and its Effects on Living Resources and Fishing.* FIR: MP/70/E-40. 6 November 1970.

Stroud, R.K. and A.V. Nebeker. 1974. A study of the pathogenesis of gas bubble disease in steelhead trout *(Salmo gairdneri).* pp 66-71 *in* Fickeisen, D.H. and M.J. Schneider (Eds.), *Gas Bubble Disease,* Technical Information Center, USERDA, Conference No. 741033. Oak Ridge, Tenn.

Sutterlin, A.M. and N. Sutterlin. 1970. Taste responses in Atlantic salmon *(Salmo salar)* parr. *J. Fish. Res. Bd. Canada* 27:1927-1942.

Swift, D.J. and R. Lloyd. 1974. Changes in urine flow rate and haematocrit value of rainbow trout, *Salmo gairdneri* (Richardson), exposed to hypoxia. *J. Fish. Biol.* 6:379-387.

Tamura, Tamotsu and Hiroshi Niwa. 1967. Spectral sensitivity and color vision of fish as indicated by S-potential. *Comp. Biochem. Physiol.* 22:745-754.

Thomas, A.E., et al. 1964. *A device for stamina measurement of fingerling salmonids.* U.S. Dept. Interior, Fish and Wildlife Service, Bureau of Sport Fisheries and Wildlife, Research Report 67.

Tomita, T. 1971. Vision: electrophysiology of the retina. pp 33-57 in Hoar, W.S. and D.J. Randall (Eds.), *Fish Physiology,* Vol. V. Academic Press, N.Y.

Vibert, Richard. 1963. Neurophysiology of electric fishing. *Trans. Am. Fish. Soc.* 92 (3):265-275.

Vibert, Richard (Ed.), 1969. *Fishing with electricity, its application to biology and management.* Fishing News Book, Ltd. 276 pp.

Watters, K.W. and L.S. Smith. 1973. Respiratory dynamics of the starry flounder *Platichthys stellatus* in response to low oxygen and high temperature. *Marine Biol.* 19 (2):133-148.

Webb, P.W. 1975. Hydrodynamics and energetics of fish propulsion. *Bulletin No. 190 of the Fisheries Research Board of Canada.* Ottawa.

Webb, P.W. and J.R. Brett. 1972a. Respiratory adaptations of prenatal young in the ovary of two species of viviparous seaperch, *Rhacochilus vacca* and *Embiotoca lateralis. J. Fish. Res. Bd. Canada* 29:1525-1542.

Webb, P.W. and J.R. Brett. 1972b. Oxygen consumption of embryos and parents, and oxygen transfer characteristics within the ovary of two species of viviparous seaperch, *Rhacochilus vacca* and *Embiotoca lateralis. J. Fish. Res. Bd. Canada* 29:1543-1553.

Wedemeyer, Gary. 1970. The role of stress in the disease resistance of fishes. *Am. Fish. Soc. Symposium on the Diseases of Fish and Shellfish. Spec. Publ. No. 5.*

Wedemeyer, G.A. and W.T. Yasutake. 1977. Clinical methods for the assessment of the effects of environmental stress on fish health. *Technical Papers of the U.S. Fish and Wildlife Service,* No. 89, 18 pp.

Weiss, R.F. 1970. The solubility of nitrogen, oxygen and argon in water and seawater. *Deep-Sea Research* 17:721-735.

Wittenberg, J.B., and R.L. Haedrich. 1974. The choroid rete mirable of the fish eye. II. Distribution and relation to the pseudobranch and to the swimbladder rete mirable. *Biol. Bull.* 146:137-156.

Wydoski, R.S., G.A. Wedemeyer and N.C. Nelson. 1976. Physiological response to hooking stress in hatchery and wild rainbow trout *(Salmo gairdneri). Trans. Am. Fish. Soc.* 105 (5):601-606.

Yokote, N. 1970. Sekoke disease, spontaneous diabetes in carp found in fish farms. III. Response to mammalian insulin. *Jap. Soc. Sci. Fish.* 36:1219-1223.

INDEX